Basudev Nag Chowdhury

Quantum and Brain-Inspired Device Modeling

De Gruyter Series in Mathematics and Life Sciences

Volume 16

Basudev Nag Chowdhury

Quantum and Brain-Inspired Device Modeling

Non-Equilibrium Green's Function Simulator for Qubits to Neuromorphic Computing

DE GRUYTER

Author
Basudev Nag Chowdhury
Principal Research Scientist
IIT Kharagpur
Kharagpur, West Bengal
India
and
Research & Innovation Head
QSciT Research Private Limited
Kolkata, West Bengal
India

ISSN 2195-5530
ISBN 978-3-11-154580-6
ISBN 978-3-11-154595-0 (PDF)
ISBN 978-3-11-154611-7 (EPUB)
DOI https://doi.org/10.1515/9783111545950

Library of Congress Control Number: 2026938084

Bibliographic information published by the Deutsche Nationalbibliothek
The Deutsche Nationalbibliothek lists this publication in the Deutsche Nationalbibliografie;
detailed bibliographic data are available on the internet at http://dnb.dnb.de.

De Gruyter and Walter de Gruyter GmbH are part of De Gruyter Brill.
www.degruyterbrill.com

Questions about General Product Safety Regulation:
productsafety@degruyterbrill.com

Printing: CPI books GmbH, Leck

In memory of my teachers
Late Swapan Kumar Datta
and
Late Dipanjan Rai Chaudhuri

Acknowledgements

It is really not possible to acknowledge all of them who I owe that I could write this book. Even then I cannot, but express my deepest gratefulness to Dr Sandeep Chaubey for the relentless debates and discussions over various topics of science and their methodologies. I have also been continuously enriched through discussions with Dr Manasvi Lingam and Mr Baijayanta Roy. I cannot convey my gratitude enough to my teachers Late Dr Swapan Kumar Datta and Late Prof. Dipanjan Rai Chaudhuri, who taught me the fundamentals of physics and motivated me toward dedicating for physics. I am also greatly thankful to Prof. Sanatan Chattopadhyay, who guided me learning classical MOSFET technology. Finally, I must express my thanks to all my colleagues, friends, and family members for their unconditional support and help.

 | https://doi.org/10.1515/9783111545950-202

Preface

Never judge a book by its cover, so as by its preface. I am writing the preface of this book to describe neither its content nor its attractive features, if even any, rather to confess the limitations of this book from my perspective. In fact, my own perspectives are possibly the major limitation of this book, since I think that I have not been able to restrict myself from throwing some random glimpses of my ideas. One such idea is to collate two different subjects, i.e. quantum properties of matter and the functionality of human brain, together to construct the core base of this book. It has definitely got some key inspiration from R. Penrose, who has claimed quantum collapse to be an important factor behind the functioning of consciousness of human brain.

From my personal view, however, I believe that the creation of life on Earth is a natural consequence of the evolution of non-living systems since the birth of universe, and apparently so are the creation of succeeding animal life and human beings with their extraordinary body parts, especially the brain. I further believe that a more evolved system may have the capacity to extract and analyse all information from a less evolved system (i.e. valid also for non-living systems), at least within the range of communication, but not the *vice versa*. While communicating to others, there can be no adiabatic evolution in this sense, so that the more evolved system must have something neither extractable nor analysable by a less evolved system. Penrose has possibly upheld this concept using Gödel's incompleteness theorem to discard the expectation of making an ideal artificial intelligence. I have introduced an idea of two fundamentally different kinds of numbers (which I call *C-numbers* and *V-numbers*), almost arbitrarily in this book, that may have some resemblance with this concept in the sense that *C-numbers* are close to represent Gödel's transcend, while *V-numbers* are not. Nevertheless, such concept is very premature, and still only my perspective.

Furthermore, the intersystem communications along with their time development, including that between two quantum systems, quantum and classical world, system and apparatus or its surroundings, as well as human brain and non-living things as its constituents or as its projection on circuit planes, had engrossed my mind while preparing the content of this book. Possibly this is why I found the non-equilibrium Green's function (NEGF), especially in Keldysh formalism, to be my best expressive weapon, not only due to its beautiful mathematical structure, but rather owing to its fascinating concept of describing a system in its *retarded* and *advanced* forms.

While describing a system in terms of its communication with others, i.e. on the basis of information exchange, an apparent subjectivity may appear. For instance, quantum theory is widely interpreted as a model to subjectively describe (based on randomness during measurements) a system without having any objective reality of its own. Debate between A. Einstein and N. Bohr on this is still so charming, which, in turn, has significantly motivated me to argue in this book that, information with maximum entropy is indeed objective and provides us with the *truth* of the system concerned, beyond the set of all individual *facts*. This has got some inspiration from

 | https://doi.org/10.1515/9783111545950-203

E. T. Jaynes and R. Tagore. Thus, I argue that information about a system is not only measureable as *bits/qubits* or *action potential* in human brain nerves, but it may represent, when entropy maximized, the characteristic property describing a system objectively.

In this context, this book describes some of the basic concepts of quantum devices and quantum properties and human brain functionalities from the perspective of theoretical modelling on the basis of NEGF formalism, which however includes, as burden, some of my unconventional ideas, such as on quantum-classical complementarity in terms of measurement paradox, harvesting the so-called dissipative energy utilizing non-equilibrium processes, charge-mass equivalence principle leading to curved surface, spin-charge-phonon trialectics to explain superconductivity, describing brain-neuronal medium as a moderate-temperature dense plasma to theorize signal generation and transmission, and exploration of fractional dimension as the missing link between the theories of quantum and gravity. Although the formal development of NEGF theory and its applications for device simulation may give the readers some taste, they may get annoyed with such half-cooked or uncooked foodstuffs served as complimentary side dish. Thus, this preface ends here with begging the readers' pardon . . .

B. Nag Chowdhury

Contents

1 Some ideas on modern quantum and brain-inspired devices

The most appealing 'words' in the title of this book, i.e. 'Quantum' and 'Brain', may make the readers skeptical about the content of this book due to their adjacent placements in the title. Remembering Sigmund Freud's indication that *"Words have a magical power* being *capable of arousing the strongest emotions and prompting all men's actions"*, let us first start, rather in a defensive perspective, with clarifying that we are not going to propose or criticize (positively or negatively) any quantum description of brain functionality in terms of consciousness in this book. However, the consecutive placement of these two 'words' (more precisely to say 'concepts') is not mere juxtaposition; rather, the astonishing similarity between them remains in the fact that the key processes associated with both quantum phenomena and brain functionalities are non-equilibrium in general.

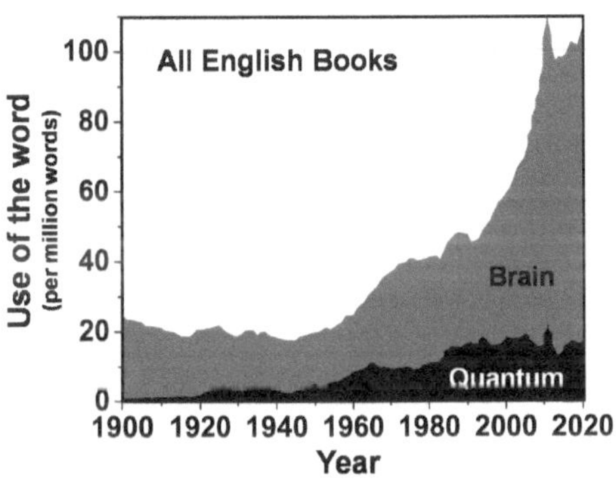

Fig. 1.1: Plots of the use of the words 'Quantum' and 'Brain' (per million words) in all the books written in English each year since 1900 (data source: *Google Books Ngram Viewer*).

Further, the appeal of these words cannot now be ignored if we look at the extensive growth in the use of these words, 'Quantum' and 'Brain', in literatures, both fiction and non-fiction. Figure 1.1 plots the use of these words in all the books written in English since the last century to the current years, where 'Quantum' shows a consistent growth since the 1920s (for evident reason, i.e. the birth of quantum physics). However, 'Brain' is exhibiting an exponential escalation in the last two decades, which cannot be explained by its conventional inflated use in literatures, and indeed is attributed to the large scale research initiatives taken in this period: for instance, the Cuban Human Brain Mapping Project [1] and the European Human Brain Project [2].

It is apparent, therefore, that it is not possible to even briefly cover such an enormous area of the diverse fundamental aspects, as well as applications of these two fields, in a single book. Hereby, we must be apologetic to limit the scope of the present

 | https://doi.org/10.1515/9783111545950-001

book to a very small but extremely important subsection of these fields, quite itself like a quantum object or a human brain in its size and activities. Here, we will border us only to walking around the non-equilibrium processes related to quantum phenomena and brain functionalities. And our focus will be to describe how to model such processes in a structured and consistent mathematical framework, so that one can simulate such phenomena on a virtual platform. We will show that the non-equilibrium Green's function (NEGF) formalism can provide a fascinating route towards such development.

1.1 Which shall we call a 'Quantum device' to be?

It is now familiar to us that what we call 'quantum phenomena', such as wave-particle duality, uncertainty, superposition, entanglement, etc., cannot be explained by classical laws of physics. Thus, one can simply define a 'quantum device' by whether its operating principle is effectively based on quantum mechanics. However, it is fundamentally paradoxical that all the classical objects which do not exhibit quantum phenomena empirically are made up of tiny objects that essentially follow quantum mechanical principles (i.e. the unsolved Schrödinger's cat paradox [3]). Then, depending on such causes of action, each device would be a quantum device, at least in principle! Clearly, this cannot be an identifying definition of a quantum device. Thus, people started to look at the effects instead of concentrating on the causes. And in this regard, recently, physicists have almost reached a consensus that a device which shows an effect in practice that cannot be achieved by any device which follows classical laws of physics, and can solely be explained by quantum theory, is said to be a 'quantum device'. For instance, a sensor that offers sensitivity beyond its theoretical classical limit, or a photodetector providing a resolution that surpasses the classical limit-of-detection, or any device violating Bell or Bell-type inequalities, is identified as a quantum device (for a comprehensive understanding, see Peres [4]).

Evidently, to realize such quantum devices in practice is difficult; however, surprisingly, not only due to their quantum properties, but rather more for the loss of such quantum properties during measurement. It is still an unsolved and open problem what a quantum measurement process is, indeed. Nevertheless, that all the measured outcomes must be compatible with our (i.e. humans') sense perception and thus must be classical (not quantum) in nature cannot be denied. Thus, it would not be wrong to say that quantum measurement is merely a communication between the quantum and classical worlds. Consequently, to theoretically model quantum devices, it is not sufficient to work only with the quantum equations of motion, but it is also necessary to take into account the quantum-classical interaction.

At this point, it is worth explaining that another eccentric difficulty arises in this context: a number of properties that are obvious even to our every second feeling are not at all defined for quantum systems. A classic example is the 'Temperature'. It is such an apparent property of any substance that we can touch it and feel it. However, we need to recall that when such a subjective feeling is taken away, the temperature of a system, from the point of view of physics, is an objective measure of the average kinetic/vibrational energy of its atoms/molecules. Interestingly, 'average' is a statistical quantity that is meaningfully defined only for a 'large' sample size. Here, 'large' is described by a 'quantity' that eventually leads to a probability distribution function (e.g. Gaussian distribution). For instance, if the number of atoms/molecules is of the order of Avogadro's number ($\sim 10^{23}$), the 'average' is described by a Boltzmann distribution. Without such a 'large' number of atoms/molecules, neither any distribution nor any average is at all meaningfully defined. Thus, for instance, a quantum dot (QD) with a few (~10's) numbers of atoms/molecules does not have any defined temperature (!). However, the measuring apparatus is generally a classical object with a definite temperature. And we need to also model the interaction between two such systems (one classical, the other quantum), one having a definite temperature and the other without any defined temperature. Evidently, such interaction must be a non-equilibrium process in general. Similar unorthodoxy appears for the Fermi level (i.e. chemical potential) or stress/pressure. Such unconventional stuffs are necessary to consider for developing a realistic model to describe a quantum device.

1.2 The concept of 'Quantum' and criticality for practical quantum devices

Assuming that the readers have basic knowledge of quantum mechanics, we will now introduce a new line of thought here on 'Numbers' to discuss the concept of 'Quantum' for practical quantum devices. We want to draw the attention of our readers to the fact that we are indeed not much aware that we, in practice, generally deal with two very different kinds of numbers. For instance, suppose in a restaurant we are observing the volume of water in a bowl placed on a running induction oven and the number of customers having food. Obviously, the former number (i.e. the volume of water) will change continuously, including through its fractional numbered values, while the latter number (i.e. of the customers) will change discretely only with integer values. However, that is not the mere difference that we are discussing here. In fact, if we look into it a little closer, the former number represents something (e.g. the volume of water in this case) that changes, preserving its own existence (here for simplicity, we neglect the evaporation of water). However, the latter number corresponds to something that can only change by creating or destructing its own existence (e.g. through the entry of a new customer or exit of an existing customer in the restaurant). Consequently, the former number represents a 'value' assigned to it, and therefore, can be origin-shifted or scaled as per subjective perspective (we will call it *V-number*). On the other hand, the latter number characterizes a 'count' that is completely objective (we will call it *C-number*). And it is imperative to note that, for such reasons, irrespective of whether they are fractional or integral, these two kinds of numbers, although often symbolized by the same 'Numbers', are indeed completely different in their inherent nature. For instance, if some particle number in some region of space is 'zero', it means there is no particle existing at all (i.e. a *C-number*); whereas a 'zero' value of the position of a particle in some region of space does not mean that its position does not exist, it is rather just a 'value' assigned to it (i.e. a *V-number*): the two 'zeros' have completely different meaning!

In practical quantum devices, we need to work with both kinds of such 'Numbers' or 'Mathematical elements'. For instance, if a voltage is applied on a QD in such a manner (e.g. using logic gates) that an electron is constrained to move inside it, the motion of the electron is described by a wave function (which itself, as well as the variables describing it, are *V-numbers* in this case) following the Schrödinger equation. Here, the existence of the electron, as well as the existence of its properties within the QD, is preserved (commonly called unitary evolution). On the other hand, if a voltage is applied to the QD that drives out the electron into the probing metal, or an electron enters the QD from the probe, then the electron's existence within the boundary of the QD is destroyed or created, respectively; and we require, according to standard quantum field theory, the second quantization filed operators corresponding to its creation and annihilation (i.e. the *C-numbers* in these cases) to appropriately describe such motion. Thus, in fact, a qubit 'manipulation' can be characterized by the *V-numbers* while the 'preparation' or 'measurement' needs *C-numbers* to describe it. Similarly, a MOS capacitor that accumulates (both for 'accumulation' and 'inversion' regions) or depletes electrons or holes can be well described by *V-numbers* . However, in a photodetector/solar-cell where a photon is absorbed (i.e. destroyed), resulting in the generation of an electron-hole pair, or in a laser where photon numbers are amplified (i.e. created), we must deal with *C-numbers*.

The NEGF framework, which is the focus area of the present book, accommodates both kinds of numbers, i.e. the *V-numbers* and *C-numbers*, in its general formalism. The NEGF for a quantum system contains information about the system (through *V-numbers*), as well as about the other surrounding systems that it interacted with (through *C-numbers*) within a certain period of time; and thereby it comprises of all the information about the evolution of a quantum system through the continuous changes of its preserved elements of properties along with the discrete changes of its created/destructed elements of properties. We will elaborate on such foundational and applicative aspects of the NEGF framework in the succeeding chapters of this book.

1.3 Digitized world: MOSFET as a *bit* and its evolution towards the quantum scale

It would not be an overstatement to say that the concept of the modern digital world is practically built upon the seminal paper by C. E. Shannon in 1948 that introduced a quantifiable definition of information in terms of probability [5]. It provided the idea of measurable information in the amounts of *bits*; and the contemporary invention of semiconductor field-effect-transistor by Shockley, Bardeen, and Brattain [6] for electronically realizing such *bits* in practice, followed by the continuously progressing metal-oxide-semiconductor field-effect-transistor (MOSFET)-technology, made an unstoppable path towards digitization of the entire society, that is driven by the relentless demand of storage and processing of information. In this tremendous drive of accommodating larger amounts of information along with their various processing tasks, the semiconductor technology has found its evolution through making the physical space for each *bit* smaller, i.e. (down) scaling of MOSFETs, which has inevitably led us to enter the quantum domain in recent years. *Bits* are now being replaced by *Qubits* (i.e. quantum *bits*) that claim to exponentially speed up the computation processes.

The MOSFET is operationally a three-terminal electronic device with source, drain, and gate terminals (although, practically conventional planar MOSFETs have one more terminal for ground), with a gate-voltage controlled internal channel for current flow (see Fig. 1.2(a)). The operating principle of MOSFET is quite like the water flowing from one reservoir to another through a siphon, controlled by a faucet, as shown in Fig. 1.2(b). Just as water flows from a high-pressure reservoir (at a higher height) to a low-pressure reservoir (at a lower height) surmounting the siphon-barrier when the faucet lets it flow through, in a MOSFET, electron/hole flows from source (at zero bias) to drain (at a positive/negative bias) through the channel when gate-voltage allows it to transport. Thus, below or above a certain gate voltage, known to be the *threshold voltage*, the device current in MOSFET is (qualitatively) *low* or *high*, respectively, thereby providing the opportunity to exploit it as a digital switch with a corresponding OFF or ON state, accordingly assigned with a '0' or '1' binary value. And the switching speed of this jump from '0' to '1' is measured in terms of *sub-threshold swing*, i.e. the voltage required for raising the off-current by a decade order (which has the classical limiting value of 60 mV/dec at room temperature (300 K)). It is also important to note that, depending on whether electrons or holes are flowing through the channel, the MOSFET is known to be n-type or p-type, respectively; and an adjacent combination of both is known as the complementary MOSFET, which is the applicative unit of the famous CMOS technology as a whole (for details on MOSFETs, see Ng and Sze [7]).

It is worthy to note at this point that MOSFET is a two-field device, where the gate voltage generates the switching electric field for creating the path for electron/hole flow, while the drain bias produces the driving electric field for directional transport of such electrons/holes. It is also imperative to note that the gate operates as a MOS capacitor, and thus, except for practical leakages, predominantly it does not dissipate power; whereas drain-bias directly drives out the electrons/holes, thereby resulting in various resistive losses. Thus, in order to reduce the power dissipation while keeping the driving field intact, the channel length must be shortened accordingly as the operating drain voltage is needed to decrease. This, in turn, has led to adopting the route of scaling of MOSFET channel length (and the gate length as well) for the sustained improvement of its performance.

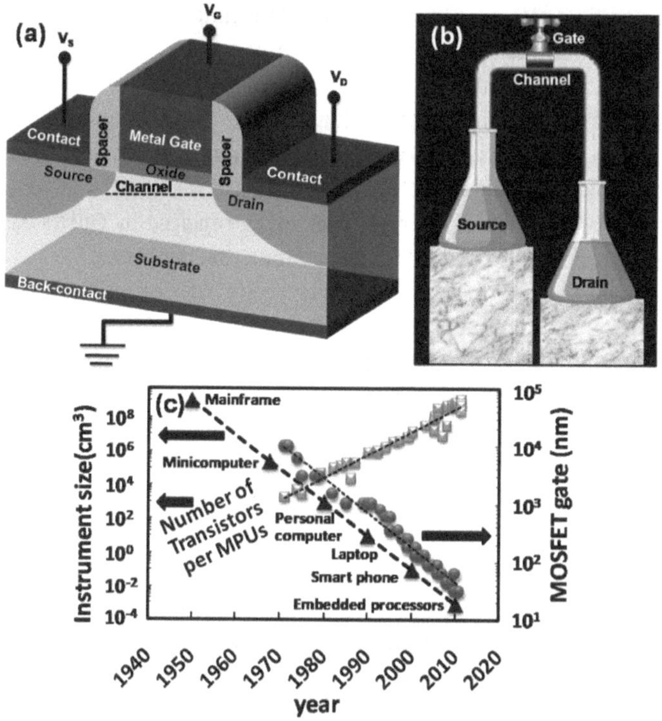

Fig. 1.2: (a) The schematic of a conventional planar MOSFET architecture. The channel is formed underneath the oxide layer (which electrically isolates it from the gate) upon the application of a suitable gate voltage, and transports current when a source/drain bias is applied; (b) an analogical illustration of a siphon that transports water from one reservoir to another using a controlling faucet; (c) left axis: plot of the volume of electronic instruments, i.e. decreasing from approximately a cube of 10 m to a cube of 1 mm, and the number of transistors per micro-processing units (MPUs), i.e. increasing from thousands to billions; and right axis: the MOSFET gate length, i.e. decreasing from about 10 μm to 10 nm over the years.

In 1965, Gordon Moore identified that the logarithm (at base 2) of the number of components of an integrated electronic function linearly increases per year (known as Moore's law) [8], thereby indicating a preferred direction of sustained technology development through the (down) scaling approach for all the component devices (e.g. MOSFET). Although Moore predicted such a trend for the next 10 years, the very trend of utilizing the advantages of miniaturization is still continuing even after six decades have passed (see Fig. 1.2(c)). It can be seen from Fig. 1.2(c) that in the last 50 years, the size of computing systems has shrunk from ~ m to ~ mm order along with the MOSFET gate length reducing from the order of ~10 μm to ~10 nm, whence the number of transistors per MPU has increased by more than an order of ten each decade [9].

Nevertheless, such continuous miniaturization of device size up to the order of a few nanometres has compelled us to enter the quantum world. The electrons/holes do not show their classical averaged-out behaviour at this scale. Now the nanostructure channel surrounded by a gate is a quantum device coupled to two classical reservoirs, i.e. source and drain. The question is whether such ultra-scaled devices will be able to provide us with beyond classical limit outputs to perform as superior *bits*, or they will exhibit only the quantum effects that are detrimental to the device performance. This is still an open problem. Meanwhile, *qubits* are already in action!

1.4 Qubits

Let us start with a famous (or rather, to be called 'infamous') illustration borrowed (not in exact form) from Schrödinger (i.e. the Schrödinger's cat paradox) [3]. If a poisonous gas is triggered towards an unfortunate cat confined in a room, it will be either *dead* or *alive*, and such information can be stored accordingly in two *bits*, representing the cat's states as '0' and '1', respectively. Quantum theory tells us that the cat may be more unfortunate to be in a *coma* under ventilator, i.e. something between *dead* and *alive* (see the illustration of Fig. 1.3(a)). Such information thus can be stored in a *qubit* through coherent superposition of quantum states, mathematically described as $|\psi\rangle = a|0\rangle + b|1\rangle$ ('$|\rangle$' symbolically represents a state). Here, a, b are two complex numbers in general, so that $|a|^2$ and $|b|^2$ are the respective probabilities of finally getting the cat as *dead* ($|0\rangle$) or *alive* ($|1\rangle$) once the ventilator is removed. It indicates that the information about the possibility of two states (e.g. *dead* or *alive*) requires 2 *bits* to be deterministically stored in, whereas only 1 *qubit* to get probabilistically stored. Or in other words, 1 *qubit* can probabilistically store the information that 2 *bits* can deterministically store; thus similarly, 2 *qubits* for 4 *bits*, 3 *qubits* for 8 *bits*,

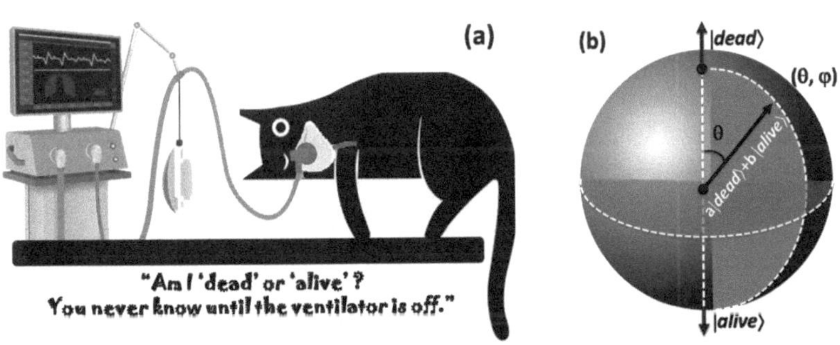

Fig. 1.3: (a) A virtual illustration of an unfortunate cat (known as Schrödinger's cat) at *coma* under ventilator, in a state something between *dead* and *alive*, mathematically represented as $|\psi\rangle = a|\text{dead}\rangle + b|\text{alive}\rangle$, with $|a|^2$ and $|b|^2$ being the probabilities that the cat would be *dead* or *alive* once the ventilator is removed; (b) Bloch sphere representation of the cat's quantum state.

and in general, *n qubits* for 2^n *bits*. Now, suppose one needs to work with the information about each of the molecules of a ~1 mole gas; it needs ~10^{23} numbers of deterministic *bits* to store such information, whereas only ~76 non-deterministic *qubits* for it (since $2^{76} \sim 10^{23}$).

It may be interesting to note that any such state ($|\psi\rangle = a|0\rangle + b|1\rangle$) of a qubit can be expressed, without loss of generality, as $|\psi\rangle = \cos(\theta/2)|0\rangle + e^{i\phi}\sin(\theta/2)|1\rangle$, since the total probability of the two states is $|a|^2 + |b|^2 = |\cos(\theta/2)|^2 + |e^{i\phi}\sin(\theta/2)|^2 = 1$. Thus, any of the information stored in a qubit can be represented by the points in terms of the (θ, ϕ)-coordinates on a sphere of unit radius (known as the Bloch sphere) as shown in Fig. 1.3(b). Note that the north pole (i.e. $\theta \equiv 0^\circ$) of the Bloch sphere represents $|0\rangle$-state while its south pole (i.e. $\theta \equiv 180^\circ$) indicates the $|1\rangle$-state. All other (θ, ϕ)-coordinates represent some kind of a superposed state of the qubit. Evidently, if measured once, the qubit state will be either $|0\rangle$ or $|1\rangle$ (just like the cat to be *dead* or *alive* after the removal of the ventilator); however, if measured repeatedly but identically for a significantly large number (N) of times, one would get $|0\rangle$-state for $\approx N\cos^2\theta$ times while $|1\rangle$-state for $\approx N\sin^2\theta$ times (known as the Born rule). It may be noted that the phase difference ϕ between $|0\rangle$ and $|1\rangle$ cannot be estimated from single-qubit measurements and may require multi-qubit operations. Nevertheless, the process of such large numbers of identical single-qubit measurement runs is not an easy job though, since once measured, the qubit state collapses to either $|0\rangle$ or $|1\rangle$, thereby destroying all the remaining information about the superposed state. Thus, for the next run of measurement, we need to prepare its state, first to $|0\rangle$ (or $|1\rangle$); then to coherently manipulate it (in a *deterministic* and *reversible* way) to attain the previous superposed state identically by relevant logic-gate operation(s); and subsequently, to again measure its state. Hence, one qubit measurement run can be described by three consecutive steps, i.e. *preparation* (or *initialization*), *manipulation*, and *measurement* (or *read-out*). All such successive steps need to be repeated identically each time for all the measurement runs. It can be done by sequential measurements (including three steps) on the same qubit or simultaneous measurements on many replicas of a single qubit (for detailed study, see Peres [4] and Nielsen and Chuang 10]). Further, it is imperative to note that although the *manipulation* step of the qubit-states can be exactly described by linear Schrödinger evolution, the *initialization* and *read-out* cannot, since these are non-equilibrium processes in general. In this context, NEGF framework has been utilized to theoretically describe and simulate such phenomena [11].

Suppose now our previous cat (name her as Alice) is in some kind of a platonic love relationship, i.e. an existing bonding without any intimacy, with another cat (name him as Bob). While Alice got poisoned and is in a *coma* under ventilator, Bob, being even at a large distance from Alice, without being informed by any messenger sent by her, is also in a *dead-dead* or *alive-alive* situation! What would happen when the ventilator is removed from Alice's neck? If Alice is *dead*, Bob is *dead* instantly, whereas if Alice is *alive*, Bob will also be *alive*. We really don't know whether this is possible for our cats or any other Alice and Bob; however, this '*spooky*' (as termed by Einstein himself, who discovered it but was not happy with it [12]) phenomenon, i.e. quantum entanglement, is indeed possible for two quantum systems (such as two qubits A and B). Such an entangled state of two qubits can thus be represented mathematically as $|\psi\rangle = a|0\rangle_A \otimes |0\rangle_B + b|1\rangle_A \otimes |1\rangle_B$. Consequently, if a quantum *measurement* (i.e. *read-out*) is performed on A in such a localized manner that cannot anyhow influence B (which may be at a very large distance from A), the resultant state will be collapsed to either $|0\rangle_A \otimes |0\rangle_B$ or $|1\rangle_A \otimes |1\rangle_B$ with the outcome of *read-out* of A to be '0' or '1', respectively. Now, if such an outcome is '0', the resultant state would be $|0\rangle_A \otimes |0\rangle_B$; and in such a case, if a quantum *measurement* (i.e. *read-out*) is performed on B, the outcome can only be '0' (there is no possibility to get a '1' *read-out* value even of B since the state has already turned out to be $|0\rangle_A \otimes |0\rangle_B$). Similarly, if the outcome of *measurement* on A reads '1', the result of *measurement* on B will definitely be '1', although A and B may be several light-years apart. *Isn't it really spooky!*

Then, naturally the question emerges: can information be transported instantaneously and inherently to an indefinite non-local distance (i.e. *spacelike*, a distance that can be covered only with a speed more than that of light in free-space) through local operations (e.g. *measurements*)? At this point it is imperative to note that, our cat's state could be represented, instead of looking in terms of *dead* and *alive* (i.e. $|0\rangle$ and $|1\rangle$, which we call as basis states), in the basis of *coma* and *conscious* (i.e. $1/\sqrt{2}(|0\rangle + |1\rangle)$ and $1/\sqrt{2}(|0\rangle - |1\rangle)$), or similar many others depending on the '*preparation*' step. Now, once the state of A is measured, it collapses to one of the prepared basis states. However, which basis state of A has been prepared in for such a measurement can only be known to B if a classical messenger carries such information, obviously with a speed not greater than that of light in vacuum respecting the relativistic restriction, unless a cloning of any qubit-state would be made by B arbitrarily so as to get the precise value from a large number of measurement runs, as described previously. Unfortunately, it has been theoretically proved that such 'perfect cloning' is not allowed (the 'No-cloning theorem') [13]. Nevertheless, quantum entanglement can be utilized for efficient quantum computing, as well as teleportation and cryptography by using several correction mechanisms. For instance, researchers have shown that if entangled qubits can be utilized appropriately, quantum computing for several tasks such as prime number factorization using Shor's algorithm or searching with Grover's algorithm can outperform classical computation with exponentially enhanced computing efficiency. However, both quantum superposition and quantum entanglement are highly fragile due to decoherence effects [14], appearing from the interactions between quantum systems and their environment that destroy the coherence between superposed states or the one-to-one correlated states. Thus, to combat the effects of decoherence is one crucial job in developing practical quantum computers.

It has already been described that any two-state quantum system can be used as a qubit, which is the basic unit of quantum computers. Interestingly, depending on the availability of two such states in real systems, qubits can be classified into three categories. **1) Intrinsic two-state qubits:** those having naturally two quantum states that can describe any of their possible states by superposition (i.e. mathematically, the complete set of orthonormal basis states in Hilbert space). The examples are spin qubits developed on the platforms of semiconductor QDs, nuclear magnetic resonance molecules, phosphorus donors in Si, liquid helium, Paul trap etc.; charge qubits using double QDs; photon polarization qubits, etc.; **2) two-state subset qubits:** those which have more than two quantum states that form the complete basis set in Hilbert space and thus two of them needs to be selected for qubit, such as, trapped ions based fine-structure qubits, optical qubits, Zeeman qubits, and hyperfine qubits; trapped Rydberg atom qubits; nitrogen vacancy-centre qubits in Si platform; topological qubits, etc.; and **3) engineered two-state qubits:** where the circuits are engineered to select two quantum states out of many to form the computational subspace. This is exactly what is done by introducing a Josephson junction in a superconducting circuit to add an anharmonicity to a quantum harmonic oscillator. We will briefly describe some of these schemes of practical qubits in Chapter 8.

1.5 Quantum optoelectronic devices

The journey of light as quantum (i.e. photons) started in the field of physics merely through light-matter interaction, or more precisely through electron-photon interaction, due to the incredible theories proposed by Planck (black-body radiation) and Einstein (photoelectric effect; and the absorption, spontaneous emission, and stimulated emission of photons) at the beginning of twentieth century, thereby giving birth to the quantum epoch. However, it took two decades to invent the first photodetector, i.e. the photomultiplier tube, four decades to develop the modern semiconductor-junction based solar cell (although photovoltaic effect was experimentally observed by Becquerel in 1839) and phototransistor, and more than 50 years to practically realize the first light-emitting-diode and LASER (i.e. Light Amplification by Stimulated Emission of Radiation). Nevertheless, all such optoelectronic devices could have thrived after entering the quantum domain, especially in the last two decades, to offer efficacies (e.g. efficiency, resolution, limit-of-detection, etc.) beyond their classical limits.

Conceptually, a photodetector is an optoelectronic device that factually transforms photons of specific wavelengths into electron-hole pairs, which can offer to provide us with an objective measure of the colour of light. Further, if such photogenerated electron-hole pairs can be separated out in opposite directions along an in-built potential slope in the device so that it exhibits a current (i.e. photocurrent) without any applied voltage bias, then the device can act as a solar cell. Conversely, a light-emitting-diode is a device that can emit light of a colour (i.e. the wavelength that generally depends on the material bandgap) as a result of the radiative recombination of electron-hole pairs. On the other hand, LASERs are sources of coherent light with a precise value of a particular wavelength (i.e. almost monochromatic), which absorbs light to excite its ground-state atoms, followed by a reverse transition to ground-state from a highly populated metastable state, thereby resulting in a stimulated emission of light, being amplified through resonating loops. Thus, it is apparent that all such optoelectronic devices operate on the basis of either the conversion of photons into electron-hole pairs or *vice versa*.

The effectiveness of such devices has been enhanced (and efforts are still going on) by utilizing the discreteness of energy states of electrons and holes in quantum structures that promise to surpass the statistically averaged-out limit (usually called classical limits) of the outputs, and therefore identified as quantum optoelectronic devices [15]. However, we argue at this point that the basic mechanism of converting a photon into an electron-hole pair, or *vice versa*, is inherently beyond the limit of classical laws of physics. In order to explain this, for instance, let us think of the photon, which is obviously associated with an electric ($\vec{E}$) and a magnetic field ($\vec{B}$) described by Maxwell's equations, and constrained by the relativistic Lorentz invariance of the quantity $I_{\text{ph}} = c^2B^2 - E^2$ (c being the speed of light in vacuum). Owing to the property $E = vB = (c/\eta)B$ of an electromagnetic wave in a medium of refractive index $\eta > 1$, such a Lorentz invariant $I_{ph} = (\eta^2 - 1)E^2$ is apparently a positive quantity. Interestingly, when such a photon is converted into an electron-hole pair (initially as an exciton), they have an electric field originating from an electric dipole but almost zero magnetic field due to opposite electron/hole current loops. Thus, the Lorentz invariant

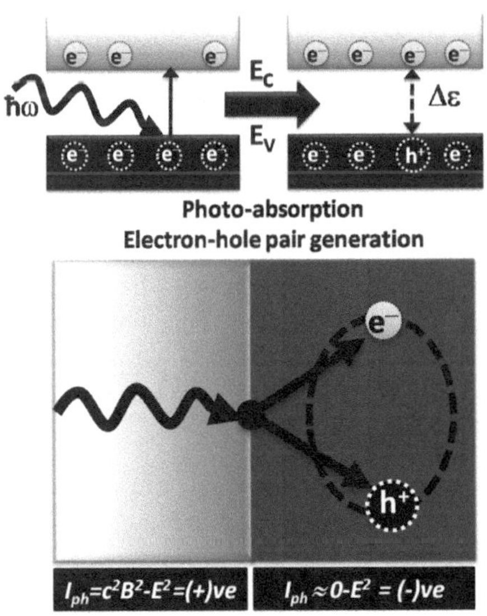

Fig. 1.4: The schematic representation of excitation of an electron from the valence band to the conduction band of a material by a photon being absorbed, thereby creating an electron-hole pair (top); the change in Lorentz invariant from (+)ve to (−)ve in the transition of EM-wave to charges (bottom).

quantity for the photogenerated electron-hole pair is now $I_{e-h} = -E^2$, which is evidently negative! (see Fig. 1.4).

Hence, we argue that (not only quantum entanglement but also) the process of photogeneration or radiative recombination violates classical relativistic invariance and thus cannot be described by any classical theory of physics, and we mandatorily need quantum theory for it. One can argue that such a process is not a mere change of inertial reference frame and thus Lorentz invariance may not be preserved. However, it is worth noting that no other external force or agent (other than the photon and the electrons in the material potential) is acting here that can change the reference frame from inertial to non-inertial. In fact, the speed of the process of any such photon↔electron conversion or transition cannot be explained by classical electromagnetic wave theory constrained by the relativistic limit.

Further, even the conservative Schrödinger (or Dirac equation) is not capable of modelling such phenomena since *C-numbers* (e.g. the numbers of electrons and photons here) change in such processes while *V-numbers* (e.g. energy or momentum) remain constant. Thus, a quantum field theoretic approach is necessary for this purpose. Moreover, such processes are inherently non-equilibrium in nature, although the initial and final particles may be in thermodynamic equilibrium, which indicates the NEGF framework to be a potential method for theoretically describing the basic mechanisms of optoelectronic devices, especially at the quantum scale.

1.6 Human brain and brain-inspired devices

It may be quite surprising for the readers of this book how their brain could be thought to allow such a quantum kind of state transition from a brief description of counterintuitive operability of quantum devices to understanding brain functionality, and the consequent imitation of such behaviour into new-generation devices! In justification, we would only like to highlight, firstly, one common feature of the quantum devices and human brain, as pointed out earlier in this chapter, to be the 'non-equilibrium processes' in their functionalities. The other most striking commonality between them must be the fact that both systems, principally and in principle, deal with some kind of 'information'.

1.6.1 Information: what do we mean by it?

The term 'information' used to be considered as much as a kind of an abstract concept until the revolutionary work by Shannon was published in 1948, which argued that information can be measured, as described earlier, in terms of *bits*. The second great contribution of the work was to identify information as a message that *is "one selected from a set of possible messages"* [5], thereby finding out the inherent probabilistic nature of any information. Thus, it can also be said that only if something belongs to a class or group (making itself one out of many similar ones), only then does it bear some information. It indicates that if we know every exact detail about some object that is completely unique, we actually have no information about it! For example, consider any measurable property, like position or momentum, of a particle; it can take on a value out of many possible values of similar nature and thus it bears some information; whereas, a universal constant contains really no information in itself; it becomes meaningful only in some interrelation of two or more properties amongst many and thus can carry some information in that regard only.

The third, and most intriguing, outcome of Shannon's work, which, we argue, was in fact upheld consequently by Jaynes, is the possible objectivity of information (which itself is inherently probabilistic and significantly depends on its receiver), in relation to the information-entropy owing to the 'maximum entropy principle' [16]. Conventionally, entropy was (and in fact still 'is', in a major school of thought) considered to be a measure of the randomness (or disorderedness) of a system. However, it is imperative to note that randomness is fundamentally subjective and depends on our knowledge and/or analyzing skill. For instance, the series 1, 2, 4, 8, 16, . . . is definitely not random for us who know at least the multiples of two, whereas the sequence 0, 1, 1, 2, 3, 5, 8, 13, . . . looks quite random to anyone until she/he finds a pattern in it. Then what would be the entropy values associated with two such series? For all of us, the entropy related to the former is 0 (i.e. the assigned minimum value) since it is not at all random, whereas that for the latter has a positive value (w.r.t. the assigned 0) for those who cannot find any pattern in it. However, the entropy for such latter set of values is 0 for those who have come to know that it is a sequence $t_n = t_{n-1} + t_{n-2}$ with $t_0 = 0$; $t_1 = 1$ (called the Fibonacci sequence). In fact, even if a single person who didn't know before and just got to know it after reading the above lines of this text, the entropy of the sequence, in general (that has no relation with the reader or this text even), suddenly changes from positive to zero. In that case, the entropy of a system is not at all its own property, but rather depends principally on the observer's knowledge and/or analyzing skill. However, the entropy of a system holistically governs the nature of its dynamics, especially the time-dynamics. Thus, such a subjective school of thought, in turn, leads us to Einstein's illusionary assumption that the arrow of time is an illusion. Jaynes showed, on the other hand, considering Shannon entropy, that the maximum entropy of a system provides us with the least biased information about the entropy of the system, so that the thermodynamic entropy is equal to Shannon entropy (note that Jaynes had proposed many other conclusions in his work, which we are not considering here or otherwise). We argue here, thus, that such least-biased information has indeed least to do anything with our subjective knowledge, and it carries the real property of a system that we can estimate most objectively.

1.6.2 Brain: an information processing system

The above point of view is more 'realistic' (in contrast to Einstein's definition of 'reality' [12]) for the quantum systems, where all the properties and dynamics of any physical object can be (to be more precise, 'must be') represented in terms of the associated information. In this context, one can corroborate von Neumann entropy with the maximum Shannon entropy for a quantum system. We argue, thus, information in this sense of least-biasness is an objective physical property of a quantum system, even if we consider its observational contextuality. On the other hand, for the human brain, or brain in general, information is the exclusive physical property that we deal with. Thus, the brain can be described (alongside its physical, physiological, and psychological aspects) fundamentally as an information processing system, the aspect that here in the two chapters of this book (Chapters 9 and 10), we will be briefly focusing on.

As mentioned earlier, information can be measured as *bits* and practically be realized, using MOSFETs in terms of electric current, or as qubits in terms of charge, spin, flux, or even photon numbers, and many others. Similarly, in the human brain, it is believed that the oscillatory propagating voltage pulses are used to realize any information (for details, see Augustine et al. [17]). In response to any external stimuli, in our brain, such electrical pulses are generated and transmitted through about 86 billion neurons that are interconnected to each other via nearly 10,000 synapses per neuron, which are packed in the volume of around only a thousand cc. It is apparent from such orders of values how complex the arrangement of the neural networks is, which are functioning inside the human brain for performing its entire information processing tasks. However, an even more fascinating fact is that all these are being done only at the expense of ~20 W of power consumption [18] (which is $\sim 10^6$ times energy efficient than the present-day supercomputers)! Such extraordinary energy-efficiency is attributed to something most special about a human brain as an information processing system, i.e. its learning ability through its memorizing capacity. It is believed that such memory-based learning by the human brain is carried out through its gigantic neural networks owing to the plasticity of neural memory.

1.6.3 Algorithmic imitation of the human brain: artificial intelligence

Algorithmically, such fundamental mechanisms of learning based information processing in human brains are translated into a mathematical framework, known as an artificial neural network, which is the basic founding stone of artificial intelligence (AI). Architecturally, the artificial neural networks imitate layered input-output structures of neural circuits of the human brain, as schematically depicted in Fig. 1.5, and optimize the weights of different inputs at each level through feedback from outputs [19], the method being commonly known as 'training'. Thereafter, it can identify anything belonging to its trained patterns and offer us extraordinary results. It is apparent that depending on the computing capacity of a practical device, AI can accommodate appropriate training with gigantic information from a large population of human beings and provide us humanly outcomes with a superman's efficiency.

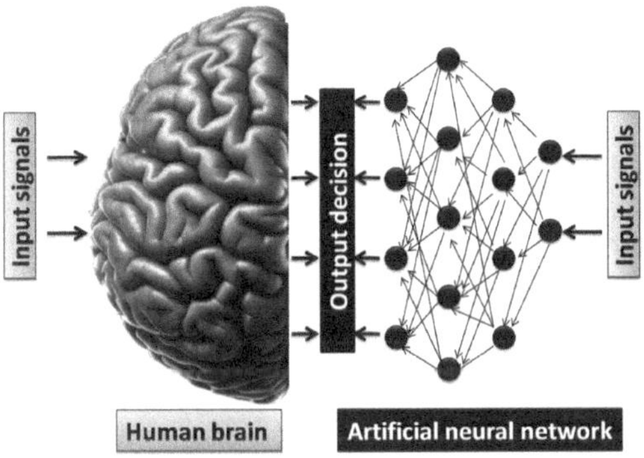

Fig. 1.5: A schematic illustration of architectural imitation of neural circuits of the human brain in the ANN framework, comprising multi-layer neural connectivity that can offer automated average human intelligence for several extraordinary works to be done.

However, it is imperative to note that, unlike what is often thought to be, the intelligence referred to in AI is merely the 'average' human intelligence, acquired only through training to memorize a pattern and act accordingly, and not even close to an 'innovative' intelligence like that of Einstein or Picasso. For instance, if many images of various cats are fed into the AI framework for training, and a new image of a cat (other than the training inputs) is asked to be identified, it can do that with excellent accuracy. It can also create the image of a new cat that is some kind of a superposition of all its training values. The method can be advanced extraordinarily to predict many scientific properties, even those that many good researchers may fail to do within a finite time. However, it cannot discover the theory of relativity that Einstein did in his first paper on it in 1905 without citing a single reference (i.e. no training from references). Or, it cannot paint a 'Guernica' (the great painting by Picasso) even if all the information about a war is fed into it. To give another virtual example, suppose we are training an AI system with all information from before the first life (i.e. prokaryotes) came on the earth (~3.8 billion years past). Can it predict there will be 'life' on the earth!

We believe, at least on the basis of the state-of-the-art concept of AI, this is impossible, since it is solely based on what is already there and thus cannot predict or create something entirely new, e.g. 'life' for the first time. However, at the same time, we believe that the human brain has a capacity to propose something really new. This optimism has been upheld by Penrose through a mathematical logic based on Gödel's Incompleteness Theorem that offers the ability to 'transcending' to out-of-the-limit of an existing framework, at least in principle [20]. Can we conceptualize and distinguish such 'average' and 'innovative' intelligence in terms of *V-numbers* and *C-numbers*, respectively? Nevertheless, the state-of-the-art AI models based on the 'average' human intelligence, without such 'transcending' or 'innovative' capacity, are already offering us outcomes that were unbelievable even a few years back.

1.6.4 Mimicking the human brain onto device platforms: brain-inspired computers

The algorithmic developments of such AI models that are not only continuing till date but also speeding up every day, however, are suffering challenges from the burden of high computational cost, especially in terms of energy requirement. Getting motivation from the fact that human brain can compute the same tasks as a supercomputer at the expense of its one millionth energy, the state-of-the-art progress of AI necessitates to look for alternative computing devices and systems: in contrast to the conventional ones known as the von Neumann computers, such devices have to operate analogous to human brain in order to perform the huge computing tasks with minimum energy requirement [18]. The fundamental difference of such systems from the von Neumann computers must be in unifying their processor and storage, just as in human brains. Therefore, we are now tasked with the goal of mimicking the neural networking topology along with the neuronal/synaptic functionality of the human brain onto electronic or some other kind of device platforms, called brain-inspired computers.

References

[1] Valdes-Sosa, P. A., Galan-Garcia, L., Bosch-Bayard, J., Bringas-Vega, M. L., Aubert-Vazquez, E., Rodriguez-Gil, I., Das, S., et al. "The cuban human brain mapping project, a young and middle age population-based EEG, MRI, and cognition dataset." *Scientific Data* 8, no. 1 (2021): 45.

[2] Amunts, K., Ebell, C., Muller, J., Telefont, M., Knoll, A., and Lippert, T. "The human brain project: Creating a European research infrastructure to decode the human brain." *Neuron* 92, no. 3 (2016): 574–581.

[3] Schrödinger, E. "The present status of quantum mechanics." *Die Naturwissenschaften* 23, no. 48 (1935): 1–26.

[4] Peres, A., ed. *Quantum Theory: Concepts and Methods*. Dordrecht: Springer Netherlands, 2002.

[5] Shannon, C. E. "A mathematical theory of communication." *The Bell System Technical Journal* 27, no. 3 (1948): 379–423.

[6] Bardeen, J. and Brattain, W. H. "The transistor, a semi-conductor triode." *Physical Review* 74, no. 2 (1948): 230.

[7] Ng, K. K. and Simon, M. S. *Physics of Semiconductor Devices*. Hoboken, NJ: Wiley-Interscience, 2007.

[8] Moore, G. E. *Cramming More Components Onto Integrated Circuits*. 1965.

[9] Bohr, M. "A 30 year retrospective on Dennard's MOSFET scaling paper." *IEEE Solid-State Circuits Society Newsletter* 12, no. 1 (2009): 11–13.

[10] Nielsen, M. A. and Isaac, L. C *Quantum computation and quantum information*. Cambridge University Press, 2010.

[11] Nag Chowdhury, B. and Chattopadhyay, S. “Dual-Gate GaAs-Nanowire FET for room temperature charge-qubit operation: A NEGF approach.” *Advanced Quantum Technologies* 6, no. 4 (2023): 2200072.

[12] Einstein, A., Podolsky, B., and Rosen, N. “Can quantum-mechanical description of physical reality be considered complete?.” *Physical Review* 47, no. 10 (1935): 777.

[13] Wootters, W. K. and Zurek, W. H. “A single quantum cannot be cloned.” *Nature* 299, no. 5886 (1982): 802–803.

[14] Zurek, W. H. “Decoherence, einselection, and the quantum origins of the classical.” *Reviews of Modern Physics* 75, no. 3 (2003): 715.

[15] Jiang, W., Chen, S., Seeds, A., and Liu, H. “Quantum dot optoelectronic devices: Lasers, photodetectors and solar cells.” *Journal of Physics D: Applied Physics* 48, no. 36 (2015): 363001.

[16] Jaynes, E. T. “Information theory and statistical mechanics.” *Physical Review* 106, no. 4 (1957): 620.

[17] Augustine, G. J., David, F., William, H., Anthony-Samuel, L., Dale, P., and Léonard, W. *Neurosciences*. (ed.) Purves, D. and et al. (3rd ed.), ISBN 0-87893-725-0. De Boeck Superleur, (2015).

[18] Marković, D., Mizrahi, A., Querlioz, D., and Grollier, J. “Physics for neuromorphic computing.” *Nature Reviews Physics* 2, no. 9 (2020): 499–510.

[19] Hinton, G. “Nobel lecture: Boltzmann machines.” *Reviews of Modern Physics* 97, no. 3 (2025): 030502.

[20] Penrose, R. “The emperor’s new mind.” *RSA Journal* 139, no. 5420 (1991): 506–514.

2 Fundamental concepts of non-equilibrium Green's function (NEGF)

Although Green's function was explored by the British mathematical physicist George Green in 1828 (astonishingly before his formal university education), the emergence of non-equilibrium Green's function (NEGF) framework for modelling and simulation of quantum scale devices as well as molecular electronics has commenced precisely in the past two and a half decades only after the seminal paper by Datta in 2000 [1]. However, the concept of NEGF was developed in the 1960s due to the pioneering works by Martin and Schwinger [2], Konstantinov and Perel [3], Kadanoff and Baym [4], and finally through the most elegant formalism by Keldysh [5]. In the present chapter, we will try to briefly present the concept of such mathematical formalism and the physical perspectives of such concepts. This may require the readers to have a basic knowledge of the standard formulation of quantum mechanics, relativistic quantum mechanics, and quantum field theory.

2.1 The basic mathematical concept of Green's function

Not going into the details of abstract mathematics, here, we will briefly explain what fundamentally the Green's function is and how it works. Consider an equation given by

$$\hat{O}(x)f(x) = s(x) \tag{2.1}$$

where $\hat{O}(x)$ is an operator, that itself is a function of the variable x, and operates on the function $f(x)$ giving rise to the function $s(x)$. Now consider another specific function $G(x,\bar{x})$ (which, in general, is dependent on $\hat{O}(x)$), so that:

$$\hat{O}(x)G(x,\bar{x}) = \delta(x-\bar{x}) \tag{2.2}$$

where $\delta(x-\bar{x})$ is known to be the Dirac delta function, a continuous function representation of the discrete Kronecker Delta (δ_{ij}). Thus, one can write:

$$\begin{aligned}
&\hat{O}(x)G(x,\bar{x})s(\bar{x}) = \delta(x-\bar{x})s(\bar{x}) \\
&\Rightarrow \int_{\bar{x}} \hat{O}(x)G(x,\bar{x})s(\bar{x})d\bar{x} = s(x) \\
&\Rightarrow \hat{O}(x)\left[\int_{\bar{x}} G(x,\bar{x})s(\bar{x})d\bar{x}\right] = s(x)
\end{aligned} \tag{2.3}$$

 | https://doi.org/10.1515/9783111545950-002

Comparing eqs. (2.1) and (2.3), the function $f(x)$ is obtained to have a solution:

$$f(x) = \int_{\bar{x}} G(x, \bar{x}) s(\bar{x}) d\bar{x} \tag{2.4}$$

Thus, Green's function, mathematically, is an integral kernel that provides a generic way to find the solutions of all the inhomogeneous equations which can be expressed as eq. (2.1). It may be interesting to note that the operator of such an equation does not necessarily need to be a linear or differential operator but may be more general in its nature. Furthermore, this entire operation can be carried out numerically using the matrix method, which may be useful for developing a simulator.

2.2 Green's function: the physical insight

In order to elaborate on Green's function from the point of view of physics, let us start with an example that is quite familiar to us. Consider the electrostatic potential ($\varphi(\vec{r})$) at a point ($\vec{r}$) in free space (permittivity being ε_0), which can be described by the famous Poisson's equation,

$$\nabla^2 \varphi(\vec{r}) = -\rho(\vec{r})/\varepsilon_0 \tag{2.5}$$

where $\rho(\vec{r})$ is the charge density at that particular point ($\vec{r}$). Thus, Poisson's equation shows the relation between the potential at a point (say, the *field point*) and the charge density at exactly that point (i.e. the same *field point*). However, if we are interested in finding the potential at a point, i.e. the *field point* at $\vec{r}$, due to a charge at some other point, say, the *source point* at $\vec{r}'$, we can use Green's function as an integral kernel technique as described in Section 2.1. For instance, if we consider $\hat{O} \to \nabla^2$; $f(x) \to \varphi(\vec{r})$; $s(x) \to -\rho(\vec{r})/\varepsilon_0$, then using eq. (2.4) it is obtained to be

$$\varphi(\vec{r}) = \int_{\vec{r}'} G(\vec{r}, \vec{r}') \left[-\rho(\vec{r}')/\varepsilon_0\right] d^3\vec{r}' \tag{2.6}$$

where $\nabla^2 G(\vec{r}, \vec{r}') = \delta^3(\vec{r} - \vec{r}')$ according to eq. (2.2) and $d^3\vec{r}'$ is the elemental volume in coordinate space of $\vec{r}'$. Now, differential geometry tells us that $\nabla^2 (1/|\vec{r} - \vec{r}'|) = -4\pi\, \delta^3(\vec{r} - \vec{r}')$, which, applying on eq. (2.6) leads to

$$\varphi(\vec{r}) = \int_{\vec{r}'} \lfloor 1/4\pi\, |\vec{r} - \vec{r}'| \rfloor \lfloor -\rho(\vec{r}')/\varepsilon_0 \rfloor d^3\vec{r}'$$

$$\Rightarrow \varphi(\vec{r}) = \frac{1}{4\pi\,\varepsilon_0} \int_{\vec{r}'} \frac{\rho(\vec{r}')}{|\vec{r} - \vec{r}'|} d^3\vec{r}' \tag{2.7}$$

For a point charge located at $(\vec{r}'=\vec{r}_0)$, the charge density is distributed as $\rho(\vec{r}')=q\delta^3(\vec{r}'-\vec{r}_0)$, which, since eq. (2.7), gives rise to

$$\varphi(\vec{r})=\frac{1}{4\pi\varepsilon_0}\frac{q}{|\vec{r}-\vec{r}_0|} \tag{2.8}$$

Equation (2.8) (and also eq. (2.7) in general) is the famous Coulomb's law, which indeed tells us about the electrostatic potential at a *field point* $(\vec{r})$ due to a charge at a distant *source point* $(\vec{r}_0)$, with a source-field distance of $|\vec{r}-\vec{r}_0|$.

Now, if we look closely into eq. (2.6), it can be considered that as if the Green's function, in fact, represents the propagation of information on the existence of a charge density at the *source point* $(\vec{r}_0)$ to the *field point* $(\vec{r})$ to create a corresponding potential out there. It would be more accurate if we write:

$$\varphi(\vec{r},t)=\int_{\vec{r}',t'} G(\vec{r},t;\vec{r}',t')\left[-\rho(\vec{r}',t')/\varepsilon_0\right]d^3\vec{r}'dt' \tag{2.9}$$

that tells us about the potential at a *field point* $(\vec{r})$ at time t due to a charge density placed at the *source point* $(\vec{r}_0)$ at any earlier time t'. Thus, Green's function is nothing but the *propagator* of information between the *source* and *field* in this case, as shown in Fig. 2.1, which can also be considered as a generic concept. It may be worthy to note that in the particular case of electrodynamics, the speed of information propagation is the speed of light (c), i.e. $|\vec{r}-\vec{r}'|=c(t-t')$. Such a concept was first consistently developed by Feynman in two consecutive papers published in 1949 [6, 7].

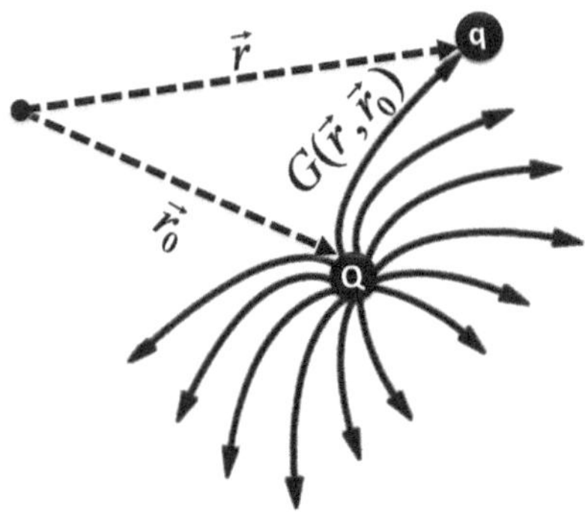

Fig. 2.1: A schematic depiction of the electric field applied on a charge at a *field point* due to another charge at a *source point* represented by Green's function as the field propagator.

2.3 Feynman diagram and the Green's function as propagator

2.3.1 Basic foundation

Suppose $\hat{H}$ is a constant (in time) Hamiltonian operator of a single particle with energy eigenvalues E_n with the corresponding eigenstates $|\varphi_n\rangle$, given by

$$\hat{H}|\varphi_n\rangle = E_n|\varphi_n\rangle \tag{2.10}$$

Then the position wave function of the particle at state $|\alpha(t)\rangle$ (α may be any physical quantity, i.e. observable, such as momentum, spin, etc.) can be written as

$$\psi_\alpha(\vec{r}, t) = \langle\vec{r}\,|\alpha(t)\rangle \tag{2.11}$$

Now the Schrödinger equation

$$[i\hbar\partial/\partial t - H]\psi(\vec{r}, t) = 0 \tag{2.12}$$

leads to the evolution of such a wave function in time as

$$\psi(\vec{r}, t') = e^{-\frac{i}{\hbar}(t' - t)\hat{H}}\psi(\vec{r}, t) \tag{2.13}$$

On the other hand, the superposition principle can be utilized to write:

$$|\alpha(t)\rangle = \sum_j c_j(t)|\varphi_j\rangle \tag{2.14}$$

where $c_j(t) = \langle\varphi_j \mid \alpha(t)\rangle$ are the coefficients of superposition that represent the corresponding probability amplitudes of eigenstates $|\varphi_j\rangle$ when the particle is in state $|\alpha(t)\rangle$. Then, since eq. (2.13), the wave function at a position $\vec{r}_2$ at some later time t_2 (with respect to time $t_1 \leq t_2$) can be expressed as

$$\begin{aligned}
\psi_\alpha(\vec{r}_2, t_2) &= e^{-\frac{i}{\hbar}(t_2-t_1)\hat{H}}\psi_\alpha(\vec{r}_2, t_1) \\
&= e^{-\frac{i}{\hbar}(t_2-t_1)\hat{H}}\langle\vec{r}_2\,|\alpha(t_1)\rangle && \text{[using Eq. (2.11)]} \\
&= e^{-\frac{i}{\hbar}(t_2-t_1)\hat{H}}\sum_j c_j(t_1)\left\langle\vec{r}_2\,\middle|\varphi_j\right\rangle && \text{[using eq. (2.14)]} \\
&= \sum_j c_j(t_1)\left\langle\vec{r}_2\,\middle|\varphi_j\right\rangle e^{-\frac{i}{\hbar}(t_2-t_1)E_j} && \text{[using eq. (2.10)]} \\
&= \sum_j \left\langle\varphi_j\,\middle|\alpha(t_1)\right\rangle\left\langle\vec{r}_2\,\middle|\varphi_j\right\rangle e^{-\frac{i}{\hbar}(t_2-t_1)E_j} && [\text{since } c_j(t) = \langle\varphi_j \mid \alpha(t)\rangle]
\end{aligned}$$

$$= \int_{\vec{r}_1} \sum_j \left\langle \varphi_j \middle| [|\vec{r}_1\rangle d^3\vec{r}_1 \langle \vec{r}_1|] |\alpha(t_1)\rangle \left\langle \vec{r}_2 \middle| \varphi_j \right\rangle e^{-\frac{i}{\hbar}(t_2-t_1)E_j} \qquad \left[\text{since} \int_{\vec{r}_1} |\vec{r}_1\rangle d^3\vec{r}_1 \langle \vec{r}_1| = 1\right]$$

$$= \int_{\vec{r}_1} d^3\vec{r}_1 \sum_j \left\langle \varphi_j \middle| \vec{r}_1 \right\rangle \left[\langle \vec{r}_1 | \alpha(t_1)\rangle\right] \left\langle \vec{r}_2 \middle| \varphi_j \right\rangle e^{-\frac{i}{\hbar}(t_2-t_1)E_j}$$

$$= \int_{\vec{r}_1} d^3\vec{r}_1 \sum_j \varphi_j{}^*(\vec{r}_1)[\psi_\alpha(\vec{r}_1, t_1)]\varphi_j(\vec{r}_2) e^{-\frac{i}{\hbar}(t_2-t_1)E_j} \qquad \text{[using eq. (2.11)]}$$

$$\Rightarrow \psi_\alpha(\vec{r}_2, t_2) = \int_{\vec{r}_1} \left[\sum_j \varphi_j(\vec{r}_2)\varphi_j{}^*(\vec{r}_1) e^{-\frac{i}{\hbar}(t_2-t_1)E_j}\right] \psi_\alpha(\vec{r}_1, t_1) d^3\vec{r}_1 \tag{2.15}$$

It is apparent that eq. (2.15) may be considered as a general expression of information propagation from $(\vec{r}_1, t_1)$ to $(\vec{r}_2, t_2)$, provided $t_2 \geq t_1$, if eq. (2.15) is generalized to

$$\psi_\alpha(\vec{r}_2, t_2) = \int_{\vec{r}_1} \left[\sum_j \varphi_j(\vec{r}_2)\varphi_j{}^*(\vec{r}_1)\vartheta(t_2-t_1) e^{-\frac{i}{\hbar}(t_2-t_1)E_j}\right] \psi_\alpha(\vec{r}_1, t_1) d^3\vec{r}_1 \tag{2.16}$$

where $\vartheta(t_2-t_1)$ is the Heaviside step function (i.e. 0 for $t_2 < t_1$, and 1 otherwise). It may be interesting to note that the term:

$$\sum_j \varphi_j(\vec{r}_2)\varphi_j{}^*(\vec{r}_1) = \sum_j \left\langle \vec{r}_2 \middle| \varphi_j \right\rangle \left\langle \varphi_j \middle| \vec{r}_1 \right\rangle$$

$$= \langle \vec{r}_2| \left[\sum_j \left|\varphi_j\right\rangle \left\langle \varphi_j\right|\right] |\vec{r}_1\rangle$$

$$= \langle \vec{r}_2 | \vec{r}_1\rangle \qquad \left[\text{since} \sum_j \left|\varphi_j\right\rangle \left\langle \varphi_j\right| = 1\right]$$

$$= \delta^3(\vec{r}_2 - \vec{r}_1)$$

$$\Rightarrow \sum_j \varphi_j(\vec{r}_2)\varphi_j{}^*(\vec{r}_1) = \delta^3(\vec{r}_2 - \vec{r}_1) \tag{2.17}$$

and thus,

$$[i\hbar\partial/\partial t_2 - \hat{H}_2]\left(\sum_j \varphi_j(\vec{r}_2)\varphi_j{}^*(\vec{r}_1)\vartheta(t_2-t_1) e^{-\frac{i}{\hbar}(t_2-t_1)E_j}\right) = i\hbar\delta(t_2-t_1)\delta^3(\vec{r}_2 - \vec{r}_1) \tag{2.18}$$

It is imperative to note that eq. (2.17) cannot be used in eq. (2.16), since $\sum_j \left|\varphi_j\right\rangle\left\langle\varphi_j\right| = 1$ but $\sum_j e^{-\frac{i}{\hbar}(t_2-t_1)E_j}\left|\varphi_j\right\rangle\left\langle\varphi_j\right| \neq 1$, in general, however, $\sum_j \left|\varphi_j\right\rangle\left\langle\varphi_j\right| = 1$ could be used to derive eq. (2.18) since $\delta(t_2-t_1)e^{-\frac{i}{\hbar}(t_2-t_1)E_j} = \delta(t_2-t_1)$ [note that, $\delta(t_2-t_1)=0$ for $t_2 \neq t_1$, while $e^{-\frac{i}{\hbar}(t_2-t_1)E_j} = 1$ for $t_2 = t_1$].

Now, the term $\left(\sum_j \varphi_j(\vec{r}_2)\varphi_j{*}(\vec{r}_1)\vartheta(t_2-t_1)e^{-\frac{i}{\hbar}(t_2-t_1)E_j}\right)$ can be considered as the Green's function $G(\vec{r}_2, t_2; \vec{r}_1, t_1)$, which describes (through eq. (2.16)) the evolution of the quantum state of a particle from $(\vec{r}_1, t_1)$ to $(\vec{r}_2, t_2)$, interestingly, as generally as the Hamiltonian does:

$$\psi_a(\vec{r}_2, t_2) = \int_{\vec{r}_1} G(\vec{r}_2, t_2; \vec{r}_1, t_1)\psi_a(\vec{r}_1, t_1)d^3\vec{r}_1 \tag{2.19}$$

Further, the total transition amplitude to find such a particle in some state at time t_2, where at time t_1 it might be in another state, can be computed, in general, using:

$$T(t_2, t_1) = \int_{\vec{r}_2} \psi_a{*}(\vec{r}_2, t_2)\psi_a(\vec{r}_2, t_2)d^3\vec{r}_2 = \int_{\vec{r}_2}\int_{\vec{r}_1} \psi_a{*}(\vec{r}_2, t_2)G(\vec{r}_2, t_2; \vec{r}_1, t_1)\psi_a(\vec{r}_1, t_1)d^3\vec{r}_1 d^3\vec{r}_2 \tag{2.20}$$

Equations (2.19) and (2.20) can be expanded further for the theoretical modelling of the scattering phenomena as discussed in the subsequent section.

2.3.2 How to model scattering events

Consider that a particle, moving freely, is suddenly scattered by the potential $\hat{V}_{Sc}$ at time t_{Sc}, acting on the particle for a small duration of time Δt_{Sc}. If the particle was initially at $(\vec{r}_1, t_1)$ with the free Hamiltonian $\hat{H}_1$, suffered the scattering at $(\vec{r}_{Sc}, t_{Sc})$ for a duration Δt_{Sc} when the Hamiltonian was $\hat{H} = \hat{H}_1 + \hat{V}_{Sc}$, and then reaches finally to $(\vec{r}_2, t_2)$ due to the post-scattering free Hamiltonian $\hat{H}_2$, then the consecutive transitions of its wave function can be symbolically shown as

$$\psi_{H_1}(\vec{r}_1, t_1) \xrightarrow{H_1} \psi_{H_1}(\vec{r}_{Sc}, t_{Sc}) \xrightarrow{\hat{H} = \hat{H}_1 + \hat{V}_{Sc}} \psi_{Sc}(\vec{r}_{Sc}, t_{Sc} + \Delta t_{Sc}) \xrightarrow{\hat{H}_2} \psi_{H_2}(\vec{r}_2, t_2)$$

Thus, since eq. (2.19), first, the free motion prior to scattering can be described by

$$\psi_{H_1}(\vec{r}_{Sc}, t_{Sc}) = \int_{\vec{r}_1} G_{H_1}(\vec{r}_{Sc}, t_{Sc}; \vec{r}_1, t_1)\psi_{H_1}(\vec{r}_1, t_1)d^3\vec{r}_1 \tag{2.21}$$

Then, the evolution during scattering can be obtained using eq. (2.13) as

$$\psi_{\mathrm{Sc}}(\vec{r}_{\mathrm{Sc}}, t_{\mathrm{Sc}} + \Delta t_{\mathrm{Sc}}) = e^{-\frac{i}{\hbar}\Delta t_{\mathrm{Sc}}\hat{H}}\psi_{H_1}(\vec{r}_{\mathrm{Sc}}, t_{\mathrm{Sc}})$$

$$\approx \left[1 - \frac{i}{\hbar}\Delta t_{\mathrm{Sc}}\hat{H}\right]\psi_{H_1}(\vec{r}_{\mathrm{Sc}}, t_{\mathrm{Sc}}) \qquad \text{[since first-order approximation for short scattering duration } \Delta t_{Sc}\text{]}$$

$$= \left[1 - \frac{i}{\hbar}\Delta t_{\mathrm{Sc}}\hat{H}_1\right]\psi_{H_1}(\vec{r}_{\mathrm{Sc}}, t_{\mathrm{Sc}}) + \left[-\frac{i}{\hbar}\Delta t_{\mathrm{Sc}}\hat{V}_{\mathrm{Sc}}\right]\psi_{H_1}(\vec{r}_{\mathrm{Sc}}, t_{\mathrm{Sc}})$$

$$\Rightarrow \psi_{\mathrm{Sc}}(\vec{r}_{\mathrm{Sc}}, t_{\mathrm{Sc}} + \Delta t_{\mathrm{Sc}}) \approx \psi_{H_1}(\vec{r}_{\mathrm{Sc}}, t_{\mathrm{Sc}} + \Delta t_{\mathrm{Sc}}) - (i/\hbar)\hat{V}_{\mathrm{Sc}}\psi_{H_1}(\vec{r}_{\mathrm{Sc}}, t_{\mathrm{Sc}})\Delta t_{\mathrm{Sc}} \tag{2.22}$$

Again applying the relation of eq. (2.19) into eq. (2.22) to get the wave function after scattering, it can be obtained:

$$\psi_{H_2}(\vec{r}_2, t_2) = \int_{\vec{r}_{\mathrm{Sc}}} G_{H_2}(\vec{r}_2, t_2; \vec{r}_{\mathrm{Sc}}, t_{\mathrm{Sc}} + \Delta t_{\mathrm{Sc}})\psi_{\mathrm{Sc}}(\vec{r}_{\mathrm{Sc}}, t_{\mathrm{Sc}} + \Delta t_{\mathrm{Sc}})d^3\vec{r}_{\mathrm{Sc}}$$

$$= \int_{\vec{r}_{\mathrm{Sc}}} G_{H_2}(\vec{r}_2, t_2; \vec{r}_{\mathrm{Sc}}, t_{\mathrm{Sc}} + \Delta t_{\mathrm{Sc}})[\psi_{H_1}(\vec{r}_{\mathrm{Sc}}, t_{\mathrm{Sc}} + \Delta t_{\mathrm{Sc}}) - (i/\hbar)\hat{V}_{\mathrm{Sc}}\psi_{H_1}(\vec{r}_{\mathrm{Sc}}, t_{\mathrm{Sc}})\Delta t_{\mathrm{Sc}}]d^3\vec{r}_{\mathrm{Sc}}$$

[since eq. (2.22)]

$$= \int_{\vec{r}_{\mathrm{Sc}}} G_{H_2}(\vec{r}_2, t_2; \vec{r}_{\mathrm{Sc}}, t_{\mathrm{Sc}} + \Delta t_{\mathrm{Sc}})[\int_{\vec{r}_1} G_{H_1}(\vec{r}_{\mathrm{Sc}}, t_{\mathrm{Sc}} + \Delta t_{\mathrm{Sc}}; \vec{r}_1, t_1)\psi_{H_1}(\vec{r}_1, t_1)d^3\vec{r}_1]d^3\vec{r}_{\mathrm{Sc}}$$

$$- (i/\hbar)\int_{\vec{r}_{Sc}} G_{H_2}(\vec{r}_2, t_2; \vec{r}_{Sc}, t_{Sc} + \Delta t_{Sc})[\hat{V}_{Sc}\left(\int_{\vec{r}_1} G_{H_1}(\vec{r}_{Sc}, t_{Sc}; \vec{r}_1, t_1)\psi_{H_1}(\vec{r}_1, t_1)d^3\vec{r}_1\right)\Delta t_{Sc}]d^3\vec{r}_{Sc}$$

[since eq. (2.21)]

$$\Rightarrow \psi_{H_2}(\vec{r}_2, t_2) = \int_{\vec{r}_{\mathrm{Sc}}}\int_{\vec{r}_1} G_{H_2}(\vec{r}_2, t_2; \vec{r}_{\mathrm{Sc}}, t_{\mathrm{Sc}} + \Delta t_{\mathrm{Sc}})G_{H_1}(\vec{r}_{\mathrm{Sc}}, t_{\mathrm{Sc}} + \Delta t_{\mathrm{Sc}}; \vec{r}_1, t_1)\psi_{H_1}(\vec{r}_1, t_1)d^3\vec{r}_1 d^3\vec{r}_{\mathrm{Sc}}$$
$$- (i/\hbar)\int_{\vec{r}_{\mathrm{Sc}}}\int_{\vec{r}_1} [G_{H_2}(\vec{r}_2, t_2; \vec{r}_{\mathrm{Sc}}, t_{\mathrm{Sc}} + \Delta t_{\mathrm{Sc}})\hat{V}_{\mathrm{Sc}}G_{H_1}(\vec{r}_{\mathrm{Sc}}, t_{\mathrm{Sc}}; \vec{r}_1, t_1)]\psi_{H_1}(\vec{r}_1, t_1)d^3\vec{r}_1 d^3\vec{r}_{\mathrm{Sc}}\Delta t_{\mathrm{Sc}} \tag{2.23}$$

The first term in the RHS of eq. (2.23) represents the evolution of the wave function from $(\vec{r}_1, t_1)$ to $(\vec{r}_2, t_2)$ through $(\vec{r}_{\mathrm{Sc}}, t_{\mathrm{Sc}})$, if there were no scattering event that took place at $(\vec{r}_{\mathrm{Sc}}, t_{\mathrm{Sc}})$. The second term, on the other hand, is attributed to the scattering phenomenon. Thus, the change in the final wave function due to such scattering is given by

$$\Delta\psi_{H_2}(\vec{r}_2, t_2) = -(i/\hbar) \int\limits_{\vec{r}_{Sc}} \int\limits_{\vec{r}_1} [G_{H_2}(\vec{r}_2, t_2; \vec{r}_{Sc}, t_{Sc} + \Delta t_{Sc})\hat{V}_{Sc} G_{H_1}(\vec{r}_{Sc}, t_{Sc}; \vec{r}_1, t_1)] \psi_{H_1}(\vec{r}_1, t_1) d^3\vec{r}_1 d^3\vec{r}_{Sc} \Delta t_{Sc} \tag{2.24}$$

If the scattering potential is assumed to exist for a long duration, it is required to integrate eq. (2.24) over time t_{Sc}, resulting in the scattering Green's function to be

$$G_{Sc}(\vec{r}_2, t_2; \vec{r}_1, t_1) = (-i/\hbar) \int\limits_{\vec{r}_{Sc}, t_{Sc}} G_{H_2}(\vec{r}_2, t_2; \vec{r}_{Sc}, t_{Sc}) \hat{V}_{Sc} G_{H_1}(\vec{r}_{Sc}, t_{Sc}; \vec{r}_1, t_1) d^3\vec{r}_{Sc} dt_{Sc} \tag{2.25}$$

Now, considering the generalized Green's function to be of 0th order for no-scattering, of 1st order for one-scattering, and so on, and using (j) in place of $(\vec{r}_j, t_j)$, and $d\tau_j$ as the elementary volume in 4D-space-time instead of $d^3\vec{r}_j dt_j$, we can rewrite eq. (2.25) for one scattering event between (1) and (2) as

$$G^{(1)}(2;1) = (-i/\hbar) \int\limits_{\mathrm{Sc}(1)} G^{(0)}(2;\mathrm{Sc}(1)) \hat{V}_{\mathrm{Sc}(1)} G^{(0)}(\mathrm{Sc}(1);1) d\tau_{\mathrm{Sc}(1)} \tag{2.26}$$

The result of eq. (2.26) can further be extended to obtain the Green's function for two successive scattering events while traversing from (1) to (2), as

$$G^{(2)}(2;1) = (-i/\hbar)^2 \int\limits_{\mathrm{Sc}(2)} \int\limits_{\mathrm{Sc}(1)} G^{(0)}(2;\mathrm{Sc}(2)) \hat{V}_{\mathrm{Sc}(2)} G^{(0)}(\mathrm{Sc}(2);\mathrm{Sc}(1)) \hat{V}_{\mathrm{Sc}(1)} G^{(0)}(\mathrm{Sc}(1);1) d\tau_{\mathrm{Sc}(1)} d\tau_{\mathrm{Sc}(2)} \tag{2.27}$$

which can, in a similar manner, be generalized to any arbitrary order of scattering phenomena. Therefore, the total Green's function can be expressed as the sum of all such orders of scattering:

$$G(2;1) = G^{(0)}(2;1) + G^{(1)}(2;1) + G^{(2)}(2;1) + \ldots \tag{2.28}$$

2.3.3 Alternative positron theory

The theory of positrons, and of the antiparticles in general, was first consistently described by the Dirac equation (i.e. the relativistic quantum equation of fermions), which exhibits equal numbers of possible solutions with both positive and negative mathematical values of energy [8]. For a relativistic particle, it implies that the particle must have possible values of kinetic energy to be greater than $+m_0c^2$ (i.e. the rest mass energy), and simultaneously, to be less than $-m_0c^2$, which, astonishingly, is allowed in the quantum case that has no classical analogue. It was shown in Dirac's theory that each of the negative kinetic energy solutions of such a quantum state corresponds to the respective positive kinetic energy solutions of another quantum state that is mathematically complex conjugate of the former one, if rest mass remains the same (i.e. $m_0{}^* = m_0$) and the charge turns out to be opposite (i.e. $q^* = -q$). Such particles with identical mass and opposite charge have been described as the antiparticles (e.g. positrons) of the former particles (e.g. electrons).

The phenomenon was explained by Dirac as all the negative energy states to be occupied by electrons and all the positive energy states to be unoccupied, together resulting in the 'vacuum'. Thus, any unoccupied negative energy state of an electron appears as a positron with a positive energy; and one needs to add an electron to fill up such unoccupied negative energy state, thereby equivalently annihilating the electron-positron pair to create the vacuum, releasing a photon of corresponding energy. Therefore, the positron (with a positive energy), i.e. the unoccupied negative energy state of an electron, is also termed as 'Dirac hole'. Converse to this annihilation process, an electron-positron pair can also be created from a vacuum through the annihilation of a photon. Interestingly, this very concept of high-energy particles (and obviously, of the antiparticles) was later used to explain the conduction band electrons and valence band holes in solid-state physics in the same manner; and consequently, the phenomena of electron-hole pair formation or annihilation can readily explain the radiative generation-recombination processes in the semiconductors, as well as the non-radiative phonon scattering phenomena.

2.3.4 Quantum electrodynamics

It is interesting to note that this can alternatively be described through Feynman's diagrammatic representation by using Green's function technique following eqs. (2.26–2.28). The only necessary modifications that are required for such a description are to replace the Schrödinger equation (e.g. eq. (2.18)) by Dirac's relativistic invariant equation and the corresponding basis state vectors (e.g. eq. (2.18)) by Dirac spinors, as well as the scattering potential (e.g. (2.26–2.28)) by the four-vector representing electromagnetic vector potential. Further, it is necessary to consider, in all such equations, only the positive energy states for time $t_2 > t_1$ and only the negative energy states for $t_2 < t_1$, in order to be consistent with Dirac's positron theory that the negative energy states are not available to the electron after scattering, i.e. the electron-positron annihilation. Thus, in fact, the alternative model of positron suggested by Feynman describes a positron conceptually equivalent to as if an electron propagates backwards in time, as illustrated in Fig. 2.2 that represents electron-positron scattering, as well as electron-photon (i.e. Compton) scattering phenomena.

In the case of quantum electrodynamics, all kinds of high-energy scatterings, including pair production or annihilation, are described in the same Feynman's diagrammatic Green's function model. The necessary modifications that are now required for such depiction are, firstly, to replace the single-particle Dirac spinors by the second-quantized field operators (creation and annihilation operators), and secondly, to consider a delayed interaction for the exchange of the positive-energy virtual particle or the propagator. For a comprehensive understanding of Feynman diagrams applicable in quantum electrodynamics, see Peskin and Schroeder [9].

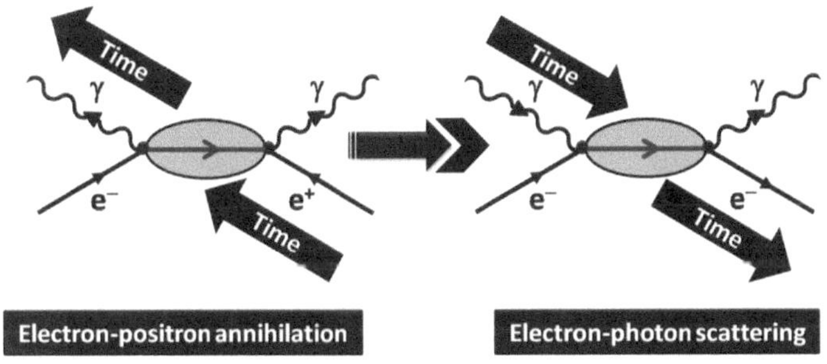

Fig. 2.2: One of the lowest-order Feynman diagrams for electron-positron annihilation and Compton scattering, by reversing the direction of time for the positron to get a scattered electron.

2.3.5 A brief discussion of Feynman's technique

The equivalence between Dirac's theory (and Hamiltonian method in general) and Feynman's diagrammatic approach in quantum theory is quite analogous to that between Gauss's law (or Poisson's equation) and Coulomb's law (as given by eq. (2.6) or in general by eq. (2.9)) in electrodynamics. The former one describes the behaviour of the electrodynamic field, while the latter elaborates the interaction between charges at a distance (although being retarded in time). Since the communication between *source point* and *field point* is accomplished through propagating light, while dealing with the distant interaction between many charges, it is almost impossible to work with the latter method, i.e. *interaction view* (since different sources can produce identical effects at a distant point); the former method, i.e. the *field view*, on the other hand, can simplify such problem by distinguishing between divergence or convergence of the field, i.e. emission or absorption of light. However, in the case of short-distance interactions between charges (e.g. scattering), it is practically impossible to distinguish between the source and absorber (i.e. where communication is made through virtual photons), thereby making the field view ambiguous; rather, the interaction view can take into account the direct interaction in such situations. Thus, one can say, as Feynman himself explained, the *field view* is more acceptable for the interactions propagating through real quanta, whereas the *interaction view* is better applicable in the cases associated with virtual quanta.

Further, it is worthy to note that, in general, the Hamiltonian method predicts the future to be developed from the present. However, to deal with a retarded interaction, the past is also needed to take into account (which is done in the Hamiltonian method by additionally considering the field oscillators along with the particle motions). In case of scattering, on the other hand, we are generally less interested in the precise time sequence of the motions within such a short duration, and rather, need to analyse the overall outcome of the process. Moreover, for relativistic motions, the Hamiltonian method, which computes the future from the instantaneous present, may be ambiguous, since the instantaneous present is subjective with respect to the observers in relative motion.

2.3.6 Martin-Schwinger non-perturbative method

A macroscopic property of a system can be considered as an integration of its microscopic behaviours. Thus, it is possible to find out particular quantities that can completely describe the local microscopic properties of a system, as well as can be utilized to predict its macroscopic behaviour. In this context, to deal with both the macroscopic and microscopic properties of a many-particle system under equilibrium in a unified approach, Martin and Schwinger developed a non-perturbative method proposing a novel Green's function technique based on imaginary parameters (e.g. complex time) [2]. Interestingly, such a technique can itself distinguish between the macroscopic and microscopic domains, respectively, in terms of the spectral densities of extensive properties and the local properties of the system in reference to the vacuum state.

They pointed out that each of the measurable quantities, such as energy, momentum, angular momentum, etc., of a system is associated with a parameter (e.g. time, space, angular coordinate) invariant under a certain group of transformations (e.g. translation, rotation, etc.). Thus, such measurable quantities can be obtained from a set of characteristic Hermitian generators related under groups, and constructed by the second-quantized field operators exhibiting the specific particle statistics. It was shown by them, using such imaginary invariant parameters, that in a many-particle system, the corresponding macroscopically measurable quantities (e.g. energy, number of particles) can be found from a generator function uniquely defined by corresponding real intensive parameters (e.g. temperature, chemical potential/Fermi-level). The microscopic properties of such many-particle systems, on the other hand, can be obtained from averaging the two-time correlation functions (i.e. two Green's functions for propagation in forward and backward time) over all the states at constant intensive parameters.

2.3.7 Konstantinov and Perel formulation

In order to represent such forward and backward propagating Green's functions, a diagram technique was then proposed by Konstantinov and Perel for describing the single-electron density matrix, leading to the transport equation even in the presence of an inhomogeneous external field, as well as the self-consistent field [3]. Such formalism was then advanced by Kadanoff and Baym [4] to develop the kinetic theory of many-particle systems. The core of this entire concept is briefly described in the following section.

2.3.8 Kadanoff and Baym approach

In terms of quantum field theory, a many-particle system can be described by the second-quantized field operators (i.e. creation and annihilation operators), so that the creation operator $\varphi^+(\vec{r},t)$ operating on a vacuum state $|0\rangle$ creates or adds a particle at the position $\vec{r}$, while the annihilation operator $\varphi(\vec{r},t)$ destroys or removes a particle from $\vec{r}$ at time t. The total number of particles at time t is thus given by $N(t)=\langle 0|\varphi^+(\vec{r},t)\varphi(\vec{r},t)|0\rangle=\langle\varphi^+(\vec{r},t)\varphi(\vec{r},t)\rangle$, where the field operators abide by the equal-time commutation (or anti-commutation) relations given by

$$\begin{cases} \varphi(\vec{r},t)\varphi(\vec{r}',t)\pm\varphi(\vec{r}',t)\varphi(\vec{r},t)=0 \\ \varphi^+(\vec{r},t)\varphi^+(\vec{r}',t)\pm\varphi^+(\vec{r}',t)\varphi^+(\vec{r},t)=0 \\ \varphi(\vec{r},t)\varphi^+(\vec{r}',t)\pm\varphi^+(\vec{r}',t)\varphi(\vec{r},t)=\delta(\vec{r}-\vec{r}') \end{cases} \tag{2.29}$$

where '$+$' are for fermions and '$-$' are for bosons due to their anti-symmetric and symmetric exchange properties, respectively.

Thus, since eq. (2.16), the Green's function representing the transport of a single particle in quantum field theory is given by, replacing the basis state vectors with field operators as

$$G(\vec{r}_1{}',t_1{}';\vec{r}_1,t_1)=-i\langle\hat{T}[\varphi(\vec{r}_1{}',t_1{}')\varphi^+(\vec{r}_1,t_1)]\rangle \tag{2.30}$$

where $\hat{T}$ is the Wick time-ordering operator that rearranges the field operators, placing the function of a later time to the left (indicating later operation) with respect to the function of a former time, preserving the symmetry/anti-symmetry property of the particles. Hence,

$$\hat{T}[\varphi(\vec{r}_1{}',t_1{}')\varphi^+(\vec{r}_1,t_1)]=\begin{cases} \varphi(\vec{r}_1{}',t_1{}')\varphi^+(\vec{r}_1,t_1) & t_1{}'>t_1 \\ \pm\varphi^+(\vec{r}_1,t_1)\varphi(\vec{r}_1{}',t_1{}') & t_1{}'<t_1 \end{cases} \tag{2.31}$$

In the exchange of operators in eq. (2.31), '+' is for bosons and '–' is for fermions, as per the last equation of eq. (2.29). Thus, it leads us to two Green's functions, for $t_1{}'>t_1$ and $t_1{}'<t_1$, respectively, as

$$G^>(1';1)=-i\langle\varphi(\vec{r}_1{}',t_1{}')\varphi^+(\vec{r}_1,t_1)\rangle \tag{2.32a}$$

$$G^{<}(1'; 1) = \mp i\langle \varphi^{+}(\vec{r}_1, t_1)\varphi(\vec{r}_1{}', t_1{}')\rangle \tag{2.32b}$$

The former Green's function $G^{>}(1'; 1)$ (the short form of $G^{>}(\vec{r}_1{}', t_1{}'; \vec{r}_1, t_1)$), given by eq. (2.32a), represents the disturbance created by addition of a particle at the position $\vec{r}_1$ at time t_1 in a many-particle system resulting in a quasi-shift from its equilibrium, followed by the propagation of such disturbance and consequent removal of a particle at another position $\vec{r}_1{}'$ at some later time $t_1{}'$, thereby bringing the system back to its equilibrium. Similarly, the latter Green's function $G^{<}(1'; 1)$, given by eq. (2.32b), describes the disturbance to the equilibrium in the many-particle system due to the removal of a particle at the position $\vec{r}_1{}'$ at time $t_1{}'$, that is propagated and counter-balanced by the addition of a particle at another position $\vec{r}_1$ at some later time t_1, returning it back the equilibrium. This can easily be extended to creation and annihilation of more than one particles.

Now, for a many-particle system in thermodynamic equilibrium, the expectation value of any operator (i.e. average measured value of a physical quantity) can be written as

$$\langle A\rangle = \frac{\sum_j \langle j|\hat{A}|j\rangle e^{-\beta(E_j-\mu N_j)}}{\sum_j e^{-\beta(E_j-\mu N_j)}} = \frac{Tr[\hat{A}e^{-\beta(\hat{H}-\mu\hat{N})}]}{Tr[e^{-\beta(\hat{H}-\mu\hat{N})}]} \tag{2.33}$$

where $\langle j|\hat{A}|j\rangle$ represents the quantum expectation value of the operator $\hat{A}$, and $e^{-\beta(E_j - \mu N_j)}$ is its thermodynamic statistical probability in the many-particle system, $\beta = 1/K_B T$ being the inverse of thermal energy at temperature T, with K_B representing the Boltzmann constant.

The remarkable finding by Kadanoff and Baym is the strong similarity between $e^{-\beta\hat{H}}$, i.e. the thermodynamic probability factor, and $e^{-\frac{i}{\hbar}t\hat{H}}$, i.e. the time-evolution operator of a quantum system (see eq. (2.13)). Such an incredible mathematical similarity between $\beta \leftrightarrow it/\hbar$ has been exploited by them to find the relation between two Green's functions, $G^{>}(1'; 1)$ and $G^{<}(1'; 1)$, and to extend the theory for non-equilibrium processes.

Now, using the formula of eq. (2.33), followed by applying the time evolution of the field operators, eq. (2.32a) leads to

$$G^{>}(1'; 1) = -i\frac{Tr[\varphi(\vec{r}_1{}', t_1{}')\varphi^{+}(\vec{r}_1, t_1)e^{-\beta(\hat{H}-\mu\hat{N})}]}{Tr[e^{-\beta(\hat{H}-\mu\hat{N})}]}$$

$$\Rightarrow G^{>}(1'; 1) = -i\frac{Tr[e^{\frac{i}{\hbar}\hat{H}t_1{}'}\varphi(\vec{r}_1{}', 0)e^{-\frac{i}{\hbar}\hat{H}(t_1{}'-t_1)}\varphi^{+}(\vec{r}_1, 0)e^{-\frac{i}{\hbar}\hat{H}t_1}e^{-\beta(\hat{H}-\mu\hat{N})}]}{Tr[e^{-\beta(\hat{H}-\mu\hat{N})}]} \tag{2.34}$$

which is an analytic function, i.e. convergent, in the limit $0 > \mathrm{Im}(t_1{}' - t_1) > \beta$. It is apparent that in a similar manner, it can be shown that $G^{<}(1'; 1)$ is analytic in the limit

$0 < \mathrm{Im}(t_1' - t_1) < \beta$. Further, using the cyclic relation in trace and considering t_1 to be the initial time, it can be written:

$$G^>(1'; 1)|_{t_1=0} = -i\frac{Tr[\varphi(\vec{r}_1', t_1')\varphi^+(\vec{r}_1, 0)e^{-\beta(\hat{H}-\mu\hat{N})}]}{Tr[e^{-\beta(\hat{H}-\mu\hat{N})}]}$$

$$= -i\frac{Tr[\varphi(\vec{r}_1', t_1')e^{\beta\mu(\hat{N}-1)}\varphi^+(\vec{r}_1, 0)e^{-\beta\hat{H}}]}{Tr[e^{-\beta(\hat{H}-\mu\hat{N})}]}$$

[Note that $\varphi^+(\vec{r}_1, 0)f(\hat{N}) = f(\hat{N}-1)\varphi(\vec{r}_1, 0)$ for any function of the number operator in order to balance the addition of a particle by $\varphi^+(\vec{r}_1, 0)$]

$$= -ie^{-\beta\mu}\frac{Tr[\varphi(\vec{r}_1', t_1')e^{-\beta(\hat{H}-\mu\hat{N})}e^{\beta\hat{H}}\varphi^+(\vec{r}_1, 0)e^{-\beta\hat{H}}]}{Tr[e^{-\beta(\hat{H}-\mu\hat{N})}]}$$

$$= -ie^{-\beta\mu}\frac{Tr[\varphi(\vec{r}_1', t_1')e^{-\beta(\hat{H}-\mu\hat{N})}\varphi^+(\vec{r}_1, -i\beta\hbar)]}{Tr[e^{-\beta(\hat{H}-\mu\hat{N})}]}$$

[assuming time – evolution to $t = -i\beta\hbar$]

$$= -ie^{-\beta\mu}\frac{Tr[\varphi^+(\vec{r}_1, -i\beta\hbar)\varphi(\vec{r}_1', t_1')e^{-\beta(\hat{H}-\mu\hat{N})}]}{Tr[e^{-\beta(\hat{H}-\mu\hat{N})}]}$$

$$\Rightarrow G^>(1'; 1)|_{t_1=0} = \pm e^{-\beta\mu}G^<(1'; 1)|_{t_1=-i\beta\hbar} \tag{2.35}$$

Interestingly, $G^{>/<}(1'; 1)$, i.e. $G^{>/<}(\vec{r}_1', t_1'; \vec{r}_1, t_1)$, can also be written as $G^{>/<}(\vec{r}_1' - \vec{r}_1, t_1' - t_1; \vec{0}, 0)$, or in short as $G^{>/<}(\vec{r}_1' - \vec{r}_1, t_1' - t_1)$, due to the translational symmetry of space-time. Then, it is apparent that the Fourier transform of such a Green's function from time to the energy domain can be written as

$$G^>(E) = \int dt_1' e^{\frac{i}{\hbar}Et_1'}G^>(\vec{r}_1' - \vec{r}_1, t_1' - t_1)|_{t_1=0} = \int dt_1' e^{\frac{i}{\hbar}Et_1'}G^>(\vec{r}_1' - \vec{r}_1, t_1')$$

$$= \pm e^{-\beta\mu}\int dt_1' e^{\frac{i}{\hbar}Et_1'}G^<(\vec{r}_1' - \vec{r}_1, t_1' - t_1)|_{t_1=-i\beta\hbar}$$

$$= \pm e^{-\beta\mu}\int dt_1' e^{\frac{i}{\hbar}Et_1'}G^<(\vec{r}_1' - \vec{r}_1, t_1' + i\beta\hbar)$$

$$= \pm e^{\beta(E-\mu)}\int d(t_1' + i\beta\hbar)e^{\frac{i}{\hbar}E(t_1'+i\beta\hbar)}G^<(\vec{r}_1' - \vec{r}_1, t_1' + i\beta\hbar)$$

$$\Rightarrow G^>(E) = \pm e^{\beta(E-\mu)}G^<(E) \tag{2.36}$$

2.3.9 Keldysh technique

It is interesting to note at this point that the entire formulation shown above is fundamentally aimed at addressing the phase accumulated by the system during its time-evolution, even if a quasi-static assumption is made so that the time-dependence is only included as an adiabatic switching ON and OFF at a non-interacting ground state in the distant past ($t \to -\infty$) and distant future ($t \to \infty$), respectively. In fact, such a phase is tackled by mathematically assuming an imaginary time that makes it equivalent to a thermodynamic equilibrium statistical distribution function (see eq. (2.35)).

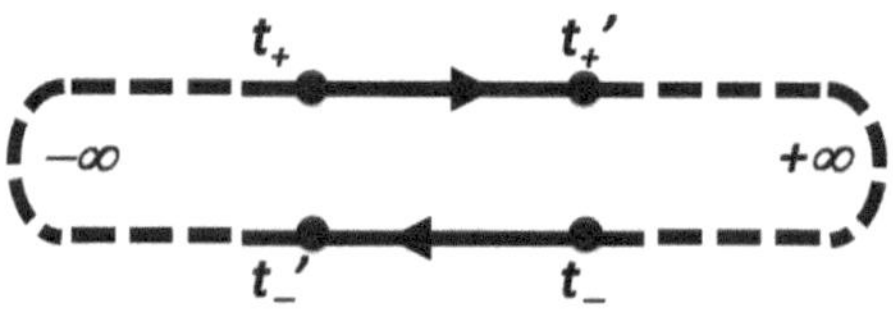

Fig. 2.3: The Keldysh contour for a complete loop over time covering negative, as well as positive infinity, so that the Green's function can also describe non-adiabatic evolutions.

In this context, Keldysh developed a diagrammatic technique [5] without considering any imaginary time, and thus not requiring the equilibrium density matrix, which is therefore capable of describing also the non-equilibrium or far-from-equilibrium processes. In fact, for a complete time loop (see Fig. 2.3) from $t \to -\infty$ to $t \to \infty$ (forward) followed by $t \to \infty$ to $t \to -\infty$ (backward), instead of imaginary time, he first introduced two times (t_+, t_-) for forward and backward evolution branches, and thus four Green's functions instead of previously shown two in eq. (2.32a) and (2.32b), as

$$G^{(-)\to(+)} = -i\langle \hat{T}[\varphi(\vec{r}, t_+)\varphi^+(\vec{r}', t'_-)]\rangle = i\langle[\varphi^+(\vec{r}', t'_+)\varphi(\vec{r}, t_-)]\rangle \tag{2.37a}$$

$$G^{(+)\to(-)} = -i\langle \hat{T}[\varphi(\vec{r}, t_-)\varphi^+(\vec{r}', t'_+)]\rangle = -i\langle[\varphi(\vec{r}, t_-)\varphi^+(\vec{r}', t'_+)]\rangle \tag{2.37b}$$

$$G^{(+)\to(+)} = -i\langle \hat{T}[\varphi(\vec{r}', t'_+)\varphi^+(\vec{r}, t_+)]\rangle = -i\langle[\varphi(\vec{r}', t'_+)\varphi^+(\vec{r}, t_+)]\rangle \tag{2.37c}$$

$$G^{(-)\to(-)} = -i\left\langle \hat{\hat{T}}[\varphi(\vec{r}', t'_-)\varphi^+(\vec{r}, t_-)]\right\rangle = -i\langle[\varphi(\vec{r}', t'_-)\varphi^+(\vec{r}, t_-)]\rangle \tag{2.37d}$$

where $\hat{\hat{T}}$ is the reverse time-ordering operator. It may be noted that the above Green's functions are not all independent and lead to a more physical representation as their linear combinations:

$$G^R = G^{(+)\to(+)} - G^{(+)\to(-)} = +\vartheta(t' - t)(G^{(-)\to(+)} - G^{(+)\to(-)}) \tag{2.38a}$$

$$G^A = G^{(+)\to(+)} - G^{(-)\to(+)} = -\vartheta(t - t')(G^{(-)\to(+)} - G^{(+)\to(-)}) \tag{2.38b}$$

$$G^K = \left(G^{(-)\to(+)} + G^{(+)\to(-)}\right) \tag{2.38c}$$

Here G^R is known to be the retarded Green's function which physically implies the net delayed response of a non-adiabatic change, while G^A represents correspondingly an (virtually) advanced response. On the other hand, G^K, the Keldysh component, characterizes the static distribution of the system. The vast applications of such formalism, especially the significant role of retarded Green's function for any non-equilibrium processes, are illustrated in the succeeding chapters of this book.

References

[1] Datta, S. "Nanoscale device modelling: The Green's function method." *Superlattices and Microstructures* 28, no. 4 (2000): 253–278.

[2] Martin, P. C. and Schwinger, J. "Theory of many-particle systems. I." *Physical Review* 115, no. 6 (1959): 1342.

[3] Konstantinov, O. V. and Perel, V. I. "A diagram technique for evaluating transport quantities." *Soviet Physics JETP-USSR* 12, no. 1 (1961): 142–149.

[4] Kadanoff, L. P. and Gordon, A. B. *Quantum Statistical Mechanics Green's Function Methods in Equilibrium Problems.* Benjamin: 1962.

[5] Keldysh, L. V. "Diagram technique for nonequilibrium processes." *Soviet Physics JETP* 20, no. 4 (1965).

[6] Feynman, R. P. "The theory of positrons." *Physical Review (US) Superseded in part by Phys. Rev. A, Phys. Rev. B: Solid State, Phys. Rev. C, and Phys. Rev. D* 76, (1949).

[7] Feynman, R. P. "Space-time approach to quantum electrodynamics." *Physical Review* 76, no. 6 (1949): 769.

[8] Dirac, P. A. M. *The Principles of Quantum Mechanics.* no. 27, Oxford university press, 1981.

[9] Peskin, M. E. and Schroeder, D. V. "An Introduction to quantum field theory." (1995).

3 General NEGF formalism: communication between the quantum and classical world

The exotic behaviours (e.g. wave-particle duality, uncertainty, superposition, entanglement) of our world that appear at the so-called quantum scale are quite counterintuitive *per se* in respect to our everyday experiences in the so-called classical world. However, this does not lead to any difficulty in theoretically modelling the behaviour of a quantum system; a simple linear differential equation, i.e. Schrodinger equation (or Dirac equation for the relativistic case), can perform extraordinarily well in this regard. In fact, the obscurity arises from the fact, how Heisenberg described it by saying, that, the *Copenhagen interpretation of quantum theory starts from a paradox*, so that any *experiment in physics, whether it refers to the phenomena of daily life or to atomic events, is to be described in the terms of classical physics*, and thus provided the range of applicability of the classical physics, one *cannot and should not try to improve them* [1]. Thus, all the counterintuitive quantum properties are manifested as classical outcomes in experimental measurements, and that is the origin of all the difficulties. Or, in other words, one can say that the quantum world communicates to the classical world through measurements performed on it.

In fact, Bohr distinguished quantum and classical systems in terms of measurements pointing out: *"that the behaviour of a physical object relative to a given system of coordinates is uniquely determined, quite independently of whether it is observed or not . . . retains its adequacy only as long as all quantities of action entering into the description are large compared to Planck's quantum*; however, for quantum systems, *it is no longer possible sharply to distinguish between the autonomous behaviour of a physical object and its inevitable interaction with other bodies serving as measuring instruments*" [2]. He further explained: *The very fact that quantum phenomena cannot be analyzed on classical lines thus implies the impossibility of separating a behaviour of atomic objects from the interaction of these objects with the measuring instruments which serve to specify the conditions under which the phenomena appear. In particular, the individuality of the typical quantum effects finds proper expression in the circumstance that any attempt at subdividing the phenomena will demand a change in the experimental arrangement, introducing new sources of uncontrollable interaction between objects and measuring instruments* [3]. Hence, presuming such inseparability of any property of a quantum system and its interaction with the measuring apparatus (assumedly classical in nature), the key question in the perspective of theoretical modelling is how to deal with such interaction. As mentioned earlier in Chapter 1, for more realistic theoretical modelling, it is also necessary to take into account the interaction between two fundamentally different systems, one (the classical system, such as an apparatus) exhibiting definite statistically averaged quantities (e.g. temperature, Fermi level, etc.) while the other (the quantum system) without any such defined quantities. In this chapter, we will focus on elaborating how the non-equilibrium

 | https://doi.org/10.1515/9783111545950-003

Green's function framework can accommodate all these requirements to describe the interaction between a quantum system and a measuring apparatus, as well as distinguish between such quantum and classical systems.

3.1 The mathematical approach to describing a quantum system interacting with an apparatus

The dynamics of an isolated quantum system can be described by the Schrodinger equation (or Dirac equation for the relativistic case) as

$$i\hbar\frac{d}{dt}|\psi\rangle - \hat{H}|\psi\rangle = 0 \tag{3.1}$$

where $\hat{H}$ is the Hamiltonian operator characteristic to the particular system that remains in the state $|\psi\rangle$, which, on the other hand, evolves in time unitarily depending on the Hamiltonian. However, when such a quantum system interacts with another system, the equation can be conceptually modified (later in this chapter, we will show how this can be done mathematically) as

$$i\hbar\frac{d}{dt}|\psi\rangle - \hat{H}|\psi\rangle - \hat{\sum}|\psi\rangle = \hat{S}|\Phi\rangle \tag{3.2}$$

It is interesting to note that in eq. (3.2), we have two additional terms: 1) self-energy $\hat{\Sigma}$; and 2) a source term $\hat{S}|\Phi\rangle$, in order to incorporate an outflow and an inflow of quantum of information carrying out between the quantum system and the apparatus.

Such incorporation of new terms is not *ad hoc* though; rather, they arise as a result of the attempt to abstract out the quantum system as a separate part from the whole of the inseparable system + apparatus. Conceptually, when the apparatus probes the system, a quantum of information inflows that disturbs the stationary state and its unitary evolution; and in response, the system reacts to it as an outflow of information to perturb the state of the apparatus, which, in turn, alters its probing interaction with the system, and this goes on in cycles. Thus, the net outflow of information from the system is not only due to such probing but also due to its own course of interaction with itself through the apparatus, and thus, we can identify it as the self-energy. The spectral density of the quantum system in energy space, which should be sharp (i.e. delta function) for an isolated system, is now broadened due to such self energy. Thus, such broadening of the spectral density is a measure of communication between the system and the apparatus, or more precisely, between the quantum and the classical world.

It is further worthy to note that in the quantum mechanical mathematical set-up, the system and the apparatus enjoy equal footing so that the apparatus also exhibits such inflow and outflow. Then how can one distinguish between the system and the apparatus, or a quantum and a classical system in general? In fact, such inflow or outflow for a classical system (or the apparatus here) is unaffected by the filling or emptying of any states due to interaction with the other (quantum) system. It will be shown in this chapter that this happens only when the interaction is negligibly small compared to the energy of the system. This idea is inspired from Dirac, who distinguished between the 'big' (i.e. so-called classical) and 'small' (i.e. so-called quantum) objects by explaining that "*science is concerned only with observable things and that we can observe an object only by letting it interact with some outside influence. An act of observation is thus necessarily accompanied by some disturbance of the object observed. We may define an object to be big when the disturbance accompanying our observation of it may be neglected, and small when the disturbance cannot be neglected*" [4].

3.2 The quantum field theoretic model: wave-particle duality

Equations (3.1) and (3.2) represent only a wave view for a quantum system or a quantum object or a collection of quantum particles. It is apparent that such descriptions disregard the particulate nature of an object that complementarily coexists with its wave nature. Further, if we consider the particulate view, for a number of particles, it is necessary to accommodate their symmetry property against exchange. For instance, over exchange of particles, the bosons (e.g. photons) remain symmetric (i.e. unaltered phase) while fermions (e.g. electrons) exhibit an anti-symmetric property (i.e. a π-phase difference) which imposes the restriction of the exclusion principle (i.e. more than one particle cannot coexist in the same state) on the latter. Thus, in order to develop a comprehensive model encompassing both the wave and particulate properties, as well as the exchange-symmetry relations, the relevant equations of motion can be obtained by considering the second-quantized field operators φ_j^+ and φ_j, representing the creation and annihilation of particles, respectively, at jth state. As mentioned in Chapter 2, $\hat{N}_j = \hat{\varphi}_j^+ \hat{\varphi}_j$ represents the number operator at jth state, which can assume only one of the two values, 0 or 1, for fermions (due to the exclusion principle), whereas any value for bosons.

Equation (3.2) can then be modified as

$$i\hbar \frac{d}{dt}\hat{\varphi}_i - \hat{H}\hat{\varphi}_i - \hat{\sum}_{ij}\hat{\varphi}_j = \hat{S}_i \tag{3.3}$$

where the field operators follow the commutation/anti-commutation relations (as mentioned in Chapter 2) for discrete states as

$$\begin{cases} \hat{\phi}_i \hat{\phi}_j \pm \hat{\phi}_j \hat{\phi}_i = 0 \\ \hat{\phi}_i^+ \hat{\phi}_j^+ \pm \hat{\phi}_j^+ \hat{\phi}_i^+ = 0 \\ \hat{\phi}_i \hat{\phi}_j^+ \pm \hat{\phi}_j^+ \hat{\phi}_i = \delta_{ij} \end{cases} \tag{3.4}$$

the '+' being for fermions and '−' being for bosons to uphold their anti-symmetry and symmetry properties, respectively. Such an anti-symmetric exchange property of fermions indeed imposes the exclusion constraint on their occupational behaviour. For instance, the square of the number operator for fermions is given by

$$\hat{N}_j^2 = \left(\hat{\phi}_j^+ \hat{\phi}_j\right)\left(\hat{\phi}_j^+ \hat{\phi}_j\right) = \hat{\phi}_j^+ \left(\hat{\phi}_j \hat{\phi}_j^+\right) \hat{\phi}_j = \hat{\phi}_j^+ \left(1 - \hat{\phi}_j^+ \hat{\phi}_j\right) \hat{\phi}_j = \hat{\phi}_j^+ \hat{\phi}_j = N_j$$

[using eq. (3.4) for fermions]

$$\Rightarrow \hat{N}_j(\hat{N}_j - 1) = 0 \tag{3.5}$$

Equation (3.5) leads to the only possible solution for $\hat{N}_j$ to be either of '0' and '1' in any given jth state, which restricts more than one fermion from occupying a single state as suggested by Pauli's exclusion principle.

Thus, the probability distribution of the occupied and empty states of fermions, or accordingly, the absorbed and emitted states of bosons, are obtained respectively by averaging the quantities:

$$\begin{cases} \left\langle \hat{\phi}_j^+ \hat{\phi}_j \right\rangle = f_j \\ \left\langle \hat{\phi}_j \hat{\phi}_j^+ \right\rangle = 1 \mp f_j \end{cases} \tag{3.6}$$

f_j being the Fermi-Dirac distribution, or Bose-Einstein distribution function of j th state at equilibrium, for fermions and bosons, respectively, with '–' and '+' signs.

It is interesting to note, as discussed in Chapter 2, that the non-equilibrium transition of such particles from ith state to jth state (i.e. occupation of fermions or absorption of bosons) between time t to t' is accompanied by the annihilation of a particle at ith state at time t followed by the creation of such particle at jth state at time t' ($t' > t$), and thus can be given by the two-time correlation function:

$$n_{ij}(t, t') = \left\langle \hat{\phi}_j^+(t') \hat{\phi}_i(t) \right\rangle \tag{3.7a}$$

Thus, such a correlation function is the quantum analogue of the classical thermodynamical distribution function. Similarly, the transition of a Dirac-hole or emission of a boson can be described by

$$p_{ij}(t, t') = \left\langle \hat{\phi}_i(t) \hat{\phi}_j^+(t') \right\rangle \tag{3.7b}$$

Thus, the net spectral density function (i.e. density of states in per energy unit) is obtained to be

$$A_{ij}(t,t') = p_{ij}(t,t') \pm n_{ij}(t,t') \tag{3.8}$$

where '+' and '−' signs are for fermions and bosons, respectively. Under steady-state conditions, such correlation functions, as well as the spectral density, depend only on the difference time coordinate $(t' - t)$ and thus can be easily Fourier transformed into the energy or frequency domain.

3.3 The equation of motion for system and apparatus: interplay between the whole and the part

We will assume both the quantum system and the apparatus to follow quantum theory, and later we will show how the classicalness (or 'bigness') of the apparatus can be retrieved for negligibly small interaction strength, as discussed earlier in this chapter. We will consider the system (S) field operators to be $(\hat{\phi}_i(t), \hat{\phi}_j^+(t))$ and that for the apparatus (A) to be $(\hat{\Phi}_r(t), \hat{\Phi}_s^+(t))$. When the system and the apparatus are isolated, their Hamiltonian (mathematically operating on their own Hilbert spaces) can be written as

$$\begin{cases} \hat{H}_{\mathrm{S}} = \sum_i \varepsilon_i^{\mathrm{S}} \hat{\phi}_i^+ \hat{\phi}_i \\ \hat{H}_{\mathrm{A}} = \sum_r \varepsilon_r^{\mathrm{A}} \hat{\Phi}_r^+ \hat{\Phi}_r \end{cases} \tag{3.9}$$

where $\varepsilon_i^{\mathrm{S}}$ and $\varepsilon_r^{\mathrm{A}}$ are energy eigenstates of the system and apparatus, respectively. Now, when such a system and the apparatus are subjected to interact with each other, their total Hamiltonian (mathematically operating on the tensor product of their Hilbert spaces) will be given by

$$\hat{H} = \sum_i \varepsilon_i^{\mathrm{S}} \hat{\phi}_i^+ \hat{\phi}_i + \sum_r \varepsilon_r^{\mathrm{A}} \hat{\Phi}_r^+ \hat{\Phi}_r + \left(\sum_{i,r} \tau_{ir}^{\mathrm{S}\leftrightarrow\mathrm{A}} \hat{\phi}_i^+ \hat{\Phi}_r + H.C. \right) \tag{3.10}$$

where $\tau_{ir}^{\mathrm{S}\leftrightarrow\mathrm{A}}$ represents the interaction between each ith state of the system and each rth state of the apparatus, while $H.C.$ signifies its Hermitian conjugate that keeps the total Hamiltonian Hermitian. Physically, it implies that although each of the quantum system and the apparatus, while interacting, becomes an open system as two mutually coupling parts, they together remain a closed system as a whole.

Thus, using the Heisenberg equation of motion for any operator, i.e. $i\hbar \frac{d}{dt}\hat{A} = [\hat{A}, \hat{H}]_{\pm}$, '+' and '−' being respectively for fermions and bosons, we have a set of equations for the system and apparatus as

$$\begin{cases} i\hbar\frac{d}{dt}\hat{\phi}_i = \varepsilon_i{}^{\mathrm{S}}\hat{\phi}_i + \sum\limits_r \tau_{ir}{}^{\mathrm{S}\leftrightarrow\mathrm{A}}\hat{\Phi}_r \\ i\hbar\frac{d}{dt}\hat{\Phi}_r = \varepsilon_r{}^{\mathrm{A}}\hat{\Phi}_r + \sum\limits_i \tau_{ri}{}^{\mathrm{S}\leftrightarrow\mathrm{A}\,*}\hat{\phi}_i \end{cases} \tag{3.11}$$

The set of equations given by (3.11) can also be written in matrix form:

$$i\hbar\frac{d}{dt}\begin{Bmatrix} \{\hat{\phi}\} \\ \{\hat{\Phi}\} \end{Bmatrix} = \begin{bmatrix} [\varepsilon^{\mathrm{S}}] & [\tau^{\mathrm{S}\leftrightarrow\mathrm{A}}] \\ [\tau^{\mathrm{S}\leftrightarrow\mathrm{A}}]^{+} & [\varepsilon^{\mathrm{A}}] \end{bmatrix}\begin{Bmatrix} \{\hat{\phi}\} \\ \{\hat{\Phi}\} \end{Bmatrix} \tag{3.12}$$

It is imperative to note that while interacting, the quantum system and the apparatus cannot be described by their isolated Hamiltonians, given by eq. (3.9), and rather are described by the equations given by (3.11), which are more than the conventional Schrodinger equation (or Dirac equation for the relativistic case) given by eq. (3.1). The equations of system field operators are now dependent on the apparatus field operators and *vice versa*. Such equations reflect the nature of open systems rather than closed systems. On the other hand, eq. (3.12) shows the field operators of both the system and apparatus together to be analogous to the Schrodinger equation (or Dirac equation for the relativistic case) given by eq. (3.1), and thereby represents a closed system as a whole.

In order to solve the set of equations given by (3.11), let us start with the equations for the apparatus $\left(i\hbar\frac{d}{dt} - \varepsilon_r{}^{\mathrm{A}}\right)\hat{\Phi}_r(t) = \sum_i \tau_{ri}{}^{\mathrm{S}\leftrightarrow\mathrm{A}\,*}\hat{\phi}_i(t)$, which can be solved using the apparatus Green's function $\left(G_{rr}{}^{\mathrm{A}}(t,t_1) = (-i/\hbar)\vartheta(t-t_1)e^{-\frac{i}{\hbar}\varepsilon_r(t-t_1)}\right)$ to obtain

$$\hat{\Phi}_r(t) = \hat{\Phi}_r(0)e^{-\frac{i}{\hbar}\varepsilon_r t} + \sum_i \int\limits_{-\infty}^{+\infty} dt_1 G_{rr}{}^{\mathrm{A}}(t,t_1)\tau_{ri}{}^{\mathrm{S}\leftrightarrow\mathrm{A}\,*}\hat{\phi}_i(t_1) \tag{3.13}$$

The first term in eq. (3.13) represents the initial state of the apparatus before interacting with the system, while the second term appears due to its coupling with the system and thus depends on the system field operators. Putting the expression of $\hat{\Phi}_r(t)$ into the system equations given in (3.11) gives rise to

$$i\hbar\frac{d}{dt}\hat{\phi}_i(t) = \varepsilon_i{}^{\mathrm{S}}\hat{\phi}_i(t) + \sum_r \tau_{ir}{}^{\mathrm{S}\leftrightarrow\mathrm{A}}\left(\hat{\Phi}_r(0)e^{-\frac{i}{\hbar}\varepsilon_r t} + \sum_j \int\limits_{-\infty}^{+\infty} dt_1 G_{rr}{}^{\mathrm{A}}(t,t_1)\tau_{rj}{}^{\mathrm{S}\leftrightarrow\mathrm{A}\,*}\hat{\phi}_j(t_1)\right)$$

$$\Rightarrow i\hbar\frac{d}{dt}\hat{\phi}_i(t) - \varepsilon_i{}^{\mathrm{S}}\hat{\phi}_i(t) - \sum_{r,j}\int\limits_{-\infty}^{+\infty} dt_1[\tau_{ir}{}^{\mathrm{S}\leftrightarrow\mathrm{A}}G_{rr}{}^{\mathrm{A}}(t,t_1)\tau_{rj}{}^{\mathrm{S}\leftrightarrow\mathrm{A}\,*}]\hat{\phi}_j(t_1) = \sum_r \tau_{ir}{}^{\mathrm{S}\leftrightarrow\mathrm{A}}\hat{\Phi}_r(0)e^{-\frac{i}{\hbar}\varepsilon_r t} \tag{3.14}$$

It is apparent that eq. (3.14) is identical to the conceptual equation given by (3.3), where the self-energy term in the two-time coordinate is obtained to be

$$[\Sigma_{\rm A}(t,t_1)]_{ij} = \sum_r [\tau_{ir}{}^{\rm S\leftrightarrow A} G_{rr}{}^{\rm A}(t,t_1)\tau_{rj}{}^{\rm S\leftrightarrow A\,*}] \tag{3.15a}$$

It is worthy to note that the self-energy appearing in the system equation is due to the system field operators interacting with the apparatus field operators (through $\tau_{rj}{}^{\rm S\leftrightarrow A\,*}$) followed by (through $G_{rr}{}^{\rm A}(t,t_1)$) the apparatus field operators coupling to the system field operators (through $\tau_{ir}{}^{\rm S\leftrightarrow A}$), and thereby, as a whole, due to an action of the system field operators on themselves through an intermediate communication with the apparatus. Under steady-state conditions, where this can be expressed as a function of the difference-time coordinate, a Fourier transform of eq. (3.15a) gives rise to

$$[\Sigma_{\rm A}(E)] = [\tau^{\rm S\leftrightarrow A}][G^{\rm A}(E)][\tau^{\rm S\leftrightarrow A}]^{+} \tag{3.15b}$$

On the other hand, the source term in eq. (3.14) equivalent to that in eq. (3.3) is given by $\hat{S}_i(t) = \sum_r \tau_{ir}{}^{\rm S\leftrightarrow A}\hat{\Phi}_r(0)e^{-\frac{i}{\hbar}\varepsilon_r t}$, which depends exclusively on the initial conditions of the apparatus and appears in the equation solely due to its probing of the system through coupling $[\tau^{\rm S\leftrightarrow A}]$. The solution of eq. (3.14) is obtained in terms of the system Green's function ($G_{ik}{}^{\rm S}(t,t_1)$) as

$$\hat{\phi}_i(t) = \sum_k \int_{-\infty}^{+\infty} dt_1 G_{ik}{}^{S}(t,t_1)\hat{S}_k(t_1) \tag{3.16}$$

where

$$\left[\sum_j \delta_{ij}\left(i\hbar\frac{d}{dt} - \varepsilon_i\right) - \int_{-\infty}^{+\infty} dt'[\Sigma_{\rm A}(t,t')]_{ij}\right] G_{jk}{}^{S}(t,t_1) = \delta_{ik}\delta(t-t_1) \tag{3.17a}$$

While applying Fourier transform on eq. (3.17a into the energy domain) under steady-state conditions, we obtain in the matrix form:

$$[G^{\rm S}(E)] = \left[E[I] - [\varepsilon^{\rm S}] - [\Sigma_{\rm A}(E)]\right]^{-1} \tag{3.17b}$$

Now, since eqs. (3.7a) and (3.16), the two-time correlation function is obtained to be

$$n_{ij}{}^{\rm S}(t,t') = \left\langle \hat{\phi}_j{}^{+}(t')\hat{\phi}_i(t)\right\rangle = \left\langle \sum_{k,k'} \int_{-\infty}^{+\infty} dt_2 G_{k'j}{}^{\rm S\,*}(t_2,t')\hat{S}_{k'}{}^{+}(t_2) \int_{-\infty}^{+\infty} dt_1 G_{ik}{}^{\rm S}(t,t_1)\hat{S}_k(t_1)\right\rangle$$

$$= \int_{-\infty}^{+\infty} dt_2 \int_{-\infty}^{+\infty} dt_1 \sum_{k,k'} G_{ik}{}^{\rm S}(t,t_1)\left\langle \hat{S}_{k'}{}^{+}(t_2)\hat{S}_k(t_1)\right\rangle G_{k'j}{}^{\rm S\,*}(t_2,t')$$

$$\Rightarrow n_{ij}{}^{\rm S}(t,t') = \int_{-\infty}^{+\infty} dt_2 \int_{-\infty}^{+\infty} dt_1 [G^{\rm S}(t,t_1)][\Sigma_{\rm A}^{In}(t_1,t_2)][G^{\rm S^{+}}(t_2,t')] \tag{3.18}$$

where $[\Sigma_A^{In}(t_1,t_2)]_{kk'} = \langle S_{k'}^{+}(t_2)S_k(t_1)\rangle$ represents the inflow of information from the apparatus to the system, and can be simplified to

$$[\Sigma_A^{In}(t_1,t_2)]_{kk'} = \langle S_{k'}^{+}(t_2)S_k(t_1)\rangle = \left\langle \sum_s \tau_{sk'}^{S\leftrightarrow A\,*}\,\hat{\Phi}_s^{+}(0)e^{\frac{i}{\hbar}\varepsilon_s t_2} \sum_r \tau_{kr}^{S\leftrightarrow A}\hat{\Phi}_r(0)e^{-\frac{i}{\hbar}\varepsilon_r t_1}\right\rangle$$

$$= \sum_{r,s} \tau_{kr}^{S\leftrightarrow A}\left\langle \hat{\Phi}_s^{+}(0)\hat{\Phi}_r(0)\right\rangle \tau_{sk'}^{S\leftrightarrow A\,*}\,e^{-\frac{i}{\hbar}(\varepsilon_r t_1-\varepsilon_s t_2)}$$

$$= \sum_{r,s} \tau_{kr}^{S\leftrightarrow A} n_{rs}^{A}(t_2,t_1)\tau_{sk'}^{S\leftrightarrow A\,*}$$

[assuming the evolution of the apparatus states to be unaffected by the interaction with the system, keeping in mind that the vice versa is not true]

$$\Rightarrow [\Sigma_A^{In}(t_1,t_2)] = [\tau^{S\leftrightarrow A}][n^A(t_2,t_1)][\tau^{S\leftrightarrow A}]^{+} \tag{3.19}$$

Thus, the Fourier transforms of eqs. (3.18) and (3.19) over the difference-time coordinate into the energy domain leads respectively to

$$[n^S(E)] = [G^S(E)][\Sigma^{in}(E)][G^S(E)]^{+} \tag{3.20a}$$

$$[\Sigma_A^{In}(E)] = [\tau^{S\leftrightarrow A}][n^A(E)][\tau^{S\leftrightarrow A}]^{+} \tag{3.20b}$$

3.3.1 Spectral density function: local density of states (LDOS)

In order to obtain the spectral density or the local density of states (LDOS) in per energy unit, we need to calculate the empty state (or emission) density:

$$p_{ij}^S(t,t') = \left\langle \hat{\phi}_i(t)\hat{\phi}_j^{+}(t')\right\rangle = \int_{-\infty}^{+\infty} dt_2 \int_{-\infty}^{+\infty} dt_1 \sum_{k,k'} G_{ik}^S(t,t_1)\left\langle \hat{S}_k(t_1)\hat{S}_{k'}^{+}(t_2)\right\rangle G_{k'j}^{S\,*}(t_2,t')$$

where

$$\langle S_k(t_1)S_{k'}^{+}(t_2)\rangle = \left\langle \sum_r \tau_{kr}^{S\leftrightarrow A}\hat{\Phi}_r(0)e^{-\frac{i}{\hbar}\varepsilon_r t_1} \sum_s \tau_{sk'}^{S\leftrightarrow A\,*}\,\hat{\Phi}_s^{+}(0)e^{\frac{i}{\hbar}\varepsilon_s t_2}\right\rangle$$

$$= \sum_{r,s} \tau_{kr}^{S\leftrightarrow A}\left\langle \hat{\Phi}_r(0)\hat{\Phi}_s^{+}(0)\right\rangle \tau_{sk'}^{S\leftrightarrow A\,*}\,e^{-\frac{i}{\hbar}(\varepsilon_r t_1-\varepsilon_s t_2)}$$

$$= \sum_{r,s} \tau_{kr}^{S\leftrightarrow A}\left\langle \delta_{rs} \mp \hat{\Phi}_s^{+}(0)\hat{\Phi}_r(0)\right\rangle \tau_{sk'}^{S\leftrightarrow A\,*}\,e^{-\frac{i}{\hbar}(\varepsilon_r t_1-\varepsilon_s t_2)}$$

['–' and '+' signs for fermions and bosons, respectively]

Now, using the relation

$$[i(G_{rr}{}^{A}(t_2,t_1) - G_{rr}{}^{A}(t_1,t_2)*)] = i[(-i/\hbar)\vartheta(t_2-t_1)e^{-\frac{i}{\hbar}\varepsilon_r(t_2-t_1)} - (i/\hbar)\vartheta(t_1-t_2)e^{\frac{i}{\hbar}\varepsilon_r(t_1-t_2)}]$$
$$= (1/\hbar)e^{-\frac{i}{\hbar}\varepsilon_r(t_2-t_1)}$$

we have

$$\langle S_k(t_1)S_{k'}{}^{+}(t_2)\rangle = \hbar\sum_r \tau_{kr}{}^{S\leftrightarrow A}[i(G_{rr}{}^{A}(t_2,t_1) - G_{rr}{}^{A}(t_1,t_2)*)]\tau_{rk'}{}^{S\leftrightarrow A}*$$
$$\mp\sum_{r,s}\tau_{kr}{}^{S\leftrightarrow A}n_{rs}{}^{A}(t_2,t_1)\tau_{sk'}{}^{S\leftrightarrow A}*$$

Thus, applying the Fourier transform in the difference-time coordinate, it is obtained to be

$$[p^S(E)] = [G^S(E)]\left([\tau^{S\leftrightarrow A}]i([G^A(E)] - [G^A(E)]^{+})[\tau^{S\leftrightarrow A}]^{+} \mp [\tau^{S\leftrightarrow A}][n^A(E)][[\tau^{S\leftrightarrow A}]^{+}\right)[G^S(E)]^{+}$$

$$= [G^S(E)]\left(i([\Sigma_A(E)] - [\Sigma_A(E)]^{+}) \mp [\Sigma_A^{In}(E)]\right)[G^S(E)]^{+} \quad \text{[since eq. (3.15b) and (3.20b)]}$$
$$= [G^S(E)](i([\Sigma_A(E)] - [\Sigma_A(E)]^{+}))[G^S(E)]^{+} \mp [n^S(E)] \quad \text{[since eq. (3.20a)]}$$

Then, using the expression of the system Green's function given by eq. (3.17b), we have

$$[p^S(E)] = i\left([G^S(E)] - [G^S(E)]^{+}\right) \mp [n^S(E)] \tag{3.21}$$

Thus, eqs. (3.8) and (3.21) lead to

$$[A^S(E)] = i\left([G^S(E)] - [G^S(E)]^{+}\right) \tag{3.22a}$$

Using the expression of the system Green's function given by eq. (3.17b), the spectral density can also be written:

$$[A^S(E)] = [G^S(E)][\Gamma_A(E)][G^S(E)]^{+} \tag{3.22b}$$

where $[\Gamma_A(E)] = i([\Sigma_A(E)] - [\Sigma_A(E)]^{+}) = -2\,\mathrm{Im}(\Sigma_A(E))$, which represents the net outflow of information from the system to the apparatus.

In order to physically explain the local density of states of the quantum system and how it changes due to probing by the apparatus, let us consider a single-energy-level ($\varepsilon_i{}^S$) quantum system. Its spectral density should be sharp and highly peaked at energy $E = \varepsilon_i{}^S$ so that it can be represented as

$$A^S(E) \equiv \delta(E - \varepsilon_i) \tag{3.23}$$

where the delta function may be multiplied by 2 for fermions or the number of bosons. However, since eq. (3.22a), it is given by

$$A^S(E) \equiv i\left(\left[E - \varepsilon_i^S - \Sigma_A\right]^{-1} - \left[E - \varepsilon_i^S - \Sigma_A{}^*\right]^{-1} \right)$$

$$\Rightarrow A^S(E) \equiv \frac{2\ \mathrm{Im}(\Sigma_A)}{(E - \varepsilon_i^S - \mathrm{Re}(\Sigma_A))^2 + (\mathrm{Im}(\Sigma_A))^2} \tag{3.24}$$

It is imperative to note that eq. (3.24) gives rise to a delta function, i.e. returns to eq. (3.23), at the limit $\Sigma_A \to 0$. It clearly shows that if the quantum system is not probed by the apparatus, its energy is precisely quantized, leading to a sharp and highly peaked spectral density, but only hypothetically, since that can never be achieved in experimental measurement. How can one measure a property of a system without probing it with an apparatus! And if not measured, how can one be sure of what is claimed hypothetically! However, one can approach such a result by arbitrarily weakening the coupling strength of the interaction between the system and the apparatus (this has close resemblance with 'weak' measurement or 'protective' measurement [5]). Interestingly, while interacting with the apparatus, the peak position of the spectral density of the system is shifted from $E = \varepsilon_i^S$ by the real part of the self-energy (Σ_A). This may be considered as the instrumental error which can be made zero through appropriate experimental arrangements. Experimentally, this is done by applying origin correction method, which, from a theoretical perspective, is carried out by the process of *renormalization*. On the other hand, the imaginary part of the self-energy cannot be set to zero (in that case, the spectral function would be undefined), and this $\mathrm{Im}(\Sigma_A)$ introduces a finite broadening into the spectral density of the system. Thus, such broadening of the energy level of the system is unavoidable due to the non-zero imaginary part of the self-energy, which is an action of the system on itself while interacting with an apparatus, whatsoever small the strength of interaction may be. It will be shown in the next section that such phenomenon has a close resemblance with natural uncertainty relation.

3.3.2 Self-energy and interaction life-time

In order to obtain an expression for the system Green's function given by eq. (3.17a), let us start with assuming

$$G_{jk}{}^S(t, t_1) = -i/\hbar a_{jk} \vartheta(t - t_1) e^{-\frac{i}{\hbar}\lambda_j(t, t_1)} \tag{3.25}$$

$$\Rightarrow \left[\sum_j \delta_{ij}\left(i\hbar \frac{d}{dt}\right)\right] G_{jk}{}^S(t, t_1) = \left[\sum_j \delta_{ij} a_{jk}\left(\delta(t - t_1) + (-i/\hbar)\vartheta(t - t_1) e^{-\frac{i}{\hbar}\lambda_j(t, t_1)} \partial_t \lambda_j(t, t_1)\right)\right]$$

$$= a_{ik}\delta(t-t_1) + \sum_j \delta_{ij}(\partial_t\lambda_j(t,t_1))G_{jk}{}^{S}(t,t_1)$$

Comparing such a result with eq. (3.17a), we have, firstly $a_{ik} = \delta_{ik}$, and secondly

$$\delta_{ij}(\partial_t\lambda_j(t,t_1)) = \delta_{ij}\varepsilon_i + \int_{-\infty}^{+\infty} dt'[\Sigma_A(t,t')]_{ij} \tag{3.26}$$

Summing over 'j' followed by integrating over time t from t_1 to t leads to

$$\lambda_i(t,t_1) = \varepsilon_i(t-t_1) + \sum_j \int_{t_1}^{t} dt \int_{-\infty}^{+\infty} dt'[\Sigma_A(t,t')]_{ij} \tag{3.27}$$

The system Green's function is obtained to be

$$G_{ik}{}^{S}(t,t_1) = -i/\hbar\delta_{ik}\vartheta(t-t_1)e^{-\frac{i}{\hbar}\left[\varepsilon_i(t-t_1) + \sum_j \int_{t_1}^{t} dt \int_{-\infty}^{+\infty} dt'[\Sigma_A(t,t')]_{ij}\right]} \tag{3.28}$$

As discussed in Chapter 2, such Green's function is a representation of the interaction between the system and the apparatus. Here, the step function physically implies the Green's function to represent the effect on the system at a delayed time t due to its interaction with the apparatus at an earlier time t_1, and thus is commonly known as the retarded Green's function ($G_{ik}{}^{S}(t)|_{Rt}$). Now, it is worth noting that the time evolution of the probability of the system states due to such interaction can be studied from (considering $t > t_1 = 0$)

$$\left|G_{ik}{}^{S}(t)\right|_{Rt}^{2} \equiv e^{-\frac{2}{\hbar}\left[-\sum_j \int_{0}^{t} dt \int_{-\infty}^{+\infty} dt'\mathrm{Im}[\Sigma_A(t,t')]_{ij}\right]} \tag{3.29}$$

It is apparent from eq. (3.29) that the imaginary part (which itself is negative in general for the cases of interaction with a measuring apparatus) represents the decay rate of the impulse of interaction due to probing the system by the apparatus. The lifetime of such an impulse of interaction can be obtained by averaging the interaction duration over the norm of retarded Green's function given in eq. (3.29), that results in

$$\langle t\rangle_{ii} = \int_{0}^{\infty} te^{-\frac{2}{\hbar}\left[-\sum_j \int_{0}^{t} dt \int_{-\infty}^{+\infty} dt'\mathrm{Im}[\Sigma_A(t,t')]_{ij}\right]} dt \tag{3.30a}$$

which, up to the first order approximation, gives rise to

$$[\langle t\rangle] \approx \frac{\hbar}{2}\left(-\mathrm{Im}[\Sigma_{\mathrm{A}}{}^{(1)}(E)]^{-1}\right) \tag{3.30b}$$

$$[\Gamma_{\mathrm{A}}(E)][\langle t\rangle] \approx \hbar \tag{3.30c}$$

It is imperative to note, as discussed in the previous section, that the $\mathrm{Im}(\Sigma_{\mathrm{A}})$ brings in a finite broadening ($[\Gamma_{\mathrm{A}}(E)]$) into the discreteness and sharpness of the energy spectral density of a quantum system due to its own action on itself through interacting with an apparatus, and eq. (3.30b) clearly suggests that such broadening of energy levels is related to the interaction duration through the energy-time uncertainty relation limited by Planck's quantum $\hbar$, so that in any experimental arrangement it is theoretically impossible to reduce the $\mathrm{Im}(\Sigma_{\mathrm{A}})$ to zero. This is in complete consonance with what Bohr in his response to the "spooky" nature of quantum properties as explored in EPR-paradox [6] stated: *Indeed the finite interaction between object and measuring agencies conditioned by the very existence of the quantum of action entails – because of the impossibility of controlling the reaction of the object on the measuring instruments if these are to serve their purpose – the necessity of a final renunciation of the classical ideal of causality and a radical revision of our attitude towards the problem of physical reality* [7]. At this point, it may also be interesting to note that if $\mathrm{Im}(\Sigma_{\mathrm{A}})$ is not negative, the impulse of interaction will not decay but will increase in continuous feedback loops of coupling, and thus the apparatus will not be able to serve its purpose, rather may induce a forced resonance effect.

3.3.3 Self-energy and classicalness of the apparatus

Equations (3.30b) and (3.30c), as discussed above, represent the quantumness of the system restricted by (or one can say, defined by) its inherent uncertainty in terms of its energy level broadening, which cannot be reduced arbitrarily to zero by any means of appropriately setting the apparatus arrangements. It is imperative to note that up to this point, we have described the apparatus also on the basis of quantum theory and not imposed any classical property onto it. Now we will investigate the classicalness of the apparatus and the required conditions for it. Interestingly, it can be observed from eqs. (3.11) and (3.12) that both the system and the apparatus have equal rights in their collective description, and no theoretical difference(s) have been considered or obtained that can distinguish between them. Then the natural question that arises is what decides which one is the system and which one is the apparatus! Is it just our subjective choice?

Let us consider, keeping in mind the equal rights of the system and apparatus, the spectral function (i.e. the density of states in per energy unit) just in the same manner as is given by eqs. (3.22a) or (3.24), where now the energy levels of the apparatus exhibit broadening due to its own course of action on itself through interacting with the system:

$$A^{\mathrm{A}}(E) \equiv \sum_r \frac{2\ \mathrm{Im}(\Sigma_{\mathrm{S}})}{(E - \varepsilon_r{}^{\mathrm{A}} - \mathrm{Re}(\Sigma_{\mathrm{S}}))^2 + (\mathrm{Im}(\Sigma_{\mathrm{S}}))^2} \tag{3.31}$$

Now, since for any isolated system (be it the quantum system concerned or be it the apparatus), the Schrödinger equation (or types of Dirac equation for the relativistic case) leads to infinitely many energy eigenvalues, so that in the entire range of energy of the system, the discrete summation can be replaced by a continuous integral of the form

$$A^{\mathrm{A}}(E) \equiv \int\limits_{\varepsilon^{\mathrm{A}}} d\varepsilon^{\mathrm{A}} D(\varepsilon^{\mathrm{A}}) \left(\frac{2\ \mathrm{Im}(\Sigma_{\mathrm{S}})}{(E - \varepsilon^{\mathrm{A}} - \mathrm{Re}(\Sigma_{\mathrm{S}}))^2 + (\mathrm{Im}(\Sigma_{\mathrm{S}}))^2} \right) \tag{3.32}$$

where $D(\varepsilon^{\mathrm{A}})$ is the number of eigenstates per unit elementary energy range of $d\varepsilon^{\mathrm{A}}$. It is imperative to note at this point that such a notion of changing the summation to an integral is solely attributed to the classicalness of a system by Datta, only on the basis of closely spaced eigenvalues in a classical system [8]. However, firstly, such assumption, in fact, distinguishes between classical and quantum systems on the basis of size only (i.e. with increasing size, gap between the eigenvalues reduces), which cannot explain the complete quantumness observed in large-size systems like superconductors; and secondly, any quantum system that is isolated, such as, an electron confined in a hydrogen atom potential, always gives rise to infinitely many eigenstates that can always be considered closely spaced with respect to the entire energy range. The gap between eigenvalues is thus a measure of discreteness but not solely the measure of quantumness/classicalness. Therefore, rather recalling Dirac, as noted earlier in this chapter, that "*We may define an object to be big* (i.e. so-called classical) *when the disturbance accompanying our observation of it may be neglected, and small* (i.e. so-called quantum) *when the disturbance cannot be neglected*", the classicalness of the apparatus can be attributed to its negligible impact on itself due to interacting with the system by simply putting the limit $\mathrm{Im}(\Sigma_{\mathrm{S}}) \to 0$ arbitrarily with respect to the range of energy of the apparatus (note that this could not be done with the quantum system in question, i.e. $\mathrm{Im}(\Sigma_{\mathrm{A}}) \to 0$, as discussed above). Thus, eq. (3.32) leads to, for the apparatus to be a classical system assuming instrumental error to be theoretically zero (i.e. $\mathrm{Re}(\Sigma_{\mathrm{S}}) = 0$ as explained previously),

$$\begin{aligned} A_{Cl}{}^{A}(E) &\equiv \int\limits_{\varepsilon^{A}} d\varepsilon^{\mathrm{A}} D(\varepsilon^{\mathrm{A}}) \underset{\mathrm{Im}(\Sigma_S)\to 0}{Lt} \left(\frac{2\ \mathrm{Im}(\Sigma_S)}{(E - \varepsilon^{\mathrm{A}})^2 + (\mathrm{Im}(\Sigma_S))^2} \right) \\ &= \int\limits_{\varepsilon^{\mathrm{A}}} d\varepsilon^{\mathrm{A}} D(\varepsilon^{\mathrm{A}}) \delta(\varepsilon^{\mathrm{A}} - E) \\ \Rightarrow A_{Cl}{}^{A}(E) &\equiv D(E) \end{aligned} \tag{3.33}$$

which neither depends on the discrete energy eigenvalues nor is limited by the self-energy related to inherent quantum uncertainty, and thus represents the classicalness of the apparatus. This is how the non-equilibrium Green's function framework can distinguish between a classical and a quantum system on the basis of its relative value of the imaginary part of the self-energy in comparison to its energy.

3.3.4 Self-energy and dynamic band structure of a quantum system

The system and apparatus together constitute an isolated system described by eq. (3.12) and thus Green's function of system+apparatus can be written as

$$[G^{\mathrm{Tot}}(E)] = \begin{bmatrix} [G^{\mathrm{S}}(E)] & [\ldots] \\ [\ldots]^{+} & [G^{\mathrm{A}}(E)] \end{bmatrix} = \begin{bmatrix} E[I]-[\varepsilon^{\mathrm{S}}] & [\tau^{\mathrm{S}\leftrightarrow\mathrm{A}}] \\ [\tau^{\mathrm{S}\leftrightarrow\mathrm{A}}]^{+} & E[I]-[\varepsilon^{\mathrm{A}}] \end{bmatrix}^{-1} \tag{3.34}$$

where the off-diagonal terms in the Green's function block matrix (not in the inverse matrix at the RHS) are not of our concern, since we are interested in its traced-off parts for the system ($[G^{\mathrm{S}}(E)]$) and/or the apparatus ($[G^{\mathrm{A}}(E)]$). These can be obtained from the inverse of block matrix as

$$[G^{\mathrm{S}}(E)] = \left[E[I]-[\varepsilon^{\mathrm{S}}]-[\tau^{\mathrm{S}\leftrightarrow\mathrm{A}}][E[I]-[\varepsilon^{\mathrm{A}}]]^{-1}[\tau^{\mathrm{S}\leftrightarrow\mathrm{A}}]^{+}\right]^{-1} \tag{3.35}$$

Thus, the system's self-energy due to its action on itself through the interaction with the apparatus can be calculated from

$$[\Sigma_{\mathrm{A}}(E)] = [\tau^{\mathrm{S}\leftrightarrow\mathrm{A}}][E[I]-[\varepsilon^{\mathrm{A}}]]^{-1}[\tau^{\mathrm{S}\leftrightarrow\mathrm{A}}]^{+} \tag{3.36}$$

which is consistent with the expression of eq. (3.15b), taking into account the classicalness of the apparatus as discussed in the above section. Now, if we write the apparatus matrix in discrete position coordinates, the classicalness of the apparatus leads us to the translational symmetry (like Bloch wave) so that:

$$[E[I]-[\varepsilon^{\mathrm{A}}]]^{-1} = \begin{bmatrix} E[I]-[\varepsilon^{\mathrm{A}}] & [\tau^{\mathrm{A}\leftrightarrow\mathrm{A}}] & \ldots & \ldots \\ [\tau^{\mathrm{A}\leftrightarrow\mathrm{A}}]^{+} & E[I]-[\varepsilon^{\mathrm{A}}] & [\tau^{\mathrm{A}\leftrightarrow\mathrm{A}}] & \ldots \\ \ldots & [\tau^{\mathrm{A}\leftrightarrow\mathrm{A}}]^{+} & E[I]-[\varepsilon^{\mathrm{A}}] & \ldots \\ \ldots & \ldots & \ldots & \ldots \end{bmatrix}^{-1} \tag{3.37}$$

where the elements at (1,1) represent the nodes of the apparatus that directly couple with the system, while the nodes at (2,2) couple with nodes at (1,1), nodes at (3,3) with those at (2,2), and so on. $[\tau^{\mathrm{A}\leftrightarrow\mathrm{A}}]$ represents the coupling matrices between adjacent nodes. Since the system directly couples with the apparatus only through the nodes at (1,1), the coupling matrices between the system and all other nodes (i.e. (2,2), (3,3), . . .

) of the apparatus are zero. Thus, mathematically, we are interested in the element ($[g]_{11}$) of the inverse block matrix of eq. (3.37) at (1,1) only, which, due to the translational symmetry of the apparatus matrix, is given by

$$[g]_{11} = \left(E[I] - [\varepsilon^{A}] - [\tau^{A\leftrightarrow A}][g]_{11}[\tau^{A\leftrightarrow A}]^{+}\right)^{-1}$$

$$\Rightarrow [\tau^{A\leftrightarrow A}][g]_{11}[\tau^{A\leftrightarrow A}]^{+}[g]_{11} - (E[I] - [\varepsilon^{A}])[g]_{11} + I = 0 \tag{3.38}$$

The simplest solutions of eq. (3.38) can be obtained by factorizing it as

$$\left([\tau^{A\leftrightarrow A}][g]_{11} - [I]e^{i\vec{k}\cdot\vec{\Delta}}\right)\left([\tau^{A\leftrightarrow A}]^{+}[g]_{11} - [I]e^{-i\vec{k}\cdot\vec{\Delta}}\right) = 0 \tag{3.39a}$$

provided,

$$E[I] - [\varepsilon^{A}] = 2\mathrm{Re}\left([\tau^{A\leftrightarrow A}]e^{-i\vec{k}\cdot\vec{\Delta}}\right) \tag{3.39b}$$

Although $\vec{k}\cdot\vec{\Delta}$ is an arbitrary phase appearing mathematically in the solutions given by eq. (3.39a), it physically arises due to the translational symmetry of the apparatus as described in Bloch theory. Here, $\vec{\Delta}$ is the minimum unit in position coordinate that translates the elements of the apparatus into itself, which is analogous to the lattice constant of a crystal in Bloch theory, while $\vec{k}$ is the Bloch wave vector.

Now, since the system couples with the apparatus only through the nodes at (1,1), eqs. (3.36) and (3.39a) lead to obtaining the self-energy of the system due to its interaction with the apparatus as

$$[\Sigma_{A}(E)] = [\tau^{S\leftrightarrow A}][g]_{11}[\tau^{S\leftrightarrow A}]^{+} = [\tau^{S\leftrightarrow A}][\tau^{A\leftrightarrow A}]^{-1}[\tau^{S\leftrightarrow A}]^{+}e^{i\vec{k}\cdot\vec{\Delta}} \tag{3.40}$$

Now, eq. (3.24) shows that the spectral density is maximum at $E = \varepsilon_i^{S} + \mathrm{Re}(\Sigma_{A})$, which represents each of the band extrema. Thus, near band extrema, the dynamic band structure of the system while interacting with the apparatus can be obtained from

$$E[I] - [\varepsilon^{S}] - \mathrm{Re}\left([\tau^{S\leftrightarrow A}][\tau^{A\leftrightarrow A}]^{-1}[\tau^{S\leftrightarrow A}]^{+}e^{i\vec{k}\cdot\vec{\Delta}}\right) = [0] \tag{3.40}$$

where the effective Hamiltonian of the system, while coupling with the apparatus, is eventually given by

$$[H^{S}] = [\varepsilon^{S}] + \mathrm{Re}([\Sigma_{A}(E)])e^{i\vec{k}\cdot\vec{\Delta}} \tag{4.40a}$$

Accordingly, the band velocity and effective mass tensor near band extrema are obtained to be

$$[v_j] = \frac{1}{\hbar}\frac{\partial E[I]}{\partial k_j} = \Delta_j\left(\frac{2}{\hbar}\mathrm{Im}[-\Sigma_{A}(E)]\right) = \Delta_j[\langle t\rangle]^{-1} \tag{3.41a}$$

$$[m_{ij}*]^{-1} = \frac{1}{\hbar^2}\left(\frac{\partial^2 E[I]}{\partial k_i \partial k_j}\right) = \frac{\Delta_i \Delta_j}{\hbar^2}\mathrm{Re}[-\Sigma_\mathrm{A}(E)] \tag{3.41b}$$

Thus, the band velocity of the system given by eq. (3.41a) is proportional to the imaginary part of its self-energy (i.e. energy level broadening due to natural uncertainty) while interacting with the apparatus. Such a result physically implies how motion is indeed generated from the uncertainty inherent to a system. On the other hand, the expression for effective mass given by eq. (3.41b) may seem ambiguous since it has already been discussed that the real part of the self-energy can be made zero by appropriate settings of the experiment, which would then result in the effective mass being infinite. However, that is not the case; in fact, even then, the equivalent real part of self-energy can be obtained from the Hilbert transform of its imaginary part. Thus, the (inverse of) effective mass is in a convoluted relation with the band velocity, or more precisely, with the natural uncertainty of the system.

Further, it is imperative to note at this point that eqs. (3.40) and (3.41) are more general results than they apparently look. For instance, two adjacent atoms in the unit cell of a crystal can be represented by 'S' and 'A' of the above model, where $[\varepsilon^\mathrm{S}]$ and $[\varepsilon^\mathrm{A}]$ are the electronic energy levels of the two atoms, respectively, while $[\Sigma_\mathrm{A}(E)]$ represents the interaction impact of 'A' on 'S'. If we consider all such primitive adjacent atoms (i) of one in a unit cell, each of which has the minimum translational repetition $\vec{\Delta}_i$ (i.e. the lattice constant vector), then eq. (3.40a) can be written as

$$[H^\mathrm{S}] = [\varepsilon^\mathrm{S}] + \sum_i \mathrm{Re}\left([\Sigma_\mathrm{A}(E)]_i\right)e^{i\vec{k}\cdot\vec{\Delta}_i} \tag{3.42}$$

The relation between energy eigenvalues (E_n) of such a Hamiltonian of eq. (3.42) and the Bloch wave vector $\vec{k}$ gives rise to the band structure of the crystal. This method of band-structure calculation is nothing but the well-known linear combination of atomic orbitals (LCAO) approach.

3.4 System↔apparatus information inflow and outflow

The rate of inflow/outflow of system information from/to the apparatus can be obtained by calculating the two-time flux:

$$\Pi(t,t') = \left(\frac{d}{dt} - \frac{d}{dt'}\right) n_{ij}{}^\mathrm{S}(t,t') \tag{3.43}$$

which, since eq. (3.18) leads to

$$[\Pi(t,t')] = \left(\frac{d}{dt} - \frac{d}{dt'}\right) \int_{-\infty}^{+\infty} dt_2 \int_{-\infty}^{+\infty} dt_1 [G^\mathrm{S}(t,t_1)][\Sigma_\mathrm{A}^\mathrm{In}(t_1,t_2)][G^{\mathrm{S}+}(t_2,t')]$$

$$= \left((-i/\hbar) \int_{-\infty}^{+\infty} dt_2 [\Sigma_A^{In}(t,t_2)][G^{S+}(t_2,t')] \right) - \left((-i/\hbar) \int_{-\infty}^{+\infty} dt_1 [G^S(t,t_1)][\Sigma_A^{In}(t_1,t')] \right)$$

$$+ (-i/\hbar) \int_{-\infty}^{+\infty} dt_2 \int_{-\infty}^{+\infty} dt_1 \int_{-\infty}^{+\infty} d\tilde{t} \left([\Sigma_A(t,\tilde{t})] - [\Sigma_A(\tilde{t},t')]^+ \right) [G^S(t,t_1)][\Sigma_A^{In}(t_1,t_2)][G^{S+}(t_2,t')]$$

[since eq. (3.17a)]

$$\Rightarrow [\Pi(t,t')] = (-i/\hbar) \int_{-\infty}^{+\infty} dt_0 \left([\Sigma_A^{In}(t,t_0)][G^{S+}(t_0,t')] - [G^S(t,t_0)][\Sigma_A^{In}(t_0,t')] \right)$$

$$- (1/\hbar) \int_{-\infty}^{+\infty} dt_2 \int_{-\infty}^{+\infty} dt_1 \int_{-\infty}^{+\infty} d\tilde{t}i \left([\Sigma_A(t,\tilde{t})] - [\Sigma_A(\tilde{t},t')]^+ \right) [G^S(t,t_1)][\Sigma_A^{In}(t_1,t_2)][G^{S+}(t_2,t')] \quad (3.44)$$

Equation (3.44) representing the flux of information exchanged between each state of the system and the apparatus indicates that the first term is a measure of inward flux ($[\Pi_{In}(t,t')]$) to each state of the system from all the states of the apparatus that depends directly on $[\Sigma_A^{In}(t,t_0)]$, while the second term represents the total outward flux ($[\Pi_{Out}(t,t')]$) from all the states of system to each of the apparatus states (note the broadening term $\Gamma(t,t') = \int_{-\infty}^{+\infty} d\tilde{t}i \left([\Sigma_A(t,\tilde{t})] - [\Sigma_A(\tilde{t},t')]^+ \right)$). Thus,

$$[\Pi_{In}(t,t')] = (-i/\hbar) \int_{-\infty}^{+\infty} dt_0 \left([\Sigma_A^{In}(t,t_0)][G^{S+}(t_0,t')] - [G^S(t,t_0)][\Sigma_A^{In}(t_0,t')] \right) \quad (3.45a)$$

$$[\Pi_{Out}(t,t')] = -(1/\hbar) \int_{-\infty}^{+\infty} dt_2 \int_{-\infty}^{+\infty} dt_1 \Gamma(t,t')[G^S(t,t_1)][\Sigma_A^{In}(t_1,t_2)][G^{S+}(t_2,t')] \quad (3.45b)$$

or, in the energy domain (by Fourier transform over the difference-time coordinate), these result to

$$[\Pi_{In}(E)] = -i \left([\Sigma_A^{In}(E)][G^{S+}(E)] - [G^S(E)][\Sigma_A^{In}(E)] \right) \quad (3.46a)$$

$$[\Pi_{Out}(E)] = -\Gamma(E)[G^S(E)][\Sigma_A^{In}(E)][G^{S+}(E)] \quad (3.46b)$$

Since eq. (3.46a) and (3.46b), the total inward and outward flux accompanying all the states of the system and the apparatus are accordingly obtained by taking the trace of the matrices as

$$Tr[\Pi_{In}(E)] = -iTr \left([\Sigma_A^{In}(E)][G^{S+}(E)] - [G^S(E)][\Sigma_A^{In}(E)] \right)$$

$$= -iTr \left([\Sigma_A^{In}(E)][G^{S+}(E)] - [\Sigma_A^{In}(E)][G^S(E)] \right) \qquad [\text{since } Tr([A][B]) = Tr([B][A])]$$

$$= -Tr\left([\Sigma_A^{In}(E)]i([G^{S+}(E)] - [G^S(E)])\right)$$

$$= Tr\left([\Sigma_A^{In}(E)][G^{S+}(E)][\Gamma(E)][G^S(E)]\right)$$

$$= Tr\left([\Gamma(E)][G^S(E)][\Sigma_A^{In}(E)][G^{S+}(E)]\right) \qquad \text{[since } Tr([A][B]) = Tr([B][A])\text{]}$$

$$= -Tr[\Pi_{Out}(E)] \qquad \text{[since eq. (3.46b)]}$$

$$\Rightarrow Tr[\Pi_{In}(E)] + Tr[\Pi_{Out}(E)] = 0 \tag{3.47}$$

It is highly interesting to note from eqs. (3.46a), (3.46b), and (3.47) that the inward flux to each state of the system from all the states of the apparatus are evidently quite different from the outward flux from all the states of the system to each of the apparatus states (i.e. $[\Pi_{In}(E)] + [\Pi_{Out}(E)] \neq [0]$), so that a non-zero amount of exchange of information takes place from state-to-state between the system and apparatus through their interaction at individual levels; however, i.e. total inward flux to all system states from all the apparatus states is equal and opposite to the total outward flux from all the system states to all of the apparatus states, leading to no information exchange between system and apparatus in totality (i.e. $Tr[\Pi_{In}(E)] + Tr[\Pi_{Out}(E)] = 0$). This can also be considered as Newton's third law of motion for two interacting quantum systems. Thus, if all the system states interact with all the apparatus states, no information can be obtained from such interaction. Thus, it upholds the requirement of quantum collapse of the system states into one during measurement (i.e. through interaction with the measuring apparatus) in order to extract information from the system, i.e. the purpose to be served by the measuring apparatus.

References

[1] Heisenberg, W. and MacLachlan, J. "Physics and Philosophy." *Physics Today* 11, no. 9 (1958): 36.

[2] Bohr, N. "Causality and complementarity." *Philosophy of Science* 4, no. 3 (1937): 289–298.

[3] Bohr, N. "On the notions of causality and complementarity." *Science* 111, no. 2873 (1950): 51–54.

[4] Dirac, P. A. M., *The principles of quantum mechanics*. no. 27, Oxford university press, 1981.

[5] Home, D. "Conceptual Foundations of Quantum Physics: An overview from modern perspectives." (2013).

[6] Einstein, A., Podolsky, B., and Rosen, N. "Can quantum-mechanical description of physical reality be considered complete?." *Physical Review* 47, no. 10 (1935): 777.

[7] Bohr, N. "Can quantum-mechanical description of physical reality be considered complete?." *Physical Review* 48, no. 8 (1935): 696.

[8] Supriyo, D. *Quantum transport: Atom to transistor*. Cambridge University Press, 2005.

4 Modelling quantum transport in electronic devices: MOSFET, FinFET, NanowireFET

In the present era of unprecedented facilities and services available in our everyday life in almost all spheres of modern digitized and information-based civilization, if one has to attribute it solely to some human achievement, it would possibly be nothing but the very CMOS technology, more particularly the invention and development of MOSFETs. As mentioned in Chapter 1, the unrelenting motivation for achieving low-power, higher speed, and continuously increasing packing density for improved functionality has compelled such technology to adopt the unique route of gradually miniaturizing the MOSFET device dimensions. In the pathway of such unremitting down-scaling of the physical dimension of MOSFETs by aggressively reducing the source-drain distance over the past 50 years, the device channel length has shrunk from the micrometre (μm) order to only a few nanometres (nm) [1]. Such nm-orders of the channel/gate length exhibit a shorter path traversed by the charge carriers (i.e. electrons and holes), thereby increasing the device speed, as well as offering a larger driving electric field at lower source/drain bias, which evidently reduces the power consumption per electronic function. Moreover, the downscaling of device size significantly enhances the number of functions in the chip area.

Nevertheless, advantage does not come alone; it brings with it a number of challenges. As the gate length is reduced aggressively from ~μm to a few nm ranges, the thickness of the gate oxide needed its apparent thinning up to a few atomic layers in order to maintain the strength of the switching field (in comparison to the driving field) with optimized gate control. As a result, alongside the corresponding fabrication complexities and other short-channel effects [1], quantum properties started dominating at this scale to introduce the undesired effect like quantum tunnelling of electrons through the thin oxide layer, leading to unwanted leakage current and static-power dissipation [1].

The origin of such challenges remains in the violation of the basic operating principle of MOSFET, where the switching electric field (due to the gate) must dominate the driving electric field (due to the drain). Thus, in order to overcome such challenge, two approaches have been adopted: 1) incorporating high-k insulators at the gate ('k' being the dielectric constant), which, as a linear function of its k-value can be made thick enough to reduce quantum tunnelling, and at the same time, keeping the gate-induced electric field (in terms of the 'effective oxide thickness') intact; and 2) gate engineering, by increasing the gating area to effectively increase the control of the switching field. In this latter context, the MOSFET has evolved through the exploration of several fundamentally different gate architectures: double gate (DG) [2], FinFET [3], Ω-gate [4], and gate all-around (GAA) [5–6] geometries (see Fig. 4.1(a)–(f)). Accordingly, with such gate architectures, the geometry of the channel region is also evolved, and recently reached two geometrical shapes, i.e. the cylindrical nanowire (NW) [5] and the corner-rounded nanosheet (NS) [6], as the most promising choices (Fig. 4.1(e),(f)).

 | https://doi.org/10.1515/9783111545950-004

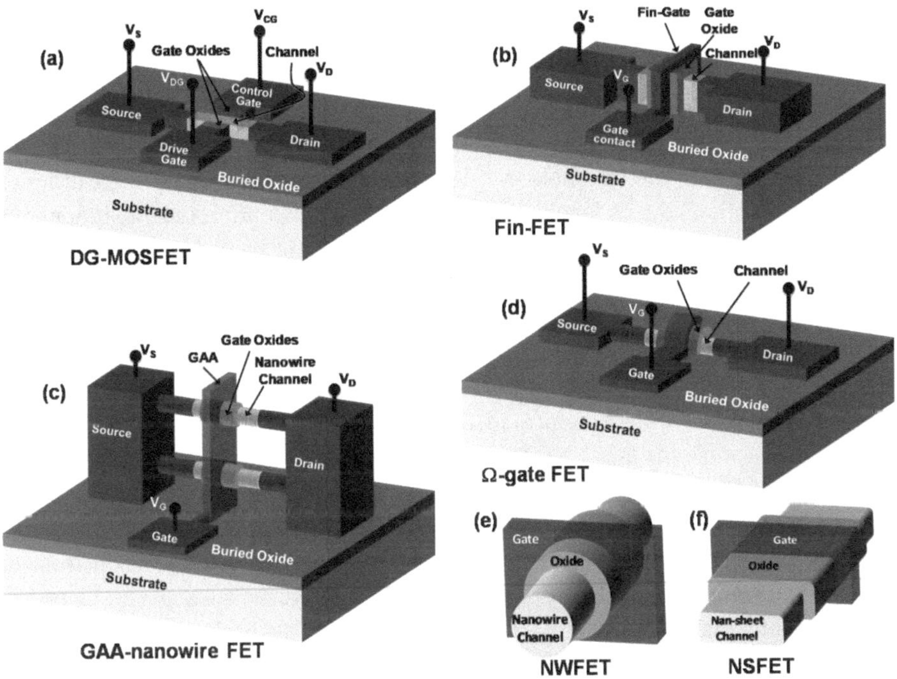

Fig. 4.1: Evolution of gate architecture of MOSFET from its planar structure: (a) double-gate (DG) MOSFET; (b) Fin-FET; (c) gate-all-around (GAA) nanowire (NW) FET; (d) Ω-gate NW FET; and state-of-the-art channel geometry: (e) cylindrical NW FET; and (f) corner-rounded nanosheet (NS) FET.

At this point, it is worth mentioning that 97% of the current CMOS technology is based upon Si substrates and also on Si as the channel material. The rationale behind this includes its natural abundance, the availability of its native oxide SiO_2, providing minimal interface traps, desired mechanical strength for larger wafer fabrication, and the crucial factor that it is non-toxic. However, alongside the aggressive scaling of Si-MOSFETs, a parallel progress is observed for improving the MOSFET performance by investigating a number of alternative channel materials with mobility higher than that of Si, such as strained-Si, SiGe, GaAs, and other III-V materials [7]. Further, in order to cross the classical limit of switching speed at room temperature, the quantum mechanical tunnelling effect has been exploited by fabricating TFETs [8].

It may be interesting to note that the large-scale manufacturers like TSMC, Intel, and Samsung are now developing 2 nm technology nodes [9] that comprise the NS-FETs with a channel length smaller than 10 nm. It has evidently escalated the device performance to an extraordinary degree, however, at the cost of the complexities from the technological aspects, as well as the fundamental limitations posed by quantum properties. It is imperative to note that in such ultra-scaled devices, quantum effects are not something that perturbs their physical properties and electrical performances, but rather, their entire behaviour is governed by the quantum rules in contrast to the semi-classical laws. Thus, the interpretations and mathematical calculations, as well as the scientific predictions of such physical and electrical characteristics of the ultra-scaled FET devices, cannot be carried out by the conventional semi-classical models, even those that include quantum effects as perturbations.

For instance, the electron/hole wavelength in such devices may be of the order of, or even larger than, the channel length itself, leading to an unprecedented input/output overlap of signals. Also, it indicates the possibility of ballistic carrier transport without any scattering, thereby making the process entirely non-equilibrium [10]. Furthermore, as discussed in Chapter 1, the parameters like 'temperature' or 'Fermi level' (chemistry people generally call it 'chemical potential') may not even be any defined property of a nanostructure-channel due to the inadequate numbers of constituting atoms to provide any meaningful statistical average. However, the source and drain that directly couple to such a channel have finite 'temperatures', as well as definite 'Fermi levels'. Therefore, a full-quantum model addressing such complicated issues is necessary to estimate the behaviour of a MOSFET device at this scale. This chapter will thus focus on developing such a mathematical model to describe the downscaled MOSFETs, in general, and their newly emerged architectures, such as FinFET or NW FET, assuming that the reader has an understanding of the basics of MOSFET device physics and operations (for details, see Ng et al. [11]).

4.1 A fundamental difference between planar and nanostructure MOSFETs

A MOSFET is a digital switching device that is operated by two counteracting orthogonal electric fields, as shown in Fig. 4.2: 1) the transverse field, due to gate voltage (V_G), that creates the semiconducting channel path beneath the metal/oxide-gate for carrier flow; and 2) the longitudinal field, due to drain voltage (V_D), that drives such carriers from source to drain through such channel. Depending on whether the carrier is an electron or a hole, the MOSFET is said to be an n-channel or p-channel device, respectively. In the case of planar MOSFETs, interestingly, an n-channel MOSFET that utilizes electrons as carriers is developed, as illustrated in Fig. 4.2, on a p-type semiconducting (e.g. p-Si) substrate with source/drain being degenerate n-type (i.e. n^+) material, and *vice versa* for a p-channel MOSFET.

Such a structure is designed for the specific purpose of operating the device at its 'inversion' condition, i.e. at a voltage bias applied to the gate of MOS so that the p-type semiconductor locally (only near the metal/oxide-gate) behaves like an n-type one (and *vice versa* for an n-type substrate). In order to understand the fundamental physical mechanism of carrier transport within the MOSFET, it is thus necessary to visualize the device in its potential energy space, where a slope or a barrier endorses or prevents, respectively, the carrier flow. The representation of the device in such a

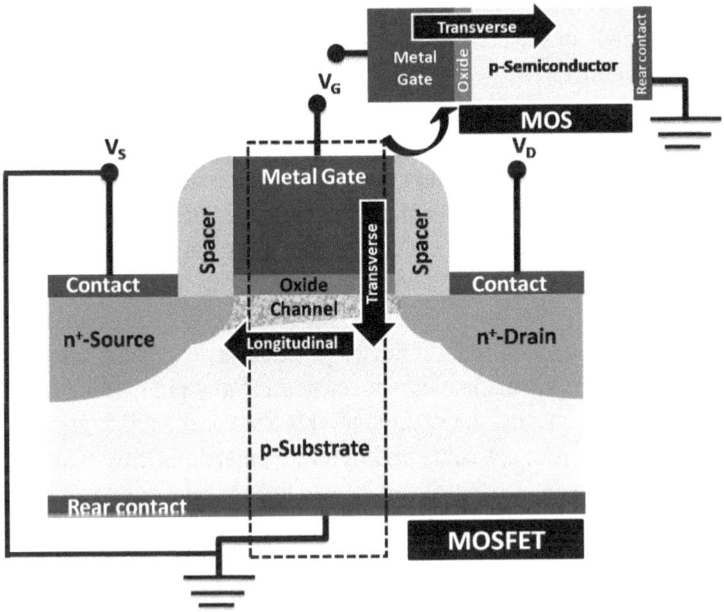

Fig. 4.2: The schematic of an n-channel planar MOSFET developed on a p-type semiconducting substrate with degenerate n-type (i.e. n^+) source/drain. The metal gate is electrically insulated from the semiconductor by the intermediate thin oxide layer, thereby creating the MOS structure (also shown separately). The direction of transverse (along MOS-gate) and longitudinal (along channel) electric fields is indicated in the figure.

space is called its band diagram (note that it should not be messed up with the band structure that depicts the energy-wave vector relations of electrons in a material). The band diagrams in MOS devices are depicted in Fig. 4.3(a)–(f) at different gate biases for a comprehensive understanding of how the inversion channel is formed in a MOSFET.

Firstly, the rear contact (see Fig. 4.2) is an MS junction that must be an ohmic contact (not a Schottky contact) for the majority carriers (e.g. holes in the p-substrate) in order to ensure uninterrupted carrier flow in the circuit. Now, when the semiconductor has not formed any junction with the metal (see Fig. 4.3(a)), it has spatially uniform conduction and valence bands. However, once the junction is formed, electrons/holes can flow from semiconductor to metal and *vice versa*, thereby forming a common Fermi level owing to the minimized Gibbs potential of the entire system. On the other hand, electron affinity (i.e. the energy required for taking an electron out of the conduction band to vacuum) of the semiconductor (χ_S) ensures the energy gap between vacuum-to-conduction band to be unchanged at the interface (note that vacuum level cannot be defined for the interior region of a material) leading to a band-bending near the MS junction depending on the relative initial values of Fermi levels of the metal (E_{FM}) and semiconductor (E_{FS}), as shown in Fig. 4.3(a). It is thus apparent that, in order to bend the band for endorsing the holes of a p-semiconductor to flow towards the metal, one needs to select a metal with a Fermi level $E_{FM} > E_{FS}$, and *vice versa* for an n-type semiconductor.

However, for the MOS junction, the case is not so, where electrons/holes cannot transport from semiconductor to metal-gate (or *vice versa*) due to the potential barrier formed by the intermediate oxide layer owing to its much smaller affinity (see Fig. 4.3 (b)), so that the corresponding Fermi levels cannot equalize (since Gibbs potential cannot be minimized without particle exchange). Now, when it is connected through an external metallic wire (without any bias), as shown in Fig. 4.3(c), carriers can flow via such a wire, thereby equalizing the Fermi levels. In the process of such Fermi level equalization, the band near the oxide-semiconductor junction bends, and a potential difference is also created within the oxide (without considering any pre-existing fixed-oxide-charges). Sometimes, accordingly, the vacuum level is also considered to shift virtually. Nevertheless, a voltage of equal value and opposite polarity is thus evidently needed to make the band flat again (see Fig. 4.3(d)), known to be the flat-band voltage.

Now, when a positive voltage is applied, as illustrated in Fig. 4.3(e), at the metal-gate of a p-semiconductor MOS device, the conduction and valence bands bend near the oxide-semiconductor junction to maintain the continuity of potential and normal component of electric displacement (i.e. physically the surface charge density) at the interface. The local electric field created here, associated with such band-bending, attracts the electrons towards the interface while pushing back the holes towards the rear-contact. Interestingly, such electrons get confined in the quantum well created along the MOS-direction by the gate-voltage-induced band-bending near the oxide-semiconductor interface (often called surface quantization).

For developing a little more clarity, the intrinsic Fermi level (E_i) (i.e. the Fermi level if the semiconductor were an intrinsic/undoped one, also known to be the mid-gap) is also depicted in the figure. It may be worthy to note that a Fermi level below the intrinsic level (i.e. closer to the valence band) indicates the semiconductor to be p-type, while above it (i.e. closer to the conduction band) refers to an n-type material. Thus, although in a p-semiconductor-based MOS, the intrinsic level remains above its Fermi level, with increasing the positive voltage applied at the metal-gate, as can be seen from Fig. 4.3(f), it goes down the Fermi level near the oxide-semiconductor interface, making the region n-type, at least in principle. This is called the 'inversion' mode of MOSFET, and now the channel is assumed to be formed to pave the path for electrons to flow (with lesser recombination loss owing to the lower density of holes), from the source to drain due to any longitudinal driving field (i.e. source-drain voltage).

It is imperative to note at this point that the FinFETs and NWFETs, or the nanostructure FET devices in general, are fundamentally different from the planar MOSFETs by the fact that their channels are isolated from the substrate by an insulating oxide layer (often called the buried oxide layer). Thus, an n-channel Fin/NW-FET device, technically speaking, does not operate at 'inversion' mode from a material point of view,

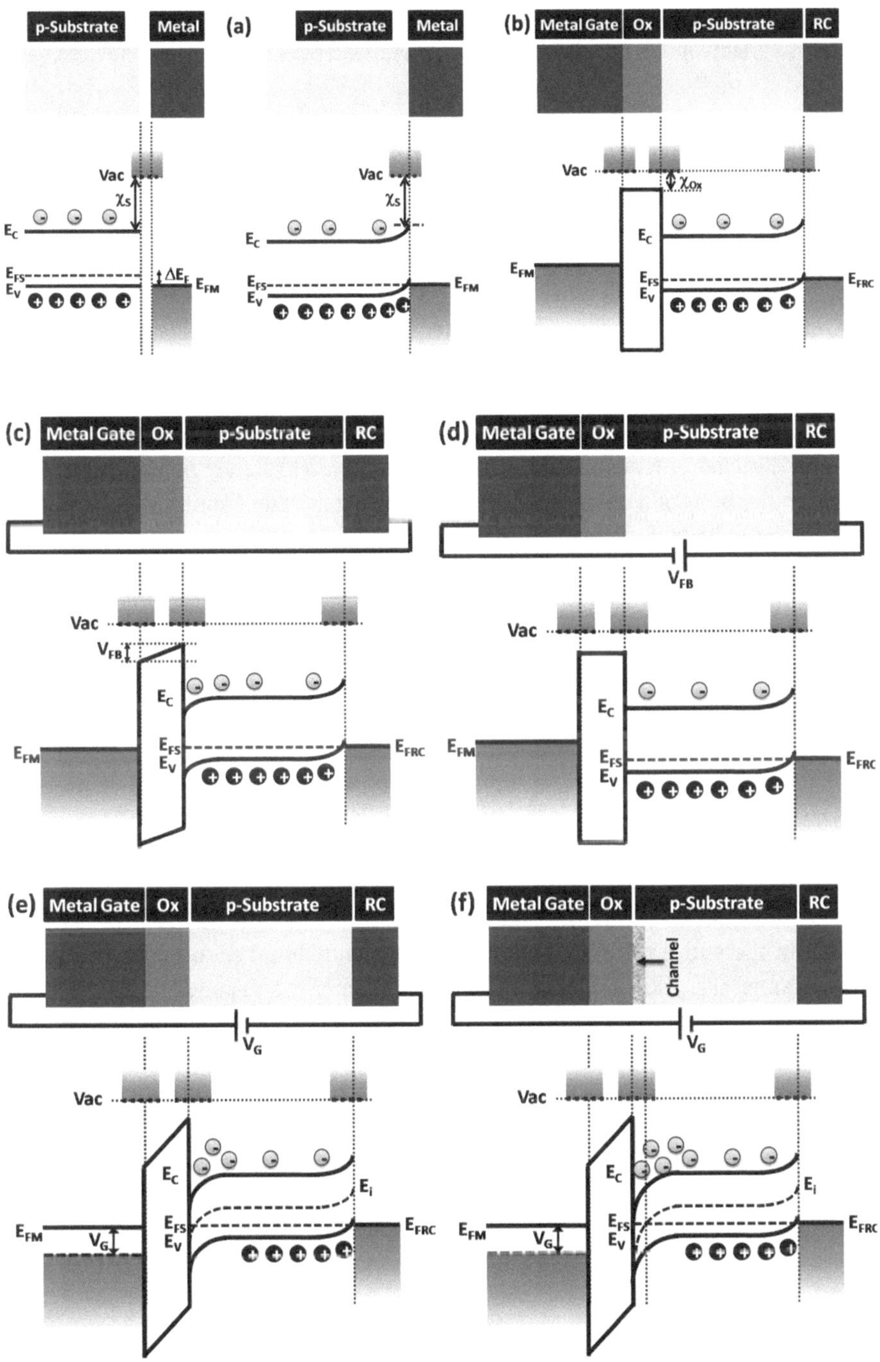

Fig. 4.3: Schematic representation of band diagram of MS junction and MOS devices illustrating a p-semiconducting substrate: (a) when p-semiconductor and metal are isolated and when they form an

since the channel material used is itself kept either intrinsic or low-doped n-type. Therefore, the 'inversion' mode is indeed a virtual or equivalent concept in such devices.

4.2 Scheme of quantum charge transport modelling in short-channel MOSFETs

In a short-channel (planar or nanostructure) MOSFET, the channel itself is a quantum system, where the motion of an electron (or a hole) is governed by electrostatic interaction, or in general by electromagnetic interaction if spin transport is considered, as well as simultaneously by the probabilities arising from quantum uncertainty. However, such a quantum channel is not isolated but rather coupled to two classical systems, i.e. source and drain. Thus, in order to develop a theoretical construct for modelling the charge transport in such a short-channel (planar or nanostructure) MOSFET on the basis of a quantum-electrostatic self-consistent framework, the following steps need to be taken up sequentially:

1. Construct the total Hamiltonian of the FET, including the channel, assuming it to be the active device, while considering the source and drain as two reservoirs coupled to it, in the second quantization field operator framework, to appropriately incorporate the wave-particle duality of electrons/holes. Then derive the equations of motion (using the Heisenberg equation) for channel and reservoirs separately, which must, on the other hand, be interdependent.
2. Note that a MOSFET channel is a semi-confined quantum system where the charges are free to move along the channel (let Z-axis) and the in-plane transverse direction (let Y-axis), as shown in Fig. 4.4(a), while confined along the MOS-gate direction (let X-axis) at a bias leading to 'inversion' condition as explained above. For nanostructure MOSFETs like FinFET or NWFET, the charges are confined in the entire transverse plane (i.e. XY-plane) and free to move only along the channel (i.e. Z-axis). Thus, in any case, the energy sub-bands of such a semi-confined active device (i.e. the channel) first need to be calculated that are not directly coupled with the reservoirs.

Fig. 4.3 (continued)
'abrupt' (not graded) junction; (b) MOS device at an open circuit condition; (c) when the circuit is shorted without any external bias, showing intrinsic band-bending; (d) recreating flat band by applying relevant opposite (here negative) bias; (e) band bending due to application of a positive bias at such flat-band condition; and (f) applying relevant positive bias to achieve 'inversion' mode in the MOS device.

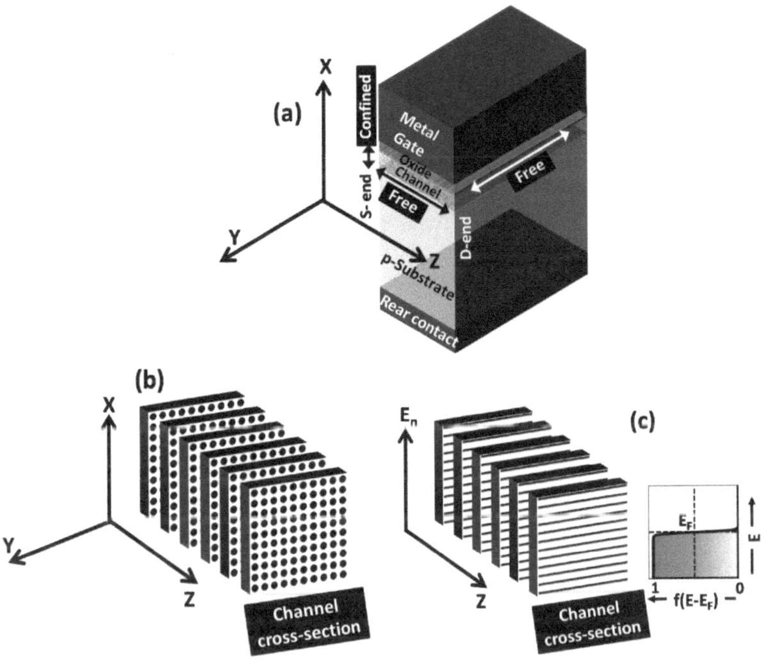

Fig. 4.4: (a) A schematic of a MOSFET channel depicting the confinement of charges in the 'inversion'-channel of MOS-gate direction along *X*-axis, and their possible free motion along the in-plane transverse direction, i.e. *Y*-axis and the channel directed towards *Z*-axis; (b) a virtual slicing of the channel into thin slabs for computing the local eigenstates at each of such slabs owing to the transverse confinement of carriers; and (c) illustrating the 'mode space' approach, where the transverse 2D positional coordinate space of (b) has been transformed into transverse 1D eigenenergy space, along with the sharp fall of their occupational probability due to the Fermi-Dirac distribution function of the reservoirs declining sharply above Fermi levels.

3. This can be done in a straight forward manner by taking the quantum state projections on the 3D position space along the three spatial coordinates, commonly called the 'real space' approach.
4. On the other hand, the channel can now be conceptually sliced into thin slabs (see Fig. 4.4(b)), resulting in the local eigenstates at each of such particular slabs due to the transverse confinement (see Fig. 4.4(c)). The degree of such confinement is defined by the gate voltage, as well as the channel material properties along with those of the oxide in planar MOSFETs; however, in nanostructure FETs, it also depends significantly on the transverse dimensions (i.e. shape and size). Such local eigenstates are solutions of the 2D Schrödinger equation in the transverse confining plane of nanostructure FETs, while for planar MOSFETs (where charges are confined only along the MOS-gate), they can be obtained by integrating over all possible momentum states along the in-plane transverse direction. At this point, it is worth mentioning that it is necessary to calculate the

local eigenstates using a Schrödinger-Poisson simultaneous solver (why it is necessary will be illustrated in Chapter 11). Such local eigenstates vary from slab-to-slab along the channel due to the variation of potential energy from source→-channel→drain, thereby leading to the corresponding semi-quantized energy sub-bands along the channel as a whole. Such sub-band variation is obtained by solving the longitudinal Poisson's equation (numerically by initially assuming an arbitrary potential and then iterating it). This method is known to be the 'mode space' approach [12–13].

This 'mode space' technique increases the computational efficiency of a quantum-scale MOSFET simulation significantly, offering a quite satisfactory degree of accuracy. It is imperative to note that in 'mode space' approach, we work, in fact, in transverse 1D energy space instead of the transverse 2D positional coordinate space, and only a few of such semi-quantized energy levels (depending on temperature) contribute to the carrier transport, owing to the sharp fall of Fermi-Dirac distribution function of the reservoirs above Fermi levels (see Fig. 4.4(c)). Thus, huge numbers of spatial grids are reduced to only a few numbers of energy 'modes' as the computational basis, and the corresponding matrix size reduces in square order.

5. Once the sub-bands of the isolated active device, i.e. the channel, are obtained, its coupling with reservoirs, i.e. source/drain, is incorporated into the channel-Hamiltonian through corresponding self-energies. Calculation of reservoir self-energies is carried out by considering these to be infinitely extended compared to the active device size, as described in Chapter 3.
6. Now, the non-equilibrium Green's function (NEGF) for the channel is obtained based on the effective Hamiltonian including self-energies, and the local density of states (LDOS) along the channel is calculated, which leads to the corresponding charge density as well.
7. Such charge density is then put into the longitudinal Poisson's equation along the channel (which was initially assumed arbitrarily) to get the potential profile.
8. Potential obtained at Step 7 is then again fed into step 4, followed by the successive steps, and such iteration loops are continued until the potential profile obtained at step 7 matches that at step 4 of the same loop (to a desired degree of computational accuracy and tolerance). The self-consistency between quantum actions and electrostatic interactions is then considered to be achieved.
9. Now the energy sub-bands, LDOS, and band potential energy profile along the channel, as well as the transmission coefficient, device current, and power dissipation, can be finally calculated on the basis of quantum-electrostatic self-consistent NEGF.
10. Scattering effects can be incorporated into the model by assuming each of the scattering centres in the channel to couple with itself as a virtual reservoir, preserving the number of carriers (note that energy may not be conserved for inelas-

tic/dephasing scattering events). This is known to be the Büttiker-probe technique, described in detail in Wang et al. [13,14] and thus not discussed here.

4.3 Mathematical framework

4.3.1 Hamiltonian of a MOSFET device

The entire operation of a MOSFET device can be comprehended on the basis of the coupling between its channel (attached to the MOS-gate and defined by it) and the source/drain. Thus, such a device is constituted of three subsystems, i.e. (MOS-gate defined) channel, source, and drain, each of which is described in its own Hilbert spaces by the corresponding Hamiltonians, while their coupling is represented by the interaction-Hamiltonians acting on the joint Hilbert space. Here we are illustrating the electron transport in an n-channel MOSFET, which can easily be mimicked for holes in a p-channel MOSFET.

The Hamiltonian of the entire n-MOSFET device can be written in terms of its second-quantized field operators as

$$\begin{aligned}\hat{H}_{\text{MOSFET}} = &\sum_i \varepsilon_i c_i^+ c_i + \sum_r \varepsilon_r{}^S c_r{}^{S+} c_r{}^S + \sum_r \varepsilon_r{}^D c_r{}^{D+} c_r{}^D \\ &+ \sum_i \left[\left(\tau_{ir}{}^S c_i{}^+ c_r{}^S + \tau_{ir}{}^D c_i{}^+ c_r{}^D\right) + \text{H.C.}\right]\end{aligned} \tag{4.1}$$

where $(c_i{}^+, c_r{}^{S+}, c_r{}^{D+})$ and $(c_i, c_r{}^S, c_r{}^D)$ are creation and annihilation operators for electrons within the channel, source, and drain of the MOSFET, respectively. The corresponding energy eigenstates and the channel-source and channel-drain interaction potentials are represented by $(\varepsilon_i, \varepsilon_r{}^S, \varepsilon_r{}^D)$ and $(\tau_{ir}{}^S, \tau_{ir}{}^D)$. It is imperative to note that the first, second, and third terms in the RHS of eq. (4.1) are acting on Hilbert spaces of the channel ($\mathscr{H}_{\text{Ch}}$), source ($\mathscr{H}_{\text{S}}$), and drain ($\mathscr{H}_{\text{D}}$), respectively, while two of the last terms (i.e. the coupling terms), along with their Hermitian conjugates (*H.C.*), are spread over the joint Hilbert spaces of channel-source ($\mathscr{H}_{\text{Ch}} \otimes \mathscr{H}_{\text{S}}$) and channel-drain ($\mathscr{H}_{\text{Ch}} \otimes \mathscr{H}_{\text{D}}$), respectively. It is apparent that they all may be considered to operate on the total joint Hilbert space $\mathscr{H}_{\text{Ch}} \otimes \mathscr{H}_{\text{S}} \otimes \mathscr{H}_{\text{D}}$.

Now the quantum state for channel-Hamiltonian in Schrödinger (i.e. non-relativistic) limit can be expressed in position coordinates (x^k) as

$$\hat{H}_{\text{Ch}}|x^k\rangle = \left[-\frac{\hbar^2}{2m_e}\partial_i[m^*]_{ij}^{-1}\partial_j + (-e)\varphi(x^k)\right]|x^k\rangle = E|x^k\rangle \tag{4.2}$$

where each of i, j, k can take the values 1, 2, or 3 for considering the 3D motion of electrons,-e is the charge and m_e is the rest mass of electron, $[m^*]$ represents relative effective mass tensor (i.e. in the unit of m_e) of the materials, and $\varphi(x^k)$ is the potential distribution within the channel bounded by the voltages applied at gate (V_G) and drain (V_D), and dependent on the material properties and geometric shape/size of the MOS constituents.

Equation (4.2) can be directly solved in a straightforward manner in 'real space' by conceptually dividing the channel into discrete grids along the three spatial directions, as mentioned earlier; however, we will elaborate here on the method of solving it in 'mode space' with the aim of getting an improved computational efficiency. Here, as mentioned earlier, the spatial coordinate axes for a planar MOSFET are considered as follows: MOS-gate→ *X*-axis, in-plane transverse direction→ *Y*-axis, and channel→ *Z*-axis, which are also the principal axes of the single-crystal (as in practical Si-MOSFETs), so that all the off-diagonal components of the effective mass tensor are zero. Now, in order to adopt the 'mode space' approach, assume the channel to be virtually sliced into thin slabs, and in such a slab at $z = z$, the local 2D Schrödinger equation (i.e. along *X*- and *Y*-axes) can be written as

$$\left[\left(-\frac{\hbar^2}{2m_e}\frac{\partial}{\partial x}\frac{1}{m_{11}*(x)}\frac{\partial}{\partial x}+(-e)\varphi(x;z)\right)-\frac{\hbar^2}{2m_e m_{22}*}\frac{\partial^2}{\partial y^2}\right]\left|n,\vec{k}_y;\,z\right\rangle=\left[\varepsilon_n(z)+\varepsilon_{\vec{k}_y}\right]\left|n,\vec{k}_y;\,z\right\rangle \tag{4.3}$$

where the electrons are free to move along the in-plane transverse direction with a continuous energy spectrum $\varepsilon_{\vec{k}_y}=\frac{\hbar^2 k_y{}^2}{2m_e m_{22}*}$ ($\vec{k}_y$ is the wave-vector along the *Y*-axis), while being confined in the *X*-direction (i.e. along the MOS-gate at 'inversion' condition) with discrete local sub-band energies $\varepsilon_n(z)$, leading to the local orthonormal sub-band states $\left|n,\vec{k}_y;\,z\right\rangle$. It is imperative to note that in nanostructure (Fin/NW) FETs, electrons are confined in the entire transverse *XY*-plane, which leads to modifying eq. (4.3) for them as

$$\left[-\frac{\hbar^2}{2m_e}\left(\frac{\partial}{\partial x}\frac{1}{m_{11}*(x,y)}\frac{\partial}{\partial x}+\frac{\partial}{\partial y}\frac{1}{m_{22}*(x,y)}\frac{\partial}{\partial y}\right)+(-e)\varphi(x,y;\,z)\right]|n;\,z\rangle=[\varepsilon_n(z)]|n;\,z\rangle \tag{4.3a}$$

where now n denotes all the 2D-confined states. Interestingly, by virtue of quantum theory (more correctly, the very nature of the quantum world, indeed), such local sub-band states ($|n,\vec{k}_y;\,z\rangle$) evidently form a complete set of basis vectors that any state of the system (*i.e.* $|x^k\rangle$) at $z = z$ can be represented with, utilizing the superposition principle, as

$$\left|x^k\right\rangle=\sum_{n,\vec{k}_y}c_n(z)\left|n,\vec{k}_y;\,z\right\rangle \tag{4.4}$$

where $c_n(z)$ are the coefficients of such a superposition. Here, $\vec{k}_y$ varies continuously and thus the summation is actually carried out by integration over all possible values. For Fin/NW-FETs, on the other hand, eq. (4.4) simply gets modified to

$$\left|x^k\right\rangle = \sum_n c_n(z)|n;\ z\rangle \tag{4.4a}$$

Now, since eqs. (4.2), (4.3), and (4.4),

$$\hat{H}_{\mathrm{Ch}} \sum_{n,\vec{k}_y} c_n(z)\left|n, \vec{k}_y;\ z\right\rangle = \left[-\frac{\hbar^2}{2m_e}\partial_i[m^*]_{ij}^{-1}\partial_j + (-e)\varphi(x^k)\right]\sum_{n,\vec{k}_y} c_n(z)\left|n, \vec{k}_y;\ z\right\rangle$$

$$= E\sum_{n,\vec{k}_y} c_n(z)\left|n, \vec{k}_y;\ z\right\rangle$$

$$\Rightarrow \sum_{n',\vec{k}'_y}\left[\left(-\frac{\hbar^2}{2m_e}\frac{\partial}{\partial z}\frac{1}{m_{33}^*(z)}\frac{\partial}{\partial z} + (-e)\varphi(z)\right) + (\varepsilon_{n'}(z) + \varepsilon_{\vec{k}'_y})\right]c_{n'}(z)\left|n', \vec{k}'_y;\ z\right\rangle$$
$$= E\sum_{n',\vec{k}'_y} c_{n'}(z)\left|n', \vec{k}'_y;\ z\right\rangle \tag{4.5}$$

Note that in eq. (4.5), the dummy indices $(n, \vec{k}_y)$ are replaced by $(n', \vec{k}'_y)$ without losing any information or generality. It may further be interesting to note that the z-component of the channel-Hamiltonian is not involved in eq. (4.3) that represents the Schrödinger equation for any transverse 2D slab of the channel at any $z = z$, whereas such component has the key operating role in eq. (4.5) where the transverse eigenstates are only forming the orthonormal basis vectors of representation. From a physical point of view, eq. (4.5), in fact, depicts the correlation between the longitudinal (along channel) and transverse (along MOS-gate) quantized motion of electrons in the device. Thus, now such a short channel MOSFET may be considered as a gate-drain correlated or two mutually orthogonal electric field correlated device. It is apparent that eq. (4.5) for Fin/NW-FETs accordingly results in

$$\Rightarrow \sum_{n'}\left[\left(-\frac{\hbar^2}{2m_e}\frac{\partial}{\partial z}\frac{1}{m_{33}^*(z)}\frac{\partial}{\partial z} + (-e)\varphi(z)\right) + \varepsilon_{n'}(z)\right]c_{n'}(z)\left|n';\ z\right\rangle = E\sum_{n'} c_{n'}(z)\left|n';\ z\right\rangle \tag{4.5a}$$

Using the orthonormal property of the basis states, i.e. $\langle n, \vec{k}_y;\ z \mid n', \vec{k}'_y;\ z\rangle = \delta_{n,n'}\delta(\vec{k}_y - \vec{k}_y{}')$ (or simply $\langle n;\ z\left|n';\ z\right\rangle = \delta_{n,n'}$ for Fin/NW-FETs), eqs. (4.5) and (4.5a) lead respectively to

$$\left[\left(-\frac{\hbar^2}{2m_e}\frac{d}{dz}\frac{1}{m_{33}^*(z)}\frac{d}{dz} + (-e)\varphi(z)\right) + (\varepsilon_n(z) + \varepsilon_{\vec{k}_y})\right]c_n(z) = Ec_n(z) \tag{4.6}$$

$$\left[\left(-\frac{\hbar^2}{2m_e}\frac{d}{dz}\frac{1}{m_{33}^*(z)}\frac{d}{dz} + (-e)\varphi(z)\right) + \varepsilon_n(z)\right]c_n(z) = Ec_n(z) \tag{4.6a}$$

It is apparent that eqs. (4.6) and (4.6a) represent the longitudinal motion of electrons in the channel of planar and nanostructure MOSFETs, respectively, which, nevertheless, is correlated with their transverse motion through the sub-bands. However, still, such equations describe the isolated channel of the corresponding MOSFET architectures, and now, these equations need to couple with the contributions from interactions of source/drain with the channel. However, before going there, we will elaborate a little more about the solutions of eqs. (4.3) and (4.3a) that lead to electron confinement along the MOS-direction, thereby generating the sub-bands along MOSFET channels.

4.3.2 Schrödinger-Poisson coupled equation for transverse confinement

The motion of an electron in a planar MOSFET device only along the MOS-direction at any (y, z) can be obtained from eq. (4.3) as

$$\left(-\frac{\hbar^2}{2m_e}\frac{d}{dx}\frac{1}{m_{11}*(x)}\frac{d}{dx}+(-e)\varphi(x)\right)|n\rangle=\varepsilon_n|n\rangle \tag{4.7}$$

coupled to Poisson's equation, given by

$$\left(\frac{d}{dx}\varepsilon_{rx}(x)\frac{d}{dx}\right)\varphi(x)=-\frac{(-e)}{\varepsilon_0 W\Delta z}\sum_{n,\vec{k}_y}\left|\langle x\,|n\rangle\right|^2 f_{FD}(\varepsilon_n+\varepsilon_{\vec{k}_y}-E_F^{\mathrm{Sub}})-\frac{(+e)}{\varepsilon_0}N_A{}^- \tag{4.8}$$

where ε_0 is the permittivity of free-space, $\varepsilon_{rx}(x)$ represents the relative permittivity of semiconductor, oxide, and metal as defined by their thicknesses along the growth direction in a practical MOS device (i.e. X-axis here), W is channel-width (along in-plane transverse direction), and Δz is the thickness of channel-slab (as assumed for computation). It is imperative to note that $|\langle x\,|n\rangle|^2$ represents the quantum probability density and $f_{\mathrm{FD}}(\varepsilon_n+\varepsilon_{\vec{k}_y}-E_{\mathrm{F}}^{\mathrm{Sub}})$ is the Fermi-Dirac distribution function (with Fermi level of the substrate being $E_{\mathrm{F}}^{\mathrm{Sub}}$) representing the thermodynamic probability density of electrons. $N_{\mathrm{A}}{}^-$ is the acceptor atom concentration of the p-type substrate. At this point, it may be noted that for a p-channel MOSFET, the sign of charges would be opposite and $N_{\mathrm{A}}{}^-$ would be replaced by $N_{\mathrm{D}}{}^+$, i.e. the donor concentration of n-type substrate.

For the Fin/NW-FETs, or the nanostructure FETs in general, such Schrödinger-Poisson coupled equations will be, respectively,

$$\left[-\frac{\hbar^2}{2m_e}\left(\frac{\partial}{\partial x}\frac{1}{m_{11}*(x,y)}\frac{\partial}{\partial x}+\frac{\partial}{\partial y}\frac{1}{m_{22}*(x,y)}\frac{\partial}{\partial y}\right)+(-e)\varphi(x,y)\right]|n\rangle=[\varepsilon_n(z)]|n\rangle \tag{4.7a}$$

$$\left(\frac{d}{dx}\varepsilon_{rx}(x,y)\frac{d}{dx}+\frac{d}{dy}\varepsilon_{ry}(x,y)\frac{d}{dy}\right)\varphi(x,y)=-\frac{(-e)}{\varepsilon_0\Delta z}\sum_n\left|\langle x,y\,|n\rangle\right|^2 \tag{4.8a}$$

Thus, in order to get the local eigenstates ($|n\rangle$) and energy eigenvalues (ε_n) of electrons confined in the (real or virtual) 'inversion'-channel MOS devices for solving eq. (4.6) (or (4.6a)), first, one needs to initially assume a potential $\varphi^{(0)}(x)$ (or $\varphi^{(0)}(x,y)$) across the MOS, put it into eq. (4.7) (or (4.7a)) to get the zeroth order iterative values of $(|n\rangle, \varepsilon_n)$, followed by putting them into eq. (4.8) (or (4.8a)) to obtain the first-order iterative value of potential $\varphi^{(1)}(x)$ (or $\varphi^{(1)}(x,y)$). Then the first order potential $\varphi^{(1)}(x)$ (or $\varphi^{(1)}(x,y)$) is put into eq. (4.7) (or (4.7a)), and the same steps are followed to get the second-order potential $\varphi^{(2)}(x)$ (or $\varphi^{(2)}(x,y)$). Such ('l_I' numbers of) iteration is carried out in loops until $\Delta\varphi^{(l_I)} = [\varphi^{(l_I)} - \varphi^{(l_I-1)}]$ can be neglected for practical purposes (i.e. to reach the desired accuracy level within tolerant runtime), and then the Schrödinger-Poisson self-consistency is considered to be achieved.

4.3.3 Non-equilibrium Green's function (NEGF) for quantum charge transport through MOSFET channel

Once the Schrödinger-Poisson self-consistent sub-band states are obtained, the contributions of source/drain coupling with the channel are incorporated as the corresponding self-energies. Using the Heisenberg equation for fermions (i.e. $(i\hbar d/dt)\hat{c} = \{\hat{c}, \hat{H}\}$), eq. (4.1) gives rise to

$$i\hbar\frac{d}{dt}\begin{pmatrix} c_r^{\mathrm{S}} \\ c_i \\ c_{r'}^{\mathrm{D}} \end{pmatrix} = \begin{bmatrix} \varepsilon_r^{\mathrm{S}} & (\tau_{ir}^{\mathrm{S}})^{+} & 0 \\ \tau_{ir}^{\mathrm{S}} & \varepsilon_i & \tau_{ir'}^{\mathrm{D}} \\ 0 & (\tau_{ir'}^{\mathrm{D}})^{+} & \varepsilon_{r'}^{\mathrm{D}} \end{bmatrix}\begin{pmatrix} c_r^{\mathrm{S}} \\ c_i \\ c_{r'}^{\mathrm{D}} \end{pmatrix} \tag{4.9}$$

Fourier transform of eq. (4.9) into the energy domain leads to the total Green's function (as discussed in Chapter 3) of the MOSFET to be

$$G_{\mathrm{MOSET}}(E) = \begin{bmatrix} E-\varepsilon_r^{\mathrm{S}} & -\tau_{ir}^{\mathrm{S}}(E)^{+} & 0 \\ -\tau_{ir}^{\mathrm{S}}(E) & E-\varepsilon_i & -\tau_{ir'}^{\mathrm{D}}(E) \\ 0 & \tau_{ir'}^{\mathrm{D}}(E)^{+} & E-\varepsilon_{r'}^{\mathrm{D}} \end{bmatrix}^{-1} \tag{4.10}$$

Then the NEGF for the channel is given by the element of the inverse matrix of eq. (4.10) at the position of $(E-\varepsilon_i)$, i.e.

$$G_{\mathrm{Ch}}(E) = [(E-\varepsilon_i) - \tau_{ir}^{\mathrm{S}}(E-\varepsilon_r^{\mathrm{S}})^{-1}(\tau_{ir}^{\mathrm{S}})^{+} - \tau_{ir'}^{\mathrm{D}}(E-\varepsilon_{r'}^{\mathrm{D}})^{-1}(\tau_{ir'}^{\mathrm{D}})^{+}]^{-1} \tag{4.11}$$

which can also be written in terms of source/drain self-energies as

$$G_{\mathrm{Ch}}(E) = [(E-\varepsilon_i) - \sum{}^{S}(E) - \sum{}^{D}(E)]^{-1} \tag{4.11a}$$

In the formulation given by eq. (4.6) (or (4.6a)), where the channel is discretized into virtual thin grids along the longitudinal direction (suppose into $l = 1, 2, \ldots, N$ grids, so that $l=1$ and $l=N$ are connected to source and drain, respectively), and the corresponding sub-bands have been obtained from transverse solutions, eq. (4.11a) can be expressed as

$$G_{\text{Ch}}(E,z_l)=[E-(E^K(z_l)+(-e)\varphi(z_l)+\varepsilon_n(z_l))-\sum{}^S(E,z_1)-\sum{}^D(E,z_N)]^{-1} \tag{4.11b}$$

where kinetic energy operator along the channel ($E^K(z)=-\frac{\hbar^2}{2m_e}\frac{d}{dz}\frac{1}{m_{33}*(z)}\frac{d}{dz}$) can be discretized as

$$E^K(z_l)=-\frac{\hbar^2}{2m_e\Delta z^2}\left[\frac{1}{m_{33}^l*(z_{l-1})}-2\left(\frac{1}{m_{33}^l*(z_l)}+\frac{1}{m_{33}^{l+1}*(z_l)}\right)+\frac{1}{m_{33}^{l+1}*(z_{l+1})}\right] \tag{4.11c}$$

The self-energy, band velocity, and effective mass, as well as the LDOS of electrons in the channel due to the source/drain, can now be separately obtained using the methods discussed in Chapter 3. Thus, the local 1D carrier concentration along the channel direction corresponding to the transverse nth sub-band is given by

$$n_{\text{1D}}^n(z_l)=\frac{1}{\Delta z}\int[A_n^{\text{S}}(E;z_l)f_{\text{FD}}(E-e_F^{\text{S}})+A_n^{\text{D}}(E;z_l)f_{\text{FD}}(E-(E_F^{\text{D}}-V_D))]dE \tag{4.12}$$

where $A_n^S(E;z_l)=\left[[G_{\text{Ch}}(E,z_l)][\Gamma^{\text{S}}(E,z_1)][G_{\text{Ch}}(E,z_l)]^+\right]_{nn}$ (with $[\Gamma^{\text{S}}(E,z_1)]=-2\text{Im}[\sum^{\text{S}}(E,z_1)]$) and $A_n^{\text{D}}(E;z_l)=\left[[G_{\text{Ch}}(E,z_l)][\Gamma^{\text{D}}(E,z_N)][G_{\text{Ch}}(E,z_l)]^+\right]_{nn}$ (with $[\Gamma^{\text{D}}(E,z_N)]=-2\text{Im}[\sum^{\text{D}}(E,z_N)]$) are the local spectral density of states of the channel due to coupling with source and drain contacts, respectively, while $f_{\text{FD}}(E-e_F^{\text{S}})$ and $f_{\text{FD}}(E-(E_F^{\text{D}}-V_D))$ are the corresponding Fermi-Dirac distribution functions, E_F^{S} and E_F^{D} being the Fermi levels of source and drain, respectively, and V_{D} being the drain voltage assuming the source to be grounded.

4.3.4 Poisson's equation in the channel of planar and nanostructure MOSFETs

The total local 3D carrier density in the channel is, therefore, obtained from the local 1D carrier concentration averaged over the weights given by transverse probability density as

$$n_{\text{3D}}(\vec{r})=\sum_{n,\vec{k}_y}n_{\text{1D}}^n(z)\left|\left\langle x,y\,|n,\vec{k}_y;\,z\right\rangle\right|^2 f_{\text{FD}}(\varepsilon_n+\varepsilon_{\vec{k}_y}-e_F^{\text{Sub}}) \tag{4.13}$$

and

$$n_{\text{3D}}(\vec{r})=\sum_n n_{\text{1D}}^n(z)\left|\langle x,y\,|n;\,z\rangle\right|^2 \tag{4.13a}$$

for planar MOSFETs and nanostructure FETs, respectively. Accordingly, the potential profile in the channel is given by Poisson's equation:

$$(\vec{\nabla}\cdot\overleftrightarrow{\varepsilon}_r\cdot\vec{\nabla})\varphi(\vec{r}) = -\frac{(-e)}{\varepsilon_0}n_{3D}(\vec{r}) \tag{4.14}$$

where $\overleftrightarrow{\varepsilon}_r$ is the relative permittivity tensor, which is diagonal if the coordinate axes are considered along the principal axes of the materials. Now, integrating eq. (4.14) over the volume of a thin channel grid of the planar MOSFET, and using Gauss' volume integral theorem (i.e. $\int dV\vec{\nabla}\cdot() = \oint d\vec{S}\cdot()$), it is obtained to be (since eq. (4.13)):

$$\oint(\overleftrightarrow{\varepsilon}_r\cdot\vec{\nabla}\varphi(\vec{r}))\cdot d\vec{S} = -\frac{(-e)}{\varepsilon_0}\int n_{3D}(\vec{r})dxdy\Delta z = -\frac{(-e)}{\varepsilon_0}\Delta z\sum_n n_{1D}^n(z)\mathrm{N}^{C(n)} \tag{4.15}$$

where the term $\mathrm{N}^{C(n)} = \sum_{\vec{k}_y}\int dy|\langle y\mid\vec{k}_y;\,z\rangle|^2 f_{FD}(\varepsilon_n+\varepsilon_{\vec{k}_y}-E_F^{Sub})$ (since closure property along X-axis) is elaborated below. On the other hand, since eq. (4.13a) for nanostructure FETs like Fin/NW-FETs:

$$\oint(\overleftrightarrow{\varepsilon}_r\cdot\vec{\nabla}\varphi(\vec{r}))\cdot d\vec{S} = -\frac{(-e)}{\varepsilon_0}\int n_{3D}(\vec{r})dxdy\Delta z = -\frac{(-e)}{\varepsilon_0}\Delta z\sum_n n_{1D}^n(z) \tag{4.15a}$$

owing to the closure property of total transverse sub-band states of such devices.

At this point, it is imperative to note that, owing to the in-plane transverse degree of freedom in planar MOSFETs, $\langle y\mid\vec{k}_y\rangle = 1/\sqrt{2\pi}\,e^{ik_y y}$, leading to the fact that the integral term $\mathrm{N}^{C(n)}$ in the RHS of eq. (4.15) appears to be $\int dk_y f_{FD}(\varepsilon_n+\varepsilon_{\vec{k}_y}-E_F^{Sub})\int dy$. Such a term cannot be expressed in a closed-form solution, and thus solved here using the following trick:

$$\begin{aligned}
\mathrm{N}^{C(n)} &= \sum_{\vec{k}_y}\int dy\left|\left\langle y\,\middle|\vec{k}_y;\,z\right\rangle\right|^2 f_{FD}(\varepsilon_n+\varepsilon_{\vec{k}_y}-E_F^{Sub})\\
&= \int_{-\infty}^{\infty} dk_y \underset{k_y'\to k_y}{Lt}\left[\int_{-\infty}^{\infty} dy\left(\frac{1}{2\pi}e^{i(k_y-k_y')y}\right)f_{FD}(\varepsilon_n+\varepsilon_{\vec{k}_y}-E_F^{Sub})\right]\\
&= \int_{-\infty}^{\infty} dk_y \underset{k_y'\to k_y}{Lt}\ \delta(k_y-k_y')f_{FD}(\varepsilon_n+\varepsilon_{\vec{k}_y}-E_F^{Sub})\\
&= 2\left|\int_{0}^{\infty} dk_y \underset{k_y'\to k_y}{Lt}\ \delta(k_y-k_y')f_{FD}(\varepsilon_n+\varepsilon_{k_y}-E_F^{Sub})\right| \qquad [\text{since } \varepsilon(\vec{k}_y)=\varepsilon(k_y^2)]\\
&= 2\left|f_{FD}(\varepsilon_n+\varepsilon_{\vec{k}_y'}-E_F^{Sub})\Big|_{k_y'\to 0}^{k_y'\to\infty}\right|
\end{aligned}$$

$$\Rightarrow \mathrm{N}^{C(n)} = \frac{2}{e^{\beta(\varepsilon_n-E_F^{Sub})}+1} \tag{4.16}$$

Here we are generalizing such a result by defining

$$\mathrm{N}^{C(a)} = \frac{2}{a+1} \tag{4.16a}$$

which offers a very good estimation of the integral $I_a = \int_{-\infty}^{+\infty} \frac{dx}{ae^{x^2}+1}$ for larger values of $|a|-1$ (see Fig. 4.5), while the limit does not exist at $a = -1$.

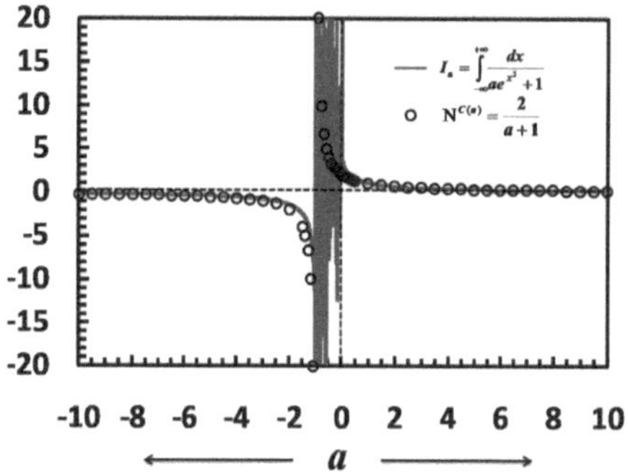

Fig. 4.5: The comparative plots of the function $I_a = \int_{-\infty}^{+\infty} \frac{dx}{ae^{x^2}+1}$ and $\mathrm{N}^{C(a)} = \frac{2}{a+1}$ with the parameter a.

Now the LHS of eq. (4.15) for planar MOSFETs can be written as

$$\begin{aligned}\oint \left(\overleftrightarrow{\varepsilon}_r \cdot \vec{\nabla}\varphi(\vec{r})\right) \cdot d\vec{S} &= \varepsilon_{rz}\left(-\frac{\partial}{\partial z}\int \varphi(\vec{r})dxdy\bigg|_{z=z}\right) + \left(\frac{\partial}{\partial z}\int \varphi(\vec{r})dxdy\bigg|_{z=z+\Delta z}\right) \\ &\quad + W\Delta z\left(\varepsilon_{rx}\frac{\partial}{\partial x}\varphi(\vec{r})\right)\bigg|_{\mathrm{Ch/Ox}}\end{aligned} \tag{4.17}$$

$$= \varepsilon_{rz}\Delta z\frac{\partial^2}{\partial z^2}\left(\int \varphi(\vec{r})dxdy\right) + W\Delta z\left(\varepsilon_{rx}\frac{\partial}{\partial x}\varphi(\vec{r})\right)\bigg|_{\mathrm{Ch/Ox}}$$

[since, to the first order approximation, $\frac{\partial}{\partial z}\Big|_{z=z+\Delta z} - \frac{\partial}{\partial z}\Big|_{z=z} = \Delta z\frac{\partial^2}{\partial z^2}\Big|_{z=z}$]

$$\Rightarrow \oint \left(\overleftrightarrow{\varepsilon}_r \cdot \vec{\nabla}\varphi(\vec{r})\right) \cdot d\vec{S} = W\Delta z\left[\varepsilon_{rz}\frac{\partial^2}{\partial z^2}\left(\int_{\mathrm{Substrate}}^{\mathrm{Ch/Ox}} \varphi(x,z)dx\right) + \left(\varepsilon_{rx}\frac{\partial}{\partial x}\varphi(x,z)\right)\bigg|_{\mathrm{Ch/Ox}}\right] \tag{4.17a}$$

since the channel potential is constant along the in-plane transverse direction (i.e. Y-axis). It may be worthy to note that while calculating the LHS of eq. (4.15) for planar MOSFET as above, only three out of six terms of the surface integral (due to six surfaces of any virtual slab) are non-zero (as can be understood from Fig. 4.4(a)–(c)), where the first two terms in eq. (4.17) correspond to the two cross-sectional surfaces of the channel slab (of thickness Δz) at $z = z$ and $z = z + \Delta z$, while the last term is due to the channel-oxide interface of the MOS-gate.

However, as shown in Fig. 4.6, for a DG-MOSFET with two MOS-gates along the Y-axis and a channel height of t_{Ch} along the X-axis, eq. (4.17a) would be modified to

$$\oint \left(\overleftrightarrow{\varepsilon}_r \cdot \vec{\nabla}\varphi(\vec{r})\right) \cdot d\vec{S} = t_{\mathrm{Ch}}\Delta z \left[\varepsilon_{rz} \frac{\partial^2}{\partial z^2}\left(\int_{\mathrm{Ch/Ox_1}}^{\mathrm{Ch/Ox_2}} \varphi(y,z)dy\right) + \left(\varepsilon_{ry} \frac{\partial}{\partial y}\bigg|_{\mathrm{Ch/Ox_1}} + \varepsilon_{ry} \frac{\partial}{\partial y}\bigg|_{\mathrm{Ch/Ox_2}}\right)\varphi(y,z)\right] \tag{4.17b}$$

For a FinFET (see Fig. 4.6), such an equation becomes:

$$\oint \left(\overleftrightarrow{\varepsilon}_r \cdot \vec{\nabla}\varphi(\vec{r})\right) \cdot d\vec{S} = \Delta z \left[\varepsilon_{rz} \frac{\partial^2}{\partial z^2}\left(\int_{\mathrm{Cross-section}} \varphi(x,y,z)dxdy\right) + \left(W\varepsilon_{rx} \frac{\partial}{\partial x}\bigg|_{\mathrm{Ch/Ox}} + 2t_{\mathrm{Ch}}\varepsilon_{ry} \frac{\partial}{\partial y}\bigg|_{\mathrm{Ch/Ox}}\right)\varphi(x,y,z)\right] \tag{4.17c}$$

while an NS-FET architecture, as schematically depicted in Fig. 4.6, gives rise to

$$\oint \left(\overleftrightarrow{\varepsilon}_r \cdot \vec{\nabla}\varphi(\vec{r})\right) \cdot d\vec{S} = \Delta z \left[\varepsilon_{rz} \frac{\partial^2}{\partial z^2}\left(\int_{\mathrm{Cross-section}} \varphi(x,y,z)dxdy\right) + 2\left(W\varepsilon_{rx} \frac{\partial}{\partial x}\bigg|_{\mathrm{Ch/Ox}} + t_{\mathrm{Ch}}\varepsilon_{ry} \frac{\partial}{\partial y}\bigg|_{\mathrm{Ch/Ox}}\right)\varphi(x,y,z)\right] \tag{4.17d}$$

Similarly, a GAA cylindrical NWFET (Fig. 4.6) leads to

$$\oint \left(\overleftrightarrow{\varepsilon}_r \cdot \vec{\nabla}\varphi(\vec{r})\right) \cdot d\vec{S} = \Delta z \left[\varepsilon_{rz} \frac{\partial^2}{\partial z^2}\left(2\pi \int_0^R \varphi(\rho,z)\rho d\rho\right) + \left(2\pi R\varepsilon_{r\rho} \frac{\partial \varphi(\rho,z)}{\partial \rho}\bigg|_{\rho=R}\right)\right] \tag{4.17e}$$

where ρ is the radial coordinate perpendicular to the cylindrical axis of the NW of radius R. Any other specific geometrical architectures of the channel cross-section can be similarly described by writing the equation equivalent to eq. (4.17) numerically in general. Interestingly, eq. (4.17), and so on, show that eq. (4.15), indeed, is not a function of the channel grid thickness (Δz), which is a numerical parameter associated with the accuracy of simulation, which upholds the generality of such an equation.

Further, since the potential on the metal gate is constant (i.e. gate-voltage) which linearly varies within the oxide (where charge density is ideally zero in absence of any fixed-oxide-charge), all the terms in expressions of $\oint \left(\overleftrightarrow{\varepsilon}_r \cdot \vec{\nabla}\varphi(\vec{r})\right) \cdot d\vec{S}$ (given by eq. (4.17) and so on) depend on the potential at the channel/oxide interface (commonly called surface potential), i.e. $\varphi(z)|_{\mathrm{Ch/Ox}}$, owing to the continuity of potential and the normal component of electric displacement vector at the interface. Such a surface potential value can be obtained for a planar MOSFET at any $z = z$ as follows:

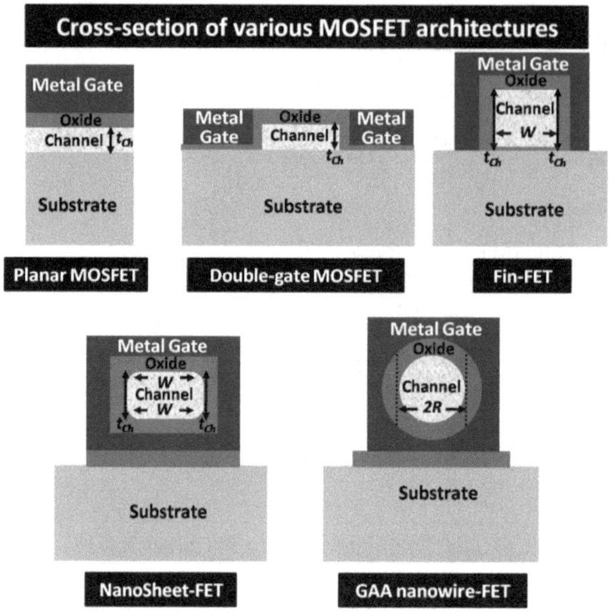

Fig. 4.6: The cross-section of any virtual slice (as illustrated in Fig. 4.4(b)) of various MOSFET architectures: planar MOSFET, DG-MOSFET, FinFET, NS-FET, and NWFET.

Assuming the channel/oxide interface at $x = 0$ (with increasing along M→O→S), the oxide being at the negative side of the X-axis with the thickness of t_{Ox}, eq. (4.8) can be rewritten as

$$\begin{cases} \varepsilon_{rx}^{\mathrm{Ox}} \frac{d^2}{dx^2} \varphi(x) = 0, & -t_{\mathrm{Ox}} \le x \le 0 \\ \varepsilon_{rx}^{\mathrm{Ch}} \frac{d^2}{dx^2} \varphi(x) = -\frac{(-e)}{\varepsilon_0 W \Delta z} \sum_n |\langle x \, | n \rangle|^2 \mathrm{N}^{C(n)} - \frac{(+e)}{\varepsilon_0} N_{\mathrm{A}}^{-}, & x \ge 0 \end{cases} \tag{4.18}$$

which exhibit the boundary conditions

$$\begin{cases} \varphi(x)|_{x \to 0-} = \varphi(x)|_{x \to 0+} = \varphi_{\mathrm{Ch/Ox}} \\ \varepsilon_{rx}^{\mathrm{Ox}} \frac{d}{dx} \varphi(x)\big|_{x \to 0-} = \varepsilon_{rx}^{\mathrm{Ch}} \frac{d}{dx} \varphi(x)\big|_{x \to 0+} \\ \varphi(x)|_{x = -t_{\mathrm{Ox}}} = V_{\mathrm{G}} \\ \varphi(x)|_{x \to +\infty} = 0 \\ \frac{d}{dx} \varphi(x)\big|_{x \to +\infty} = 0 \end{cases} \tag{4.19}$$

Hence, the solution of the potential within the oxide (i.e. $-t_{\mathrm{Ox}} \le x \le 0$) is obtained to be

$$\varphi(x) = V_{\mathrm{G}} + (\varphi_{\mathrm{Ch/Ox}} - V_{\mathrm{G}})(1 + x/t_{\mathrm{Ox}}) \tag{4.20}$$

with the normal component of electric displacement being

$$\varepsilon_{rx}^{\mathrm{Ox}} \frac{d}{dx}\varphi(x)\bigg|_{x\to 0-} = \frac{\varepsilon_{rx}^{\mathrm{Ox}}}{t_{\mathrm{Ox}}}(\varphi_{\mathrm{Ch/Ox}} - V_{\mathrm{G}}) \tag{4.21}$$

The normal component of electric displacement within the channel (i.e. $x \geq 0$), on the other hand, is obtained by integrating the corresponding eq. (4.18), as

$$\varepsilon_{rx}^{\mathrm{Ch}} \frac{d}{dx}\varphi(x)\bigg|_{x\to 0+}^{x\to +\infty} = -\varepsilon_{rx}^{\mathrm{Ch}} \frac{d}{dx}\varphi(x)\bigg|_{x\to 0+} = -\frac{(-e)}{\varepsilon_0 W\Delta z}\sum_n \mathrm{N}^{C(n)} - \frac{(+e)}{\varepsilon_0} N_{\mathrm{A}}^{-} t_{\mathrm{Ch}} \tag{4.22}$$

where $\sum_n \mathrm{N}^{C(n)} = \sum_n \int_0^\infty |\langle x\,|n\rangle|^2 xdx\mathrm{N}^{C(n)}$; and $t_{\mathrm{Ch}} = \sum_n \int_0^\infty |\langle x\,|n\rangle|^2 xdx\mathrm{N}^{C(n)} / \sum_n \int_0^\infty |\langle x\,|n\rangle|^2 xdx\mathrm{N}^{C(n)}$ is the channel thickness (at any given $z = z$), i.e. the distance from the channel/oxide interface towards the substrate, over which the electron wave spreads. Now, since eqs. (4.21), (4.22), and the boundary conditions (4.19) for continuity of normal component of electric displacement at channel/oxide interface, we have:

$$\varphi_{\mathrm{Ch/Ox}}(z) = V_{\mathrm{G}} - \frac{e}{\varepsilon_0\varepsilon_{rx}^{\mathrm{Ox}}/t_{\mathrm{Ox}}}\left[\frac{1}{W\Delta z}\sum_n \mathrm{N}^{C(n)} - N_{\mathrm{A}}^{-} t_{\mathrm{Ch}}(z)\right] \tag{4.23}$$

It is worth noting that $C_{\mathrm{Ox}} = \varepsilon_0\varepsilon_{rx}^{\mathrm{Ox}}/t_{\mathrm{Ox}}$ represents the oxide capacitance to hold the surface confined charges in the channel region, where the surface potential $\varphi_{\mathrm{Ch/Ox}}(z)$ is constituted of the contributions from two types of charges in the channel: 1) 'inversion' charge $\left[\frac{(-e)}{W\Delta z}\sum_n \mathrm{N}^{C(n)}\right]$ and 2) 'depletion' charge $[(+e)N_{\mathrm{A}}^{-} t_{\mathrm{Ch}}(z)]$. Interestingly, the 'inversion' charge is a function of V_{G} due to $\left[\sum_n \mathrm{N}^{C(n)}\right]$ through the 'implicit' dependence on the sub-band energies $\varepsilon_n(z)$, while the 'depletion' charge varies with V_{G} due to $[t_{\mathrm{Ch}}(z)]$, which, on the other hand, is 'implicitly' dependent on the sub-band energies through $\left[\sum_n \mathrm{N}^{C(n)}\right]$, and 'explicitly' depends on the sub-band states through $|\langle x\,|n\rangle|^2$.

The surface potential for DG-MOSFET or FinFET and NS-FET can be obtained by applying the same method. However, for a cylindrical GAA NWFET with an intrinsic channel of radius R, assuming ρ to increase along S→O→M, eq. (4.18) changes to

$$\begin{cases} \varepsilon_{r\rho}^{\mathrm{Ox}} \frac{1}{\rho}\frac{d}{d\rho}\rho\frac{d}{d\rho}\varphi(\rho) = 0, & R \leq \rho \leq R + t_{\mathrm{Ox}} \\ \varepsilon_{r\rho}^{\mathrm{Ch}} \frac{1}{\rho}\frac{d}{d\rho}\rho\frac{d}{d\rho}\varphi(\rho) = -\frac{(-e)}{\varepsilon_0\Delta z}\sum_n \left|\langle \rho\,|n\rangle\right|^2, & 0 \leq \rho \leq R \end{cases} \tag{4.24}$$

with the boundary conditions:

$$\begin{cases} \varphi(\rho)|_{\rho\to R+} = \varphi(\rho)|_{\rho\to R-} = \varphi_{\mathrm{Ch/Ox}} \\ \varepsilon_{r\rho}^{\mathrm{Ox}} \frac{d}{d\rho}\varphi(\rho)\big|_{\rho\to R+} = \varepsilon_{r\rho}^{\mathrm{Ch}} \frac{d}{d\rho}\varphi(\rho)\big|_{\rho\to R-} \\ \varphi(\rho)|_{\rho=R+t_{\mathrm{Ox}}} = V_{\mathrm{G}} \end{cases} \tag{4.25}$$

Now, the solution of the oxide potential (i.e. $R \le \rho \le R + t_{\mathrm{Ox}}$) is resulted to be

$$\varphi(\rho) = V_{\mathrm{G}} + (\varphi_{\mathrm{Ch/Ox}} - V_{\mathrm{G}}) \left[\frac{\mathrm{In}(\rho/(R+t_{ox}))}{\mathrm{In}(R/(R+t_{ox}))}\right] \tag{4.26}$$

since $\left(\rho \frac{d}{d\rho}\varphi(\rho)\right)$ is constant within the oxide (considering no fixed-oxide-charges). Thus, the normal component of electric displacement is obtained to be

$$\varepsilon_{r\rho}^{\mathrm{Ox}} \frac{d}{d\rho}\varphi(\rho)\bigg|_{\rho\to R+} = \frac{\varepsilon_{r\rho}^{\mathrm{Ox}}}{R\ln(R/(R+t_{\mathrm{Ox}}))}(\varphi_{\mathrm{Ch/Ox}} - V_{\mathrm{G}}) \tag{4.27}$$

Interestingly, eq. (4.27) reduces consistently to the expression for planar MOSFET given by eq. (4.21) in the limit $R \gg t_{\mathrm{Ox}}$ (i.e. when the curvature of the NW cylindrical surface approaches zero). At this point, it is imperative to note that $\varepsilon_{rx}^{\mathrm{Ox}}\left|\frac{d}{dx}\varphi(x)\right|_{x\to 0-}$ for planar MOSFET or $\varepsilon_{r\rho}^{\mathrm{Ox}}\left|\frac{d}{d\rho}\varphi(\rho)\right|_{\rho\to R+}$ for NWFET is the amount of surface charge density that the corresponding MOS-capacitor holds at the effective voltage $(V_{\mathrm{G}} - \varphi_{\mathrm{Ch/Ox}})$ across it, which shows, since eqs. (4.21) and (4.27), that the effective-oxide-capacitance of such MOS-capacitors for planar MOSFET and NWFET are, respectively,

$$C_{\mathrm{Ox}}^{\mathrm{Planar}} = \frac{\varepsilon_{\mathrm{rx}}^{\mathrm{Ox}}}{t_{\mathrm{Ox}}} \tag{4.28a}$$

and

$$C_{\mathrm{Ox}}^{\mathrm{NWFET}} = \frac{\varepsilon_{r\rho}^{\mathrm{Ox}}}{R\ln(1 + t_{\mathrm{Ox}}/R)} \tag{4.28b}$$

Thus, considering SiO_2 as the standard insulating material for MOS, the required real oxide-thickness for any other particular insulator (e.g. HfO_2, ZrO_2, La_2O_3, Al_2O_3) to get the same MOS-capacitance, provided a definite effective-oxide-thickness (EOT), is given by

$$t_{\mathrm{Ox}} = \left(\frac{\varepsilon_{rx}^{\mathrm{Ox}}}{\varepsilon_{rx}^{\mathrm{SiO_2}}}\right)\mathrm{EOT} \tag{4.29a}$$

and

$$\ln(1 + t_{\mathrm{Ox}}/R) = \left(\frac{\varepsilon_{r\rho}^{\mathrm{Ox}}}{\varepsilon_{r\rho}^{\mathrm{SiO_2}}}\right)\ln(1 + \mathrm{EOT}/R) \tag{4.29b}$$

for planar MOSFET and NWFET, respectively. In order to reduce the leakage current due to gate-tunnelling in ultra-scaled MOSFET devices, keeping the effective gate-control intact, the high-k oxides with higher thicknesses can be used following the above equations for corresponding MOSFET architectures.

Now, in a GAA NWFET, for small R (i.e. a few nm), the radial band bending is small, so that $2\pi \int_0^R \varphi(\rho, z)\rho d\rho \approx \pi R^2 \varphi_{\text{Ch/Ox}}$ and the equation along the channel can be written since eqs. (4.15a) and (4.17e) as

$$\left[\varepsilon_{rz}\frac{\partial^2}{\partial z^2}\left(2\pi\int_0^R \varphi(\rho,z)\rho d\rho\right) + \left(2\pi R\varepsilon_{r\rho}\frac{\partial\varphi(\rho,z)}{\partial\rho}\bigg|_{\rho=R}\right)\right] = -\frac{(-e)}{\varepsilon_0}\sum_n n_{1D}^n(z)$$

$$\Rightarrow \left[\left(\pi R^2\varepsilon_{rz}\frac{\partial^2}{\partial z^2}\varphi_{\text{Ch/Ox}}(z)\right) + \left(\frac{2\pi\,\varepsilon_{r\rho}^{\text{Ox}}}{\ln(R/(R+t_{\text{Ox}})}\left(\varphi_{Ch/Ox} - V_G\right)\right)\right] = -\frac{(-e)}{\varepsilon_0}\sum_n n_{1D}^n(z)$$

$$\rightarrow \left[\frac{\partial^2}{\partial z^2} - \frac{2}{R^2}\frac{\varepsilon_{r\rho}^{Ox}}{\varepsilon_{rz}}\frac{1}{\ln(1+t_{Ox}/-R)}\right]\left(\varphi_{Ch/Ox}(z) - V_G\right) = \frac{e}{\varepsilon_0\varepsilon_{rz}\pi R^2}\sum_n n_{1D}^n(z) \qquad (4.30)$$

It is apparent that eq. (4.30) is a one-dimensional second-order linear inhomogeneous differential equation that can be solved directly using boundary conditions that potential is 0 and V_D at the source and drain ends, respectively.

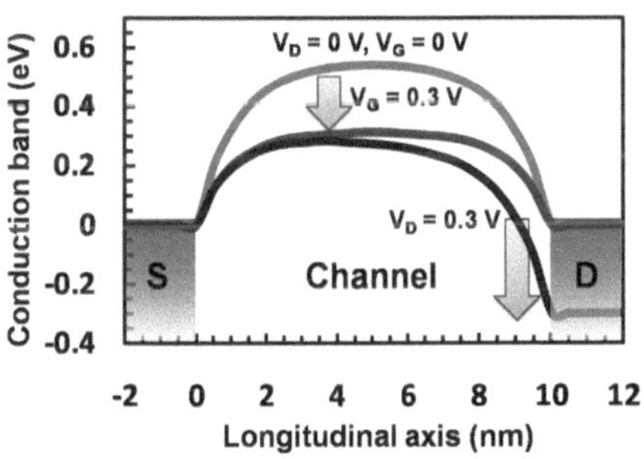

Fig. 4.7: An illustration of the longitudinal band diagram under different gate and drain biases: the plot of simulated conduction band of a Si GAA-NWFET of channel diameter of 3 nm, channel length 10 nm, oxide thickness 1 nm at unbiased condition (i.e. $V_G = 0$ V $= V_D$), at $V_G = 0.3$ V, $V_D = 0$ V, and $V_G = 0.3$ V $= V_D$ V using the NEGF formalism.

Finally, such values of the potential ($\varphi_{\text{Ch/Ox}}(z)$) along the channel obtained by solving eq. (4.30) are put into eq. (4.11b) to find out the local Green's function in the channel, followed by getting the 1D carrier density ($\sum_n n_{1D}^n(z)$) from eq. (4.12), and again put into eq. (4.30) to obtain the second-order iteration value of the channel potential. Such iteration is carried out until the difference between potential values obtained at two consecutive iteration loops is negligible and accordingly self-consistency can be considered to be achieved. Such a self-consistent conduction band potential profile along the channel of a Si GAA-NWFET, as an example, obtained using NEGF simulator is plotted in Fig. 4.7 at different gate and drain bias conditions.

4.3.5 Physics of self-consistency in the numerical computation

Unlike the conventional (semi-classical) microelectronic devices on one side, and also in contrast to the quantum particles without charge on the other, the charge transport behaviour in nanostructure semiconductor devices like quantum-scale MOSFETs cannot be appropriately described solely by either electrostatic interactions or quantum theory. For instance, the transport of charge carriers such as electrons or holes from source to drain through the channel of a nanostructure MOSFET must be modelled by solving the coupled quantum-electrostatic equations of motion simultaneously, as discussed above. It is imperative to note that Poisson's equation describes how the electrostatic force lines diverge from one point in space with higher potential energy to others at comparatively lower values of potential energy, or *vice versa,* in relation to the presence of charges at that particular point. This, in turn, shows that the more the amount of charges at such a point, the greater the relative value of potential energy at that point. However, quantum statistical properties restrict the charges to stay at such a potential energy level constrained by the condition that the more the potential energy value of the channel compared to the reservoir Fermi levels (at a definite temperature), the less is the probability of the particles to stay. Thus, the inner contradiction between such two-fold property of the charged quantum particles, i.e. 'to-stay' (due to electrostatic interactions) and 'not-to-stay' (owing to quantum statistical condition), is the fundamental origin of their motion. Therefore, in the NEGF framework, the self-consistency necessarily needs to be achieved between Schrödinger-Poisson coupled equations to describe such a quantum-electrostatic 'complementarity' relation in order to appropriately predict the transport characteristics of quantum electronic devices. This is illustrated in Fig. 4.8, which plots the difference between output and input potential energy at each iteration step that exhibits a damped oscillatory nature asymptotically reaching the self-consistent potential.

The plot of Fig. 4.8 representing the gap between output and input potential values at the centre of an NWFET channel during NEGF simulation shows an oscillatory linear decay in the semi-logarithmic scale. It is apparent that it indicates an exponential rate of convergence of such a simulation programme, upholding its computational efficiency. Further, it may be worth noting that if in any iteration loop, the assumed input potential being operated in Schrödinger equation followed by using Poisson's equation results in a lesser value of output potential (in comparison to the input), in the next iteration loop, such result it will be reversed, i.e. leading to an increased

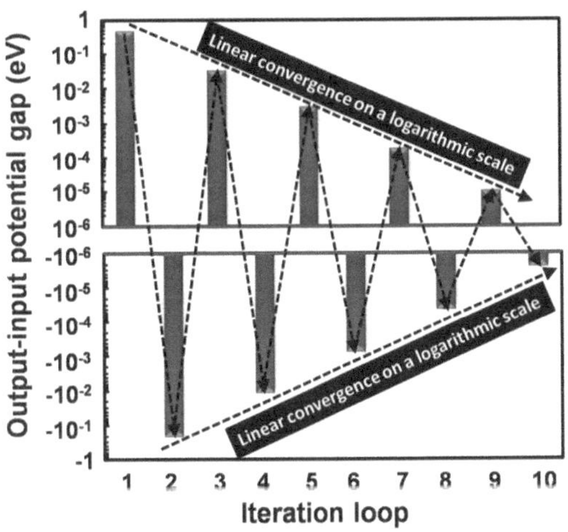

Fig. 4.8: The variation of the difference of output-input potential values, for instance, at the centre of the NWFET channel, in each consecutive iterative loops in a semi-logarithmic scale.

value of output (compared to input), although smaller than the input of previous loop. It can be analogically visualized that the iteration in NEGF simulation is quite like the damped oscillation of a pendulum around the self-consistency, which is achieved asymptotically. It seems as if the quantum-electrostatic dialectics has an inherent restoring but damping force acting unendingly towards its self-consistency that expresses the nature of the 'complementarity' relationship between quantum and electrostatic properties of nature.

4.3.6 Transmission coefficient and current

Since the charge transport in MOSFET is accompanied by the coupling of channel with two reservoirs, i.e. source and drain, the local transmission coefficient can be considered as the product of the local spectral density at any point within the channel due to source (i.e. $A_n^S(E;z_l) = \left[[G_{\mathrm{Ch}}(E,z_l)][\Gamma^S(E,z_1)][G_{\mathrm{Ch}}(E,z_l)]^+\right]_{nn}$) and the level broadening due to carrier escape to drain (i.e. $[\Gamma^D(E,z_N)]$), or just *vice versa*. Thus, once quantum and electrostatic 'complementary' self-consistency is achieved as discussed above, the net transmission coefficient ($T_{\mathrm{S}\to\mathrm{D}}$) for the transport of charge from source to drain through the MOSFET-channel can be obtained by summing up local transmission at each point in the channel as [10]

$$T_{\mathrm{S}\to\mathrm{D}}(E) = \mathrm{Trace}\left([A^{\mathrm{S}}(E;z_l)][\Gamma^{\mathrm{D}}(E,z_N)]\right) = \mathrm{Trace}\left([\Gamma^{\mathrm{S}}(E,z_1)][A^{\mathrm{D}}(E;z_l)]\right)$$

$$\Rightarrow T_{\mathrm{S}\to\mathrm{D}}(E) = \mathrm{Trace}[\Gamma_{\mathrm{S}}(E,z_1)G(E,z_l)\Gamma_{\mathrm{D}}(E,z_N)G^+(E,z_l)] \tag{4.31}$$

Recalling the concept of inflow and outflow discussed in Chapter 3, the electric current at any lth terminal (i.e. $l=1$ for source and $l=N$ for drain here) can be written as

$$I=\frac{e}{h}\int dE\sum_{l=1}^{2}\text{Trace}\left([\Pi_{\text{Out}}(E)]_l-[\Pi_{\text{In}}(E)]_l\right)\tag{4.32a}$$

$$\Rightarrow I=\frac{e}{h}\int dE\sum_{l=1}^{2}\text{Trace}\left([\Sigma_l^{\text{In}}(E)]i([G(E)]-[G^+(E)])-[\Gamma_l(E)][G(E)][\Sigma_l^{In}(E)][G^+(E)]\right)\tag{4.32b}$$

$$\Rightarrow I=\frac{2e}{h}\int T_{\text{S}\to\text{D}}(E)\left(f_{\text{FD}}(E-E_{\text{F}}^{\text{S}})-f_{\text{FD}}(E-(E_{\text{F}}^{\text{D}}-V_{\text{D}}))\right)dE\tag{4.32c}$$

where $[\Sigma_l^{\text{In}}(E)]=[\Gamma_l(E)]f_{\text{FD}}(E-E_{\text{F}}^{l})$ for coherent transport. It is interesting to note that the net MOSFET current given by eq. (4.32c) is identical to the famous Landauer formula. It may be interesting to note that if the energy is considered in 'eV' unit, the constant multiple factor in the expression of the current of eq. (4.32c) would be $G_0=\frac{2e^2}{h}$, known to be the quantum conductance, which is also equal to the fundamental quantum Hall conductance at the ground state.

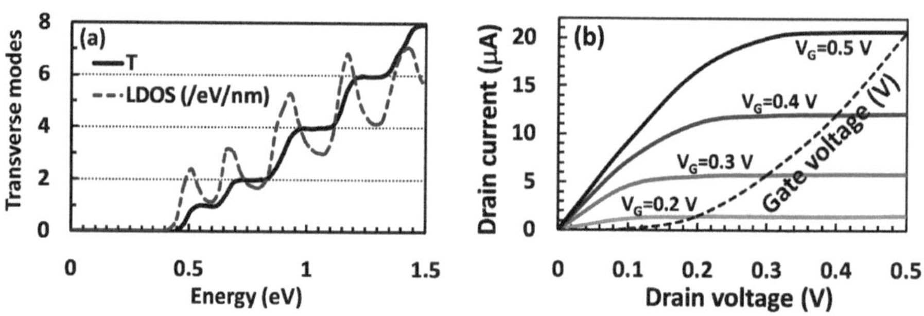

Fig. 4.9: The plots of transmission coefficient and LDOS at the centre of the channel of a Si GAA-NWFET (channel diameter of 3 nm, channel length 10 nm, oxide thickness 1 nm) with energy; (b) output characteristics of such NWFET at different gate voltages along with transfer characteristics (as a dotted line).

The results of eqs. (4.31) and (4.32) are illustrated by simulating a Si GAA-NWFET with its channel diameter to be 3 nm, channel length 10 nm, and oxide thickness being 1 nm, and plotted accordingly in Fig. 4.9(a),(b). The transmission coefficient plotted with energy in Fig. 4.9(a) shows clear steps at the transverse sub-bands owing to the corresponding quantum confinement, while the nature of LDOS ensures 1D transport of carriers, which are free only to move along the channel from source to drain. The output characteristics ($I_{\text{D}}-V_{\text{D}}$) and transfer characteristics ($I_{\text{D}}-V_{\text{G}}$) of such NWFET show typical MOSFET behaviour, although resulting in a μA order of operating currents at <1 V that offers an extremely low power device.

4.3.7 Local and average mobility of the MOSFET channel

The local channel mobility $\mu(z)$ of a MOSFET can be defined by

$$v_z(z) = \mu(z)\Xi_z(z) = \mu(z)\frac{\partial}{\partial z}\varphi(z) \tag{4.29}$$

where $v_z(z) = \frac{1}{\hbar}\frac{\partial E}{\partial k_z}$ is the local drift velocity of the electrons (or holes), $k_z(z)$ being the local wave vector, and $\Xi_z(z)$ being the local driving electric field along the channel due to applied drain voltage V_D. From the discussions of Chapter 3 on local band structure and band velocity, it is apparent that the wave vector appears as the change of phase after each lattice periodicity in the self-energy term. Thus, while getting the Green's function of two interacting systems, which may also be, for instance, any two adjacent grids of the MOSFET channel or the adjacent channel-grid and source/drain-grid, we have, in general:

$$[G(E)] = \begin{bmatrix} E-\varepsilon_1 & -\tau \\ -\tau* & E-\varepsilon_2 \end{bmatrix}^{-1} = \begin{bmatrix} [E-\varepsilon_1-\tau(E-\varepsilon_2)^{-1}\tau*]^{1} & \cdots \\ \cdots & [E-\varepsilon_2-\tau*(E-\varepsilon_1)^{-1}\tau]^{-1} \end{bmatrix} \tag{4.30a}$$

where the phases appear in the self-energy terms $\tau(E-\varepsilon_2)^{-1}\tau*$ and $\tau*(E-\varepsilon_1)^{-1}\tau$, which can be generalized as

$$[\Sigma(E)]_{ii} = [[G(E)]^{-1}]_{ii} - ([G(E)]_{ii})^{-1} \tag{4.30b}$$

Thus, the local wave vector along the channel can be obtained as

$$[k_z]_{ii} = \frac{1}{\Delta z}\arg\left([[G_{Ch}(E)]^{-1}]_{ii} - ([G_{Ch}(E)]_{ii})^{-1}\right) \tag{4.31}$$

which, since $v_z(z) = \frac{1}{\hbar}\frac{\partial E}{\partial k_z}$, leads to obtaining the local drift velocity, and also the local mobility from its relation with the local electric field using eq. (4.29). On the other hand, the average mobility $\langle\mu\rangle$ of the channel can be obtained from the net electric field $\Xi_z^{Total} = V_D/L$ across the channel of length L and the resultant wave vector $k_z = \sum_i [k_z]_{ii}$ spread over the channel.

4.3.8 Source/drain contact-resistance in MOSFETs

It is imperative to note that while modelling the charge transport in MOSFET devices on the basis of the NEGF framework, the active device region, i.e. MOSFET channel, is considered to be connected with the reservoirs, i.e. source/drain, through 'conceptually semi-infinite' contact leads. Accordingly, the Hamiltonians for such contact leads are ideally assumed to be uniform, resulting in the corresponding source/drain self-energies as discussed above and in Chapter 3. However, in practice, the local band structure in the vicinity of such contact regions, i.e. channel-source/drain junctions, must be modified with respect to the ideal ones, owing to the differences in their longitudinal effective mass that modify the transport kinetic energy, as well as the transverse effective mass components that change the transverse sub-bands, and consequently affect the corresponding self-energies. This, in turn, is manifested as the contact resistance of the device, which is thus necessary to incorporate into the model for practical device simulation. It is further worthy to note that since the effect of such contact resistances enters into the transport characteristics of MOSFET through the modified source/drain self-energies, the carrier effective mass mismatch significantly influences the local sub-band level broadenings; in fact, it reduces the allowed energy space for carriers to transport through the channel, thereby effectively degrading the channel mobility [15]. Here, the phenomenon has been illustrated by simulating an NWFET applying NEGF method in Fig. 4.10(a),(b), which clearly depict the sharpening of available energy space (in terms of LDOS) owing to the mismatch in effective mass tensor and subsequently the energy sub-bands of the active device (i.e. NWFET channel) and reservoirs (i.e. source/drain).

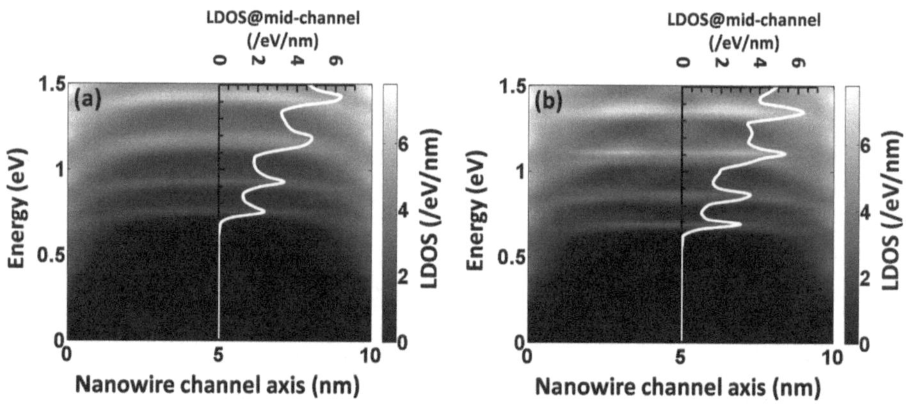

Fig. 4.10: The plots of simulated LDOS with energy along the NWFET channel, along with its variation at the channel centre: (a) without considering channel/S-D effective mass mismatch; and (b) incorporating such effective mass mismatch.

It is imperative to note at this point that while calculating the self-energy due to a reservoir (e.g. source or drain), as elaborated in Chapter 3 for the apparatus, the eigenstates or the sub-bands of the reservoir ($[\varepsilon^R]$) and the coupling matrix ($[\tau^{R\leftrightarrow R}$) were considered to be identical throughout the entire reservoir, owing to the idealistic assumption of its translational symmetry. Thus, it was trivial to assume that the inverse block matrix ($[g]$) would again give rise to the same term $[g]$ in the remaining self-energy part, leading to the characteristic equation to be $[g] = \left(E[I] - [\varepsilon^R] - [\tau^{R\leftrightarrow R}][g][\tau^{R\leftrightarrow R}]^+\right)^{-1}$, so that the net reservoir self-energy becomes $[\Sigma_A(E)] = [\tau^{\text{Ch}\leftrightarrow R}][g][\tau^{\text{Ch}\leftrightarrow R}]^+$. However, since the local band structure at the contact region of the reservoir is indeed not identical to its remaining counterpart, the characteristic equation is modified as follows:

$$[\tilde{g}] = \left(E[I] - [\varepsilon^{RC}] - [\tau^{RC\leftrightarrow R}][g][\tau^{RC\leftrightarrow R}]^+\right)^{-1} \tag{4.32}$$

with

$$[g] = \left(E[I] - [\varepsilon^{R}] - [\tau^{R\leftrightarrow R}][g][\tau^{R\leftrightarrow R}]^+\right)^{-1} \tag{4.32a}$$

where $[\varepsilon^{RC}]$ is the local sub-band matrix at the channel-contact region of the reservoir, whereas $[\varepsilon^{R}]$ represents the sub-band energies of the remaining uniform part of the reservoir; $[\tau^{RC\leftrightarrow R}]$ is the coupling matrix between the contact region and the rest of the reservoir counterparts, which couple with themselves through $[\tau^{R\leftrightarrow R}]$. Now, the reservoir self-energy matrix can be expressed as $[\Sigma_A(E)] = [\tau^{\text{Ch}\leftrightarrow R}][\tilde{g}][\tau^{\text{Ch}\leftrightarrow R}]^+$. It may be worth noting at this point that the appropriate contact-Hamiltonian and the resultant sub-band states, as well as the effective mass components, can be precisely obtained by employing sp[3]s*-type of LCAO approach or density functional theory, which can then be used to find the relevant self-energies using this method to model the impact of contact resistances in the transport behaviour of practical MOSFET devices.

References

[1] Ratnesh, Ratneshwar K., A. Goel, G. Kaushik, H. Garg, M. Singh, and B. Prasad. "Advancement and challenges in MOSFET scaling." *Materials Science in Semiconductor Processing* 134, (2021): 106002.

[2] Han, Seong-Joo, Joon-Kyu Han, Myung-Su Kim, Gyeong-Jun Yun, Ji-Man Yu, Il-Woong Tcho, Myungsoo Seo, Geon-Beom Lee, and Yang-Kyu Choi. "Ternary logic decoder using independently controlled double-gate Si-NW MOSFETs." *Scientific Reports* 11, no. 1 (2021): 13018.

[3] Weis, C. D., A. Schuh, A. Batra, A. Persaud, I. W. Rangelow, J. Bokor, C. C. Lo. et al "Mapping of ion beam induced current changes in FinFETs." *Nuclear Instruments and Methods in Physics Research Section B: Beam Interactions with Materials and Atoms* 267, no. 8–9 (2009): 1222–1225.

[4] Yang, Fu-Liang, Di-Hong Lee, Hou-Yu Chen, Chang-Yun Chang, Sheng-Da Liu, Cheng-Chuan Huang, Tang-Xuan Chung. et al "5nm-gate nanowire FinFET." In *Digest of Technical Papers. 2004 Symposium on VLSI Technology*, pp. 196–197. IEEE, 2004.

[5] Cho, Keun Hwi, Kyoung Hwan Yeo, Yun Young Yeoh, Sung Dae Suk, Ming Li, Ji Myoung Lee, M-S. Kim. et al "Experimental evidence of ballistic transport in cylindrical gate-all-around twin silicon nanowire metal-oxide-semiconductor field-effect transistors." *Applied Physics Letters* 92, no. 5 (2008).

[6] Loubet, Nicolas, S. Kal, C. Alix, S. Pancharatnam, H. Zhou, C. Durfee, M. Belyansky. et al "A novel dry selective etch of SiGe for the enablement of high performance logic stacked gate-all-around nanosheet devices." In *2019 IEEE International Electron Devices Meeting (IEDM)*, pp. 11–4. IEEE, 2019.

[7] John Chelliah, Cyril RA, and Rajesh Swaminathan. "Current trends in changing the channel in MOSFETs by III–V semiconducting nanostructures." *Nanotechnology Reviews* 6, no. 6 (2017): 613–623.

[8] Turkane, Satish M., and A. K. Kureshi. "Review of tunnel field effect transistor (TFET)." *International Journal of Applied Engineering Research* 11, 7 (2016): 4922–4929.

[9] Vaughan, Owain. “Two nanometre CMOS technology.” *Nature Electronics* 7, 12 (2024): 1063–1063.
[10] Datta, Supriyo. “Nanoscale device modeling: The Green’s function method.” *Superlattices and microstructures* 28, no. 4 (2000): 253–278.
[11] Ng, Kwok Kwok, and Simon M. S. *Physics of semiconductor devices*. Hoboken, NJ: Wiley-Interscience, 2007.
[12] Venugopal, Rajesh, Z. Ren, Supriyo Datta, Mark S. Lundstrom, and D. Jovanovic. “Simulating quantum transport in nanoscale transistors: Real versus mode-space approaches.” *Journal of Applied Physics* 92, no. 7 (2002): 3730–3739.
[13] Wang, Jing, Eric Polizzi, and Mark Lundstrom. “A three-dimensional quantum simulation of silicon nanowire transistors with the effective-mass approximation.” *Journal of Applied Physics* 96, no. 4 (2004): 2192–2203.
[14] Wang, Kuang-Chung, Roberto Grassi, Yuanchen Chu, Shree Hari Sureshbabu, Junzhe Geng, Prasad Sarangapani, Xinchen Guo, Mark Townsend, and Tillmann Kubis. “Introduction of multi-particle Büttiker probes – Bridging the gap between drift diffusion and quantum transport.” *Journal of Applied Physics* 128, no. 1 (2020).
[15] Nag Chowdhury, Basudev, and Sanatan Chattopadhyay. “Investigating the impact of source/drain doping dependent effective masses on the transport characteristics of ballistic Si-nanowire field-effect-transistors.” *Journal of Applied Physics* 115, no. 12 (2014).

5 Phonon dynamics: some unconventional ideas

Our everyday experiences tell us to believe without any scepticism that past is definite that one cannot intervene into, while future is uncertain that offers many possibilities for one to act according to her/his desire. However, in Physics, such asymmetry in the direction of time is fundamentally discarded, and all the laws of physics, ranging from classical to quantum domain, are considered to be time-symmetric, the only exception being the second law of thermodynamics. In fact, 'heat' being the uncontrolled form of microscopic motion of the atoms and molecules resulting in energy dissipation is considered to be responsible for incorporating such directionality in time. Thus, is such arrow of time somehow connected to phonons, i.e. the quanta of energy dissipated as heat owing to the inescapable quantum oscillations of atoms within any confining potential? Can this question be extended to vacuum fluctuations to a search for the arrow of time from fundamentals of space-time?

Unfortunately, these are not the issues that we are focussing here. Rather, we would like to address the question whether our fate is restricted by the natural and unavoidable dissipations regardless of whether time has an arrow or not! For instance, the curse imposed by frictional forces had been overcome by the human beings in its early stages of civilization by developing the wheels. It is not only a remarkable invention, but also an astonishing discovery of the fact that friction can be made (at least in principle) lossless. Can the same be done with phonons?

Interestingly, although the aggressive miniaturization of the electronic and optoelectronic devices, as discussed in Chapter 1, has led their physical dimensions (i.e. of the order of a few nanometres) to enter the domain of ballistic transport these days, where electrons/holes hardly face any scattering so that the medium can be thought to be perfectly elastic, they suffer from unavoidable phonon scattering usually resulting in an incoherent transport. In fact, each atom within a solid material is bonded with others and thus vibrates in such bonding potential (quite like a spring-mass system as depicted in Fig. 5.1(a)) at any non-zero finite temperature leading to several discrete quantum modes. The atoms can excite/de-excite from one such vibrational quantum state to another while interacting with the carriers (i.e. electrons/holes) by absorbing/emitting a phonon as a quantum of energy. Usually, the carriers exhibit a dissipation of their transport energy due to such phonon scattering phenomena at any finite temperature. A detail discussion on phonons is provided in [1].

5.1 Phonon interaction potential

Every atom bonded with another in a material vibrates within the potential of a nature shown in Fig. 5.1(b), which is inherently asymmetric in nature and thus cannot be described by a conventional quantum harmonic oscillator model. As shown in

 | https://doi.org/10.1515/9783111545950-005

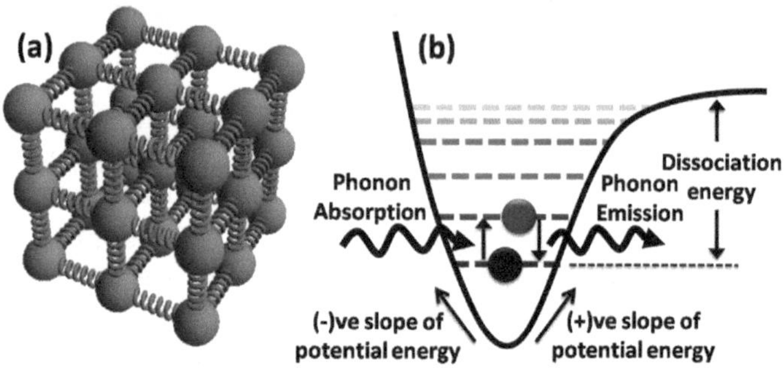

Fig. 5.1: (a) A classical view of vibrating atoms in a 3D-lattice analogically equivalent to a spring-mass system and (b) quantum view of such atoms oscillating in a confining potential (of Morse type) illustrating the absorption/emission of phonons due to atomic transitions.

Fig. 5.1(b), an atom in a lattice cannot leave the bonding with its neighbour owing to the positive potential energy slope in the furthest side, while it does not collapse to the neighbour due to negative slope in the closest side. As mentioned, such a potential is asymmetric in nature around the equilibrium position, thereby resulting in different average positions at different vibrational quantum modes. This inevitably results in a finite amount of lattice strain (Ξ) leading to modify the electron transport through an electron-phonon interaction potential (τ), that can be given in terms of the deformation potential (D) of the material as [2]:

$$\tau = D\Xi = D\vec{\nabla}\cdot\vec{u} \tag{5.1}$$

where $\vec{u}$ is the displacement of an atom (coupled to others through respective bonds) from its equilibrium position during oscillation in the bond potential and given by:

$$\vec{u} = \vec{u}_0 e^{i(\vec{\beta}\cdot\vec{r}-\Omega t)} + \vec{u}_0 \times e^{-i(\vec{\beta}\cdot\vec{r}-\Omega t)} \tag{5.2}$$

where $\vec{\beta}$ is the wave vector of such coupled oscillators (i.e. bonded atoms) and Ω is the angular frequency. It may be interesting to note that any incorporated lattice strain in a material thus can significantly modify the impact of phonons and thereby influence the electron/hole mobility (other than the strain induced band-structure modification), which we are not discussing here (for a detailed study, see [2]).

Now, the amplitude ($\vec{u}_0$) of atomic oscillation has an imaginary part in general since it can always bear a Bloch-type phase ($e^{i\phi}$) due to translational symmetry in the lattice. A crude approximation can be taken to obtain the amplitude of oscillation on an average by assuming such atomic vibrations as coupled quantum harmonic oscillations (only with respect to their amplitudes but not in terms of the asymmetric nature of potential), so that $\frac{1}{2}\rho V\Omega^2\langle|2\vec{u}_0|^2\rangle = \hbar\Omega f_{BE}(\Omega)$, where ρ is the material density, V is the normalization volume, and $f_{BE}(\Omega)$ is the Bose-Einstein (BE) distribution function for phonons (it should be noted that phonons are bosons). The phonons generally have six polarization vectors ($\hat{v}$): one longitudinal and two transverse components for each of the acoustic phonons and optical phonons, which are defined according to the relative displacement of adjacent atoms, during coupled oscillation, being either along the same direction or in the opposite direction, respectively. Thus, on an average, the amplitude of coupled oscillation can be written as [2]:

$$\vec{u}_0 = \hat{v}\left(\sqrt{\frac{\hbar f_{\mathrm{BE}}(\Omega)}{2\rho V\Omega}}\right)\frac{1}{\sqrt{2}}\begin{pmatrix}1\\ \pm 1\end{pmatrix} \tag{5.3}$$

where '+' and '–' signs, respectively, denote acoustic and optical phonons. Note that for individual nth phonon state, $f_{\mathrm{BE}}(\Omega)$ must be replaced by n in eq. (5.3).

5.2 Phonon dispersion: a proposition on a non-trivial topology

The phonons, as explained above, are the emitted or absorbed quantum of vibrational energy associated with the transition of an atom from one of its quantum state to other in the vibrating potential of other atoms in a crystal lattice. In this context, single phonon emitter, detector as well as manipulator by beam splitting exhibiting consequential interference has recently been explored [3-5]. However, the frequency dispersion of such phonons, even exhibiting quantum behaviours, can be obtained by considering the coupled oscillation of all such atoms collectively vibrating in the lattice to a satisfactory level of approximation. Thus, in order to comprehend the basic concept, we will preliminarily consider here an infinite linear chain of two types of atoms coupled through asymmetric bonds (for generalization), equivalent to an asymmetric coupled spring-mass system, with spring constant $K \in (K_1, K_2\}$ and atomic mass $m \in (m_1, m_2\}$, with a translational symmetry along the lattice (as shown in Fig. 5.2(a)). It can readily explain, at least qualitatively, the behaviour of phonons in materials that are quite important in the state-of-the-art technology, such as III–V (e.g. GaAs, GaN, InP, and InAs) and II–VI (e.g. ZnO, ZnSe, and CdTe) compound semiconductors as well as the basic semiconductors like Si and Ge (if the two atoms and two bonds are considered to be identical). It may be noted that spring constant (K) for such atomic bonds is of the order of mDyne/Å (i.e. ~10^2 N/m) and the mass of an atom is of the order of tens of amu (i.e. ~10^{-26} kg), so that the angular frequencies (considering the multiplying factors) become of the order of tens of THz and so is the phonon frequency. Thus, phonon energy ($\hbar\Omega$) in all such materials are of the order of tens of meV, i.e. comparable to thermal voltage (~26 meV) at room temperature.

It is apparent that the equation of such coupled harmonic oscillation can be given by:

$$\begin{cases} \quad\quad\quad\quad\quad\quad\quad\quad \dots \\ \frac{d^2}{dt^2}u_j = -K_1/m_1(u_j - u_{j-1}) - K_2/m_1(u_j - u_{j+1}) \\ \frac{d^2}{dt^2}u_{j+1} = -K_2/m_2(u_{j+1} - u_j) - K_1/m_2(u_{j+1} - u_{j+2}) \\ \quad\quad\quad\quad\quad\quad\quad\quad \dots \end{cases} \tag{5.4}$$

Considering $K_i/m_j = \Omega_{ij}{}^2$, the set of eq. (5.4) can be written as:

$$\begin{cases} \quad\quad\quad\quad\quad\quad\quad\quad \dots \\ \frac{d^2}{dt^2}u_j = \Omega_{11}{}^2 u_{j-1} - (\Omega_{11}{}^2 + \Omega_{21}{}^2)u_j + \Omega_{21}{}^2 u_{j+1} \\ \frac{d^2}{dt^2}u_{j+1} = \Omega_{22}{}^2 u_j - (\Omega_{22}{}^2 + \Omega_{12}{}^2)u_{j+1} + \Omega_{12}{}^2 u_{j+2} \\ \quad\quad\quad\quad\quad\quad\quad\quad \dots \end{cases} \tag{5.4a}$$

It is imperative to note that four coupled frequencies Ω_{ij}, $(i,j = 1,2)$ appear in the equations of motion of such two-atom chains in general, unlike the specific case of only two frequencies $(\Omega_{11}, \Omega_{22})$ as considered in [6] and others, due to two different spring constants owing to the directional asymmetry in the bonds. However, it is imperative to note that in the successive lattice sites, each of such two bonds or virtual springs is attached to two different atoms of two different masses, in general (e.g. III–V or II–VI materials), thereby leading to such four frequencies.

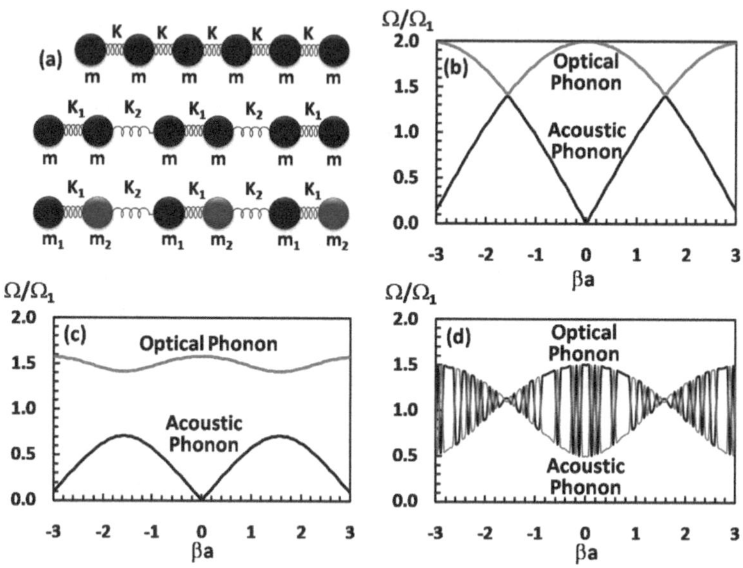

Fig. 5.2: (a) The equivalent spring-mass system schematic of infinite linear chains of identical and different atoms with symmetric and anti-symmetric inter-atomic bonds with a translational symmetry along the lattice; (b)–(d) the plots of phonon band structure: (b) $K_1 = K_2, m_1 = m_2$, i.e. $\Omega_{11} = \Omega_{21} = \Omega_{22} = \Omega_{12}$; (c) $K_1 \neq K_2, m_1 = m_2$, i.e. $\Omega_{11} = \Omega_{12}, \Omega_{21} = \Omega_{22}$; and (d) $K_1/m_1 = K_2/m_2$, $K_2/m_1 = K_1/m_2$, i.e. $\Omega_{11} = \Omega_{22}, \Omega_{21} = \Omega_{12}$.

Now, mathematically, eq. (5.4a) leads to the solutions that are eigenvalues of the block matrix:

$$\Phi = \begin{pmatrix} \begin{pmatrix} -({\Omega_{11}}^2+{\Omega_{21}}^2) & {\Omega_{21}}^2 \\ {\Omega_{22}}^2 & -({\Omega_{22}}^2+{\Omega_{12}}^2) \end{pmatrix} & \begin{pmatrix} 0 & 0 \\ {\Omega_{12}}^2 & 0 \end{pmatrix} & \cdots \\ \begin{pmatrix} 0 & {\Omega_{11}}^2 \\ 0 & 0 \end{pmatrix} & \begin{pmatrix} -({\Omega_{11}}^2+{\Omega_{21}}^2) & {\Omega_{21}}^2 \\ {\Omega_{22}}^2 & -({\Omega_{22}}^2+{\Omega_{12}}^2) \end{pmatrix} & \cdots \\ \cdots & \cdots & \cdots \end{pmatrix} \tag{5.5}$$

where eq. (5.4a) gives rise to the eigenvalue equation as:

$$\left|\Phi - (-\Omega^2)I\right| = 0 \tag{5.6}$$

Using the concept of Toeplitz block matrix, the eigenvalue equation of Φ-matrix (i.e. eq. (5.5) and (5.6)) is obtained as:

$$\left|\begin{pmatrix} -({\Omega_{11}}^2+{\Omega_{21}}^2) & {\Omega_{21}}^2+{\Omega_{12}}^2e^{-i2\phi} \\ {\Omega_{11}}^2e^{i2\phi}+{\Omega_{22}}^2 & -({\Omega_{22}}^2+{\Omega_{12}}^2) \end{pmatrix} + \Omega^2 I\right| = 0 \tag{5.6a}$$

where $\phi = (\frac{n}{N+1})\pi, n = 1, 2, \ldots, N$. Note that 2ϕ is considered in eq. (5.6a) since the angular frequency is in square, so that, owing to Bloch translational symmetry, it may be considered as $2\phi = \beta(2a)$, where a is the average lattice constant and N is the total number of atoms in the chain.

The solution of eq. (5.6a) is obtained as:

$$\Omega^2 = \frac{{\Omega_{11}}^2+{\Omega_{21}}^2}{2} + \frac{{\Omega_{22}}^2+{\Omega_{12}}^2}{2} \pm \sqrt{\left(\frac{{\Omega_{11}}^2-{\Omega_{21}}^2}{2} - \frac{{\Omega_{22}}^2-{\Omega_{12}}^2}{2}\right)^2 + 2[(\Omega_{11}\Omega_{21})^2e^{i\phi} + (\Omega_{22}\Omega_{12})e^{-i\phi}]\cos\phi} \tag{5.7}$$

It is apparent that if we would consider only the two frequencies as in [6], i.e. $\Omega_{11} = \Omega_{12} = \Omega_1$ and $\Omega_{21} = \Omega_{22} = \Omega_2$, then eq. (5.7) would give rise to:

$$\begin{aligned} \Omega^2 &= {\Omega_1}^2 + {\Omega_2}^2 \pm \sqrt{({\Omega_1}^2 - {\Omega_2}^2)^2 + 4(\Omega_1\Omega_2)^2\cos^2\phi} \\ &= {\Omega_1}^2 + {\Omega_2}^2 \pm \sqrt{({\Omega_1}^2 - {\Omega_2}^2)^2 + 2(\Omega_1\Omega_2)^2(1 + \cos 2\phi)} \\ \Rightarrow \Omega^2 &= {\Omega_1}^2 + {\Omega_2}^2 \pm \sqrt{{\Omega_1}^4 + {\Omega_2}^4 + 2{\Omega_1}^2{\Omega_2}^2\cos\beta(2a)} \end{aligned} \tag{5.7a}$$

Further, for single atomic chains like group IV semiconductors, we have $\Omega_1 = \Omega_2 = \Omega_0$, and thus, eq. (5.7a) leads to the phonon dispersion relation as:

$$\Omega^2 = 2\Omega_0{}^2 \pm \sqrt{2\Omega_0{}^4(1+\cos 2\beta a)} = 2\Omega_0{}^2 \pm \sqrt{4\Omega_0{}^4\cos^2\beta a}$$

$$\Rightarrow \Omega^2 = 2\Omega_0{}^2(1 \pm \cos\beta a) \tag{5.7b}$$

Such solutions of frequency (Ω), as a function of the wave vector (β), lead to the phonon band structure as illustrated in Fig. 5.2(b)–(d). It is to be noted that such band structures are plotted in Fig. 5.2(b)–(d), respectively, for three different cases as depicted in Fig. 5.2(a): (1) where all the atoms and their bonds are identical, i.e. $K_1 = K_2, m_1 = m_2$ and thus $\Omega_{11} = \Omega_{21} = \Omega_{22} = \Omega_{12}$; (2) where the atoms are identical but their bonds are asymmetric, so that $K_1 \neq K_2, m_1 = m_2$ and $\Omega_{11} = \Omega_{12}, \Omega_{21} = \Omega_{22}$; and (3) a hypothetical case where the atoms and bonds are such that $K_1/m_1 = K_2/m_2, K_2/m_1 = K_1/m_2$ (i.e. $\Omega_{11} = \Omega_{22}, \Omega_{21} = \Omega_{12}$).

The results show two clearly separated branching of phonons in the lower and upper strata of frequency. It can be noted from Fig. 5.2(b)–(d) that, in the equivalent first Brillouin zone for phonon wave vector, i.e. $-\pi/2 \leq \beta a \leq \pi/2$, or in any Brillouin zone in reduced representation, the phase velocity ($v_P = \Omega/\beta$) is in the same/opposite phase with the group velocity ($v_G = d\Omega/d\beta$), accordingly for such lower/upper branches, thereby indicating the acoustic and optical phonons, respectively, owing to the relative motion of successive atoms in same or opposite directions. Thus, the appearance of acoustic and optical phonons can be conceptualized as the symmetry and anti-symmetry properties, respectively, of the group and phase velocities of phonons in these materials. Interestingly, such phase symmetry can be broken, for instance, as in the last case (i.e. $\Omega_{11} = \Omega_{22}, \Omega_{21} = \Omega_{12}$) plotted in Fig. 5.2(d), where phonons can exchange their symmetry by transforming from acoustic to optical, and *vice versa* (which is different from a trivial crossover).

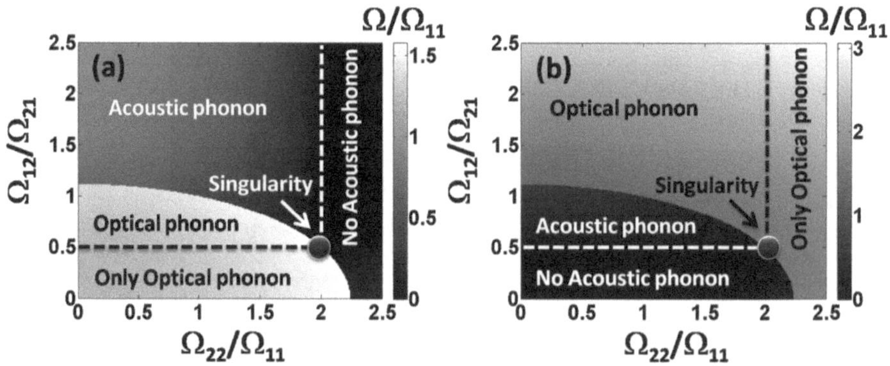

Fig. 5.3: (a) and (b) Two solutions of phonon frequency in the parameter space of $(\Omega_{22}/\Omega_{11}, \Omega_{12}/\Omega_{21})$ indicating non-trivial optical ↔ acoustic phonon transformation.

Further, the non-trivial phonon topology has been plotted in Fig. 5.3(a) and (b), where two solutions of the phonon frequencies are observed to exhibit such an optical ↔ acoustic phonon transformation. It may be interesting to note from Fig. 5.3(a) and (b) that the coupled oscillators result in an upper stratum of only optical phonon solution and lower stratum of no acoustic phonon solution for two regions in the parameter space, i.e. $\Omega_{22}/\Omega_{11} \leq 2, \Omega_{12}/\Omega_{21} \leq 0.5$ and $\Omega_{22}/\Omega_{11} \geq 2, \Omega_{12}/\Omega_{21} \geq 0.5$. Thus the trivial optical/acoustic phonons are bounded by the parameter values $\Omega_{22}/\Omega_{11} \leq 2, \Omega_{12}/\Omega_{21} \geq 0.5$. It further shows that at the point $\Omega_{22}/\Omega_{11} = 2, \Omega_{12}/\Omega_{21} = 0.5$ the acoustic and optical phonon states coalesce to generate a singularity. At such a point, the phonons may have a degeneracy of being both acoustic and optical in nature. Such a non-trivial phenomenon, we argue here, may indeed be possible, at least in principle, if one of the two bonds around an atom has a plasticity (owing to a negative value of spring constant) leading to a locally amorphous nature due to relaxed strain (e.g. in appropriately designed chalcogenides), where accordingly the effective mass (instead of the real mass) of the atoms may be considered.

5.3 NEGF model for electron-phonon scattering

5.3.1 Equation of motion describing electron-phonon scattering

It is apparent that in any electronic device or a system, in general, the motion of quasi-free electrons in the conduction band and the bound electrons through empty states of valence band (i.e. holes) are affected by the unavoidable vibrational transitions of atoms, i.e. phonons, at any finite temperature. Such electron-phonon scattering is, briefly, an inelastic process that destroys the phase-coherence of electrons (or holes) due to atomic or molecular vibrations [6]. Inside a quantum system where the electronic states are quantized as well as the atomic vibrational states are quantized within the (Morse-type) crystal potential as illustrated in Fig. 5.4(a), an electron can make a transition to such a quantum electronic state from a higher one or a lower one through phonon scattering on account of an emission or absorption of phonon associated with the corresponding atomic transition,

Such a single scattering event can be conceptualized as an electronic system coupled to itself with one excess or one less phonon, if the phonon is emitted or absorbed, respectively. In order to develop a comprehensive understanding on such electron-phonon scattering in a quantum device, e.g. a quantum dot or the channel of a nanostructure MOSFET (see Fig. 5.4(b)) with discrete energy states or semi-discrete energy sub-bands, respectively, here we are considering the case where, unlike bulk systems, such phase-breaking process cannot maintain local equilibrium for the electrons. Thus, the Hamiltonian of such a quantum electronic system (for now excluding the reservoir coupling) that exhibits electron-phonon scattering phenomena can be expressed as:

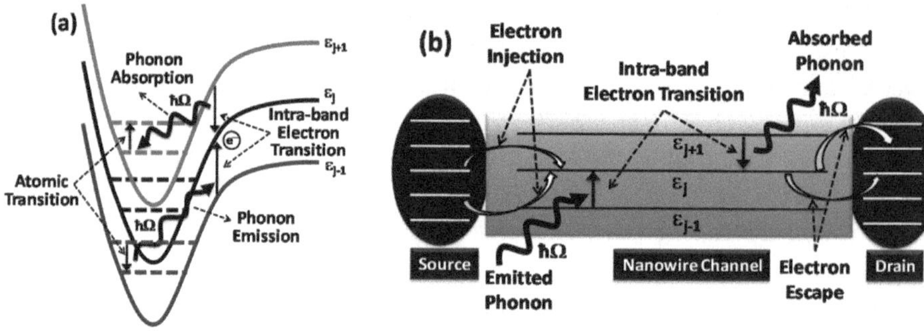

Fig. 5.4: (a) The states of electrons in a quantum system due to size quantization and atomic quantized vibration within (Morse-type) crystal potential leading to intra-band electron transition through phonon scattering and (b) schematic representation of a phonon-assisted intra-band electron transition in nanostructure FET with discrete energy sub-bands in the channel coupled to reservoir (source/drain) states.

$$\hat{H} = \sum_i \varepsilon_i c_i^+ c_i + \sum_\alpha \hbar\Omega_\alpha (b_\alpha^+ b_\alpha + 1/2) + \sum_{i,j,\alpha} \left(\tau_{ij}^\alpha c_i^+ c_j b_\alpha + \text{H.C.}\right) \tag{5.8}$$

where ε_i is the energy eigenvalue of electron at ith eigenstate, (c_i^+, c_i) are their creation and annihilation operators, respectively, and (b_α^+, b_α) accordingly are the creation and annihilation operators for phonon associated with the atomic or molecular vibrational mode α oscillating with the angular frequency Ω_α, while τ represents the electron-phonon interaction potential. The b_αs and c_is follow BE commutation and Fermi-Dirac (FD) anti-commutation relations owing to their symmetric and anti-symmetric exchange properties, respectively. As discussed in Chapter 4, the first and second terms in the RHS of eq. (5.8) act on the Hilbert spaces of electrons ($\mathscr{H}_\text{e}$) and phonons ($\mathscr{H}_\text{ph}$), respectively, whereas the coupling term (i.e. the last one) acts over their joint Hilbert space ($\mathscr{H}_\text{e} \otimes \mathscr{H}_\text{ph}$).

The equations of motion for electrons and phonons are, respectively, obtained (using Heisenberg equation) as:

$$i\hbar \frac{d}{dt} c_i = \varepsilon_i c_i + \sum_{j,\alpha} \left(\tau_{ij}^\alpha c_j b_\alpha + \tau_{ij}^\alpha{}^* c_j b_\alpha^+\right) \tag{5.9}$$

$$i\hbar \frac{d}{dt} b_\alpha = \hbar\Omega_\alpha b_\alpha + \sum_{i,j} \tau_{ji}^\alpha{}^* c_i c_j^+ \tag{5.10}$$

It may be worthy to note that the term $c_i c_j^+$ in eq. (5.10) represents the annihilation of an electron at ith state and creation of an electron jth state through the phonon interaction potential $\tau_{ji}^\alpha{}^*$. Thus, it physically implies the intra-band transition of an electron from ith state to jth state due to phonon scattering. Equation (5.10), therefore, shows the possible source of a phonon in a particular mode to be firstly from the vibrational mode itself (the first term) and secondly from the scattering of an electron (the second term). Similarly, eq. (5.9) describes the supply of electron in a particular state 'i', firstly from the ith state itself and secondly from all other states through the absorption and/or emission of phonons (see Fig. 5.4(a) and (b)).

5.3.2 Solution of the equation of motion

The method of solving the coupled equations given by eqs. (5.9) and (5.10) has been described in [6]. We are summarizing the steps here by starting with the calculation of:

$$i\hbar\frac{d}{dt}(c_j b_\alpha) = \left[i\hbar\frac{d}{dt}c_j\right]b_\alpha + c_j\left[i\hbar\frac{d}{dt}b_\alpha\right]$$

$$= \left[\varepsilon_j c_j + \sum_{k,\beta}\left(\tau_{jk}^{\beta}c_k b_\beta + \tau_{jk}^{\beta} * c_k b_\beta^{+}\right)\right]b_\alpha + c_j\left[\hbar\Omega_\alpha b_\alpha + \sum_{k,l}\tau_{kl}^{\alpha} * c_k^{+}c_l\right]$$

$$= \left[\varepsilon_j + \hbar\Omega_\alpha\right](c_j b_\alpha) + \sum_{k,\beta}\tau_{jk}^{\beta} * c_k b_\beta^{+}b_\alpha + \sum_{l,k}\tau_{lk}^{\alpha} * c_j c_l^{+}c_k$$

(using the relevant commutation and anti-commutation relations and interchanging the dummy indices in the last term):

$$= \left[\varepsilon_j + \hbar\Omega_\alpha\right](c_j b_\alpha) + \sum_{k,\beta}\sum_{l}(\tau_{lk}^{\beta} * \delta_{lj})c_k b_\beta^{+}b_\alpha + \sum_{l,k}\sum_{\beta}(\tau_{lk}^{\beta} * \delta_{\beta\alpha})c_j c_l^{+}c_k$$

$$= \left[\varepsilon_j + \hbar\Omega_\alpha\right](c_j b_\alpha) + \sum_{k,l,\beta}\tau_{lk}^{\beta} * [(b_\beta^{+}b_\alpha)\delta_{lj} + \delta_{\beta\alpha}(c_j c_l^{+})]c_k$$

$$= \left[\varepsilon_j + \hbar\Omega_\alpha\right](c_j b_\alpha) + \sum_{k,l,\beta}\tau_{lk}^{\beta} * [(b_\beta^{+}b_\alpha)\{c_j, c_l^{+}\} + [b_\alpha, b_\beta^{+}](c_j c_l^{+})]c_k$$

(using the commutation and anti-commutation relations):

$$\Rightarrow i\hbar\frac{d}{dt}(c_j b_\alpha) = \left[\varepsilon_j + \hbar\Omega_\alpha\right](c_j b_\alpha) + \sum_{k}\left[\sum_{l,\beta}\tau_{lk}^{\beta} * (b_\beta^{+}b_\alpha)(c_l^{+}c_j) + (b_\alpha b_\beta^{+})(c_j c_l^{+})\right]c_k \tag{5.11}$$

Using the Green's function ($G_{ja}^{\text{e-ph}}(t,t')$) for the electron-phonon interaction, eq. (5.11) leads to the solution:

$$c_j b_\alpha|_{t=t} = c_j b_\alpha|_{t=0}e^{-\frac{i}{\hbar}(\varepsilon_j + \hbar\Omega_\alpha)t} + \int dt' G_{ja}^{\text{e-ph}}(t,t')\sum_{k}\left[\sum_{l,\beta}\tau_{lk}^{\beta} * [N_{\alpha\beta}^{\text{ab}}(t,t')n_{jl}(t,t') + N_{\alpha\beta}^{\text{em}}(t,t')p_{jl}(t,t')]\right]c_k(t') \tag{5.12a}$$

where n, p and $N^{\mathrm{ab}}, N^{\mathrm{em}}$ are correlation functions for occupied and empty electronic states and absorption/emission of phonons, respectively, as discussed in Chapter 3. It may be interesting to note that the two terms in the last summation of eq. (5.12a) represent all possible scattering events between electrons from all states and phonons of all modes being absorbed or emitted, respectively. Similarly, we have:

$$c_j b_\alpha{}^{+}|_{t=t} = c_j b_\alpha{}^{+}|_{t=0} e^{-\frac{i}{\hbar}(\varepsilon_j - \hbar\Omega_\alpha)t} + \int dt' G_{j\alpha^{+}}^{\mathrm{e-ph}}(t,t') \sum_k \left[\sum_{l,\beta} \tau_{lk}^{\beta} [N_{\alpha\beta}^{\mathrm{ab}}(t,t') p_{jl}(t,t') + N_{\alpha\beta}^{\mathrm{em}}(t,t') n_{jl}(t,t')]\right] c_k(t') \tag{5.12b}$$

Now, since eqs. (5.9), (5.12a), and (5.12b), the equation for electron scattered by phonon can be written as:

$$\begin{aligned} i\hbar \frac{d}{dt} c_i(t) = \varepsilon_i c_i(t) &+ \sum_{j,\alpha} \tau_{ij}^{\alpha} \int dt' G_{j\alpha}^{\mathrm{e-ph}}(t,t') \sum_k \left[\sum_{l,\beta} [N_{\alpha\beta}^{\mathrm{ab}}(t,t') n_{jl}(t,t') \right. \\ &\left. + N_{\alpha\beta}^{\mathrm{em}}(t,t') p_{jl}(t,t')] \tau_{lk}^{\beta} {}^{*} \right] c_k(t') \\ &+ \sum_{j,\alpha} \tau_{ij}^{\alpha} {}^{*} \int dt' G_{j\alpha^{+}}^{\mathrm{e-ph}}(t,t') \sum_k \left[\sum_{l,\beta} [N_{\alpha\beta}^{\mathrm{ab}}(t,t') p_{jl}(t,t') \right. \\ &\left. + N_{\alpha\beta}^{\mathrm{em}}(t,t') n_{jl}(t,t')] \tau_{lk}^{\beta} \right] c_k(t') \\ &+ \sum_{j,\alpha} \left(\tau_{ij}^{\alpha} (c_j b_\alpha)|_{t=0} e^{-\frac{i}{\hbar}(\varepsilon_j + \hbar\Omega_\alpha)t} + \tau_{ij}^{\alpha} {}^{*} (c_j b_\alpha{}^{+})|_{t=0} e^{-\frac{i}{\hbar}(\varepsilon_j - \hbar\Omega_\alpha)t} \right) \end{aligned} \tag{5.13}$$

which can further be summarized in terms of the self-energy associated with phonon scattering (i.e. the second and third terms $\sum_k \int dt' \Sigma_{ik}^{\mathrm{Sc}}(t.t') c_k(t')$) and the source of inflow of electrons from other states owing to such scattering (i.e. the last term $S_i^{\mathrm{Sc}}(t)$) as:

$$i\hbar \frac{d}{dt} c_i(t) = \varepsilon_i c_i(t) + \sum_k \int dt' \Sigma_{ik}^{\mathrm{Sc}}(t.t') c_k(t') + S_i^{\mathrm{Sc}}(t) \tag{5.13a}$$

Equation (5.13a) has the steady-state solution (since it is apparent that the complementary function will decay in time due to the effect of scattering given by $\Sigma_{ik}^{\mathrm{Sc}}(t.t')$) in terms of the phonon-scattered electronic Green's function ($G_{ij}^{e}(t,t_1)$):

$$c_i(t) = \sum_j \int dt_1 G_{ij}^{e}(t,t_1) S_j^{\mathrm{Sc}}(t_1) \tag{5.14}$$

where such non-equilibrium Green's function (NEGF) is given by:

$$\sum_k \int dt' \left[\delta(t-t')\delta_{ik} \left(i\hbar \frac{d}{dt'} - \varepsilon_k \right) - \Sigma_{ik}^{Sc}(t.t') \right] G_{kj}^{e}(t', t_1) = \delta_{ij}\delta(t-t_1) \tag{5.15}$$

and the phonon scattering self-energy is:

$$\begin{aligned}\Sigma_{ik}^{Sc}(t.t') = &\sum_{j,a} \tau_{ij}^{a} \int dt' G_{ja}^{e-ph}(t,t') \sum_k \left[\sum_{l,\beta} [N_{\alpha\beta}^{ab}(t,t')n_{jl}(t,t') + N_{\alpha\beta}^{em}(t,t')p_{jl}(t,t')]\tau_{lk}^{\beta}* \right] \\ &+ \sum_{j,a} \tau_{ij}^{a}* \int dt' G_{ja^+}^{e-ph}(t,t') \sum_k \left[\sum_{l,\beta} [N_{\alpha\beta}^{ab}(t,t')p_{jl}(t,t') + N_{\alpha\beta}^{em}(t,t')n_{jl}(t,t')]\tau_{lk}^{\beta} \right]\end{aligned} \tag{5.16}$$

Now, the two-time correlation function for filled states of electrons, as discussed in Chapter 3, in the presence of phonon scattering is obtained as:

$$n_{ij}(t,t') = \sum_{k,l} \iint dt_1 dt_2 \left[[G_{ik}^{e}(t,t_1)\Sigma_{kl}^{Sc-In}(t_1,t_2)G_{lj}^{e}*(t_2,t') \right] \tag{5.17}$$

where:

$$\begin{aligned}\Sigma_{kl}^{Sc-In}(t_1,t_2) = \sum_{l,j,\alpha,\beta} \Big(&\tau_{ki}^{\alpha}[(N_{\alpha\beta}^{ab}n_{ij})\big|_{t=0} e^{-\frac{i}{\hbar}(\varepsilon_j+\hbar\Omega_\alpha)(t_1-t_2)}]\tau_{jl}^{\beta}* \\ &+ \tau_{ki}^{\alpha}*[(N_{\alpha\beta}^{em}n_{ij})\big|_{t=0} e^{-\frac{i}{\hbar}(\varepsilon_j-\hbar\Omega_\alpha)(t_1-t_2)}]\tau_{jl}^{\beta} \Big)\end{aligned} \tag{5.18}$$

The Fourier transforms of eqs. (5.15)–(5.18) into energy domain with respect to difference time coordinate finally give rise to:

$$[G^e(E)] = (E[I] - [\varepsilon] - [\Sigma^{Sc}(E)])^{-1} \tag{5.19}$$

$$\begin{aligned}[\Sigma^{Sc}(E)] = \sum_{\alpha} \big(&(N_{\Omega\alpha}+1)[\tau^{\alpha}][n(E+h\Omega_\alpha) + p(E-\hbar\Omega_\alpha)][\tau^{\alpha}]^{+} \\ &+ N_{\Omega\alpha}[\tau^{\alpha}]^{+}[n(E-\hbar\Omega_\alpha) + p(E+\hbar\Omega_\alpha)][\tau^{\alpha}] \big)\end{aligned} \tag{5.20}$$

$$[n(E)] = [G^e(E)][\Sigma^{Sc-In}(E)][G^e(E)]^{+} \tag{5.21}$$

$$[\Sigma^{Sc-In}(E)] = \sum_{\alpha} (N_{\Omega\alpha}+1)[\tau^{\alpha}][n(E+\hbar\Omega_\alpha)][\tau^{\alpha}]^{+} + N_{\Omega\alpha}[\tau^{\alpha}]^{+}[n(E-\hbar\Omega_\alpha)][\tau^{\alpha}] \tag{5.22}$$

where the first term of eqs. (5.20) and (5.22) represents phonon emission while the second term denotes absorption of phonon. $N_{\Omega\alpha}$ is the BE distribution function for phonons.

5.3.3 Power dissipation and current degradation due to electron-phonon scattering

The power dissipation due to electron-phonon scattering, i.e. the net energy outflow from the system, can be obtained, just following how the current, i.e. the net charge outflow, was calculated in Chapter 4, and is given by:

$$[P_{\mathrm{Dis}}^{\mathrm{Sc}}] = \frac{1}{h}\int E\left([\Sigma^{\mathrm{Sc-In}}(E)][i([G^e(E)] - [G^e(E)]^{+})] - [\Gamma^{\mathrm{Sc}}(E)][n(E)]\right)dE \tag{5.23}$$

Thus, the total power dissipation in all electronic states due to scattering with all possible phonon modes is:

$$P_{\mathrm{Dis}}^{\mathrm{Sc}} = Tr[P_{\mathrm{Dis}}^{\mathrm{Sc}}] = \frac{1}{h}\int Tr\left[[\Sigma^{\mathrm{Sc-In}}(E)][i([G^e(E)] - [G^e(E)]^{+})] - [\Gamma^{\mathrm{Sc}}(E)][n(E)]\right]EdE \tag{5.23a}$$

$$= \frac{1}{h}\int Tr\left[[\Sigma^{\mathrm{Sc-In}}(E)][i([G^e(E)] - [G^e(E)]^{+})] - [\Gamma^{\mathrm{Sc}}(E)][G^e(E)][\Sigma^{\mathrm{Sc-In}}(E)][G^e(E)]^{+}\right]EdE$$

[since eq. (5.21)]

$$= \frac{1}{h}\int Tr\left[[\Sigma^{\mathrm{Sc-In}}(E)][i([G^e(E)] - [G^e(E)]^{+})] - [\Sigma^{\mathrm{Sc-In}}(E)][G^e(E)]^{+}[\Gamma^{\mathrm{Sc}}(E)][G^e(E)]\right]EdE$$

(since $Tr([A][B]) = Tr([B][A])$)

$$\Rightarrow P_{Dis}^{\mathrm{Sc}} = \frac{1}{h}\int Tr\left[[\Sigma^{\mathrm{Sc-In}}(E)]\left([i([G^e(E)] - [G^e(E)]^{+})] - [G^e(E)]^{+}[\Gamma^{\mathrm{Sc}}(E)][G^e(E)]\right)\right]EdE \tag{5.23b}$$

Now, if the system is isolated and not connected to any other reservoirs, then the self-energy appears only due to phonon scattering, i.e. $i([G^e(E)] - [G^e(E)]^{+}) = [G^e(E)][\Gamma^{\mathrm{Sc}}(E)][G^e(E)]^{+}$, so that eq. (5.23b) results in:

$$P_{Dis}^{\mathrm{Sc}} = \frac{1}{h}\int Tr\left[[\Sigma^{\mathrm{Sc-In}}(E)]\left([G^e(E)][\Gamma^{\mathrm{Sc}}(E)][G^e(E)]^{+} - [G^e(E)]^{+}[\Gamma^{\mathrm{Sc}}(E)][G^e(E)]\right)\right]EdE = 0 \tag{5.23c}$$

In fact, eq. (5.23c) upholds the energy conservation principle of an isolated system, even in the presence of phase-breaking processes like electron-phonon scattering, and shows the internal consistency of the model. However, if the quantum system is coupled to other reservoirs (e.g. a nanostructure MOSFET channel connected to source/drain or a quantum dot coupled to environment), then there will be finite amount of power dissipation due to the non-zero term $i([G^e(E)] - [G^e(E)]^{+}) - [G^e(E)][\Gamma^{\mathrm{Sc}}(E)][G^e(E)]^{+}$ and the corresponding current degradation is given by:

$$I^{\mathrm{Sc}} = \frac{e}{h}\int Tr\left([\Sigma^{\mathrm{Sc-In}}(E)][i([G^e(E)] - [G^e(E)]^{+})] - [\Gamma^{\mathrm{Sc}}(E)][n(E)]\right)dE \tag{5.24}$$

5.3.4 Consequences of electron-phonon scattering on device current and power dissipation in a nanowire FET

The impact of such electron-phonon scattering phenomena, as an illustration, in a gate-all-around (GAA) Si-NWFET can be seen in the plots of its simulated transfer characteristics (i.e. drain current vs. gate voltage) and associated power dissipation in Fig. 5.5(a) and (b), respectively. As an example, the diameter of nanowire in the present case is considered to be 3 nm and its channel length to be 10 nm along [100]-crystal growth direction, surrounded by an HfO_2 layer of 3 nm thickness. Such device dimensions and material parameters are considered as per the state-of-the-art '2 nm technology node' [7]. It may be worthy to note that while simulating, the computational complexity has been minimized by not incorporating the real part of scattering self-energy matrix that has no physical contribution in such phase-breaking process except in the energy renormalization.

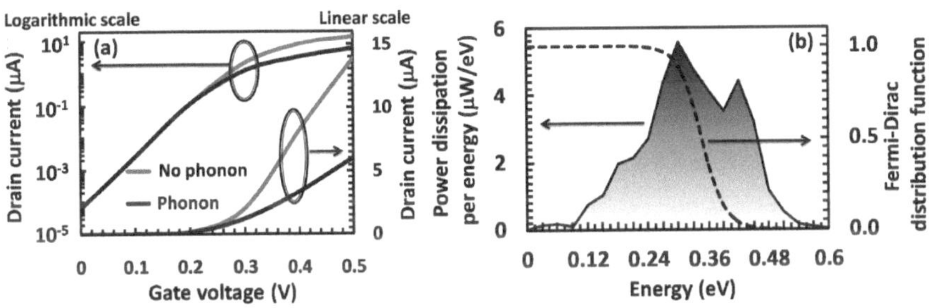

Fig. 5.5: (a) The plots of drain current with gate voltage at a drain bias of 0.6 V across the GAA Si NWFET with nanowire diameter of 3 nm and channel length 10 nm, with a gate oxide to be a 3 nm HfO_2, considering electron phonon scattering and not considering any such interaction and (b) the power dissipation per energy in such device due to electron-phonon scattering along with Fermi distribution of the reservoirs to estimate the real dissipation at room temperature.

It is apparent from the plots of Fig. 5.5(a) that the device current degrades due to phonon scattering, especially significantly, almost by 50% at the ON state (i.e. a gate voltage of 0.5 V). This is attributed to the broadening of available energy spaces around the sub-bands of nanowire channel due to phonon scattering, mathematically owing to the imaginary part of the scattering self-energy, as discussed in Chapter 3. Such a current degradation is associated with the corresponding power dissipation as evident from the results plotted in Fig. 5.5(b), where area under the curve (below Fermi level) is a measure of the net power dissipation, which is obtained to be ~1.2 μW.

5.3.5 Electron-phonon scattering lifetime

The coherence of an electron within a system is destroyed due to an electron-phonon scattering event through its impulsive action which then decays in time. The time dynamics of such phenomenon can be described in the NEGF framework as follows:

Assuming a scattering event being taken place at time $t = t_0$, the Fourier transform of eq. (5.19) into time domain gives rise to:

$$\frac{1}{2\pi\hbar}\int_{-\infty}^{\infty}\left(E[I]-[\varepsilon]-[\Sigma^{Sc}(E)]\right)[G^{e}(E)]e^{-\frac{i}{\hbar}E(t-t_0)}dE=\delta(t-t_0)[I] \tag{5.25}$$

Preliminarily, in order to understand from physical point of view, how the scattering lifetime can be described by self-energy, let us consider it to be independent of the electron energy. Then we get:

$$\frac{1}{2\pi\hbar}\left(i\hbar\frac{d}{dt}[I]-[\varepsilon]-[\Sigma^{SC}]\right)[G^{Rt}(t,t_0)]=\delta(t-t_0)[I] \tag{5.26}$$

resulting in the solution for the retarded Green's function to be:

$$[G^{Rt}(t,t_0)]=\vartheta(t-t_0)e^{-\frac{i}{\hbar}\left([\varepsilon]+[\Sigma^{Sc}]\right)(t-t_0)} \tag{5.27}$$

It may be clearly noticed from eq. (5.27) that the real part of scattering self-energy shifts the energy eigenstates of the electrons (this why it has only contribution in renormalization), while its imaginary part in the phase of retarded Green's function $(-\frac{1}{\hbar}\mathrm{Im}[\Sigma^{Sc}](t-t_0))$ represents the decay of phonon scattering impulse. Thus, the phonon scattering lifetime, as discussed in Chapter 3, can be obtained by averaging the 'time' over such impulse response (i.e. the norm of retarded Green's function) as probabilistic weight to be:

$$[\langle t_{SC}\rangle]=\frac{\int_{t_0}^{\infty} te^{-\frac{2}{\hbar}\mathrm{Im}[\Sigma^{SC}](t-t_0)}dt}{\int_{t_0}^{\infty} e^{-\frac{2}{\hbar}\mathrm{Im}[\Sigma^{SC}](t-t_0)}dt}=\frac{\hbar}{2}(-\mathrm{Im}[\Sigma^{SC}]^{-1}) \tag{5.28}$$

Now, for more general case, if the scattering self-energy is a function of electron energy, Fourier transform of the product $[\Sigma^{Sc}(E)][G^{e}(E)]$ in eq. (5.25) would be a convolution in the time domain so that:

$$\left(i\hbar\frac{d}{dt}[I]-[\varepsilon]\right)[G^{Rt}(t,t_0)]-\int\left(dt'[\Sigma^{Sc}_{Rt}(t-t')[G^{Rt}(t',t_0)]\right)=2\pi\hbar\delta(t-t_0)[I] \tag{5.29}$$

The calculation of eq. (5.29) is not straightforward and a complicated one, which however can be avoided by using the following trick:

Equations (5.25) and (5.26) can be jointly replaced by:

$$\left(i\hbar\frac{d}{dt}[I]-[\varepsilon]\right)[G^{Rt}(t,t_0)] = \int\left([I]+[\Sigma^{Sc}(E)][G^e(E)]\right)e^{-\frac{i}{\hbar}E(t-t_0)}dE$$

$$\Rightarrow [G^{Rt}(t,t_0)] = \left(i\hbar\frac{d}{dt}[I]-[\varepsilon]\right)^{-1}\int\left([I]+[\Sigma^{Sc}(E)][G^e(E)]\right)e^{-\frac{i}{\hbar}E(t-t_0)}dE$$

$$\Rightarrow [G^{Rt}(t,t_0)] = \int [G^e_{No-Sc}(E)]\left([I]+[\Sigma^{Sc}(E)][G^e(E)]\right)e^{-\frac{i}{\hbar}E(t-t_0)}dE$$

[$[G^e_{No-Sc}(E)]$ is the electron Green's function without phonon scattering]

that essentially leads to obtain:

$$[G^{Rt}(t,t_0)] = G_0 e^{-\frac{i}{\hbar}[\varepsilon](t-t_0)} + \underset{E\to(t-t_0)}{F.T.}\left([G^e_{No-Sc}(E)][\Sigma^{Sc}(E)][G^e(E)]\right) \quad (5.30)$$

Now using such expression of the retarded Green's function for phonon scattering in eq. (5.30), the scattering lifetime is calculated for the Si NWFET considered above, and the result is plotted in Fig. 5.6. The phonon scattering lifetime being in the order of ~ ns indicates the average rate of such scattering to be of the order of ~10^9/s. It can be observed from Fig. 5.6 that the scattering lifetime varies along the channel from source to drain with maximum being around the mid-channel, and significantly depend on the source/drain doping. This may be attributed to the corresponding source/drain Fermi level encompassing more energy sub-bands to contribute in phonon-associated intra-band electronic transitions.

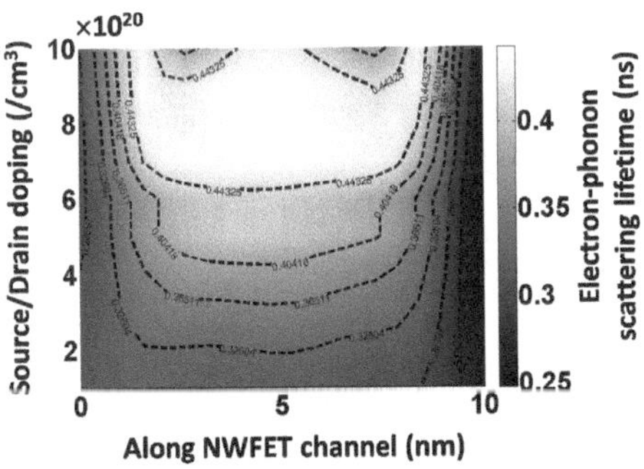

Fig. 5.6: Contour plot of electron-phonon scattering lifetime in Si NWFET (of same specifications as shown in Fig. 5.5) at the ON state (V_d = 0.4 V, V_g = 0.4 V) along the channel for various S/D doping values.

5.4 A proposition on phonon energy harvesting

In contrast to the classical heat diffusion processes in a bulk material system, heat can be conducted through a nanostructure device without resulting in any energy dissipation at the ballistic limit (i.e. up to hundreds of nm) [8]. In this section we will attempt to explore a so-called 'unconventional idea', that not only suggests whether power dissipation due to phonon scattering can be avoided but also if in any condition, rather, phonon energy can be harvested. Thus such a proposition is not to bypass or surpass electron-phonon scattering through ballistic transport, rather on the contrary, to utilize it for energy gain instead of power loss.

5.4.1 A thermodynamic perspective of dissipation: the general clue

Conventionally phonon interactions are considered (in fact, as observed in classical world) to be dissipative in its nature. Therefore, let us start with a discussion on the dissipative changes that take place in nature as a general process. One can simply notice that all such dissipative phenomena, owing to the second law of thermodynamics, are associated with an increasing amount of entropy if an isolated system is considered. This, in turn, is also directly connected to the 'arrow of time', that, although apparent from our empirical experiences that no-one can move from present to past but only to the future, is introduced in the models of physics only through this remarkable second law. Otherwise, in all the laws and equations of physics, from Newton's law to Schrodinger equation and Feynman diagrammatic approach, time is believed to be bi-directionally symmetric for any isolated system. Thus, even if in some (suppose, hypothetical) protocol phonon energy can be harvested, it is a natural question to appear, whether the second law of thermodynamics is violated in such process, that further approves Einstein's assertion that "The past, present and future are only illusions, even if stubborn ones!" It may also raise a question whether an equivalent time machine is then possible!

However, on the contrary, we argue here that energy due to phonon interaction or similar processes that usually found to get dissipated can rather be harvested utilizing the concept of 'arrow of time' that is imposed by the great second law. For any cyclic physical process where a system evolves from one state to another, the second law of thermodynamics, more precisely Carnot's theorem, asserts that:

$$\oint \frac{\bar{d}Q}{T}\bigg|_{C_1+C_2} = \int_1^2 \frac{\bar{d}Q}{T}\bigg|_{C_1} + \int_2^1 \frac{\bar{d}Q}{T}\bigg|_{C_2} \leq 0 \tag{5.31}$$

where $\bar{d}()$ mathematically represents the differential of path-dependent functions against the state functions that are represented as $d()$. Now, the heat transferred ($\bar{d}Q$) to the system through any closed path ($C_1 + C_2$) at a temperature T can further be expressed as $\bar{d}Q = dU + \bar{d}W$, dU being the change in internal energy and $\bar{d}W$ being work done by the system in the particular path $(C_1 + C_2)$ of its evolution, as per the first law of thermodynamics (for a comprehensive understanding, see [9]). The equality in eq. (5.31) is obtained only for reversible evolution defined by the two conditions: (1) the process must be 'quasi-static' (i.e. the system remains 'almost' at equilibrium throughout the evolution) and (2) there will be no 'dissipative' effects. At this point, we need to emphasize on the issue that if, at least, either of the above two conditions is violated, the process becomes irreversible leading to strict inequality in eq. (5.31). Now consider a closed path $C_1 + C_2$, as shown in Fig. 5.7(a), so that the process is 'quasi-static' (QS) but 'dissipative' (D) along the sub-path C_1, while 'non-equilibrium' (NE) with 'gain' (G) along the sub-path C_2, and hence irreversible all along. Thus, since internal energy as well as temperature both are state functions, i.e. $\oint \frac{dU}{T} = 0$, eq. (5.31) leads to the strict inequality:

$$\int_1^2 \frac{1}{T}\left(\bar{d}W_{QS} + \bar{d}W_D\right)\Big|_{C_1} + \int_2^1 \frac{1}{T}\left(\bar{d}W_{NE} - |\bar{d}W_G|\right)\Big|_{C_2} < 0$$

$$\Rightarrow \int_1^2 \frac{1}{T}\left(\bar{d}W_{QS} + \bar{d}W_D\right)\Big|_{C_1} - \int_1^2 \frac{1}{T}\left(\bar{d}W_{NE} - |\bar{d}W_G|\right)\Big|_{C_2} < 0$$

$$\Rightarrow \int_1^2 \frac{\bar{d}W_D}{T}\Big|_{C_1} + \int_1^2 \frac{\bar{d}W_G}{T}\Big|_{C_2} < \int_1^2 \frac{\bar{d}W_{NE}}{T}\Big|_{C_2} - \int_1^2 \frac{\bar{d}W_{QS}}{T}\Big|_{C_1}$$

$$\Rightarrow -\left|\int_1^2 \frac{\bar{d}W_D}{T}\right|_{C_1} + \left|\int_1^2 \frac{\bar{d}W_G}{T}\right|_{C_2} < \int_1^2 \frac{\bar{d}W_{NE}}{T}\Big|_{C_2} - \int_1^2 \frac{\bar{d}W_{QS}}{T}\Big|_{C_1} \tag{5.32}$$

It is imperative to recall at this point that the 'quasi-static' work done by an isolated system is the lowest amongst all processes of evolution resulting in a positive term in the RHS of eq. (5.32). Thus, although in the LHS, the 'dissipative' term is negative and the 'gain' term is positive, the value of the 'gain' term can be greater than that of the 'dissipative' term satisfying the inequality with a non-zero positive term at the RHS. This shows the possibility of getting a net 'gain' of energy in the complete cyclic path. It may be interesting to note that if the path C_2 would be 'quasi-static' instead, the RHS would be zero so that the maximum value of the 'gain' term must have to be less than that of the 'dissipative' term in the LHS, and we would get only energy 'loss' in a cycle. Thus, it can be said conclusively that if a non-equilibrium process is introduced in an evolution of a system, energy can be gained in a complete cycle from a process that is usually a dissipative one, interestingly, not violating but rather satisfying the second law of thermodynamics.

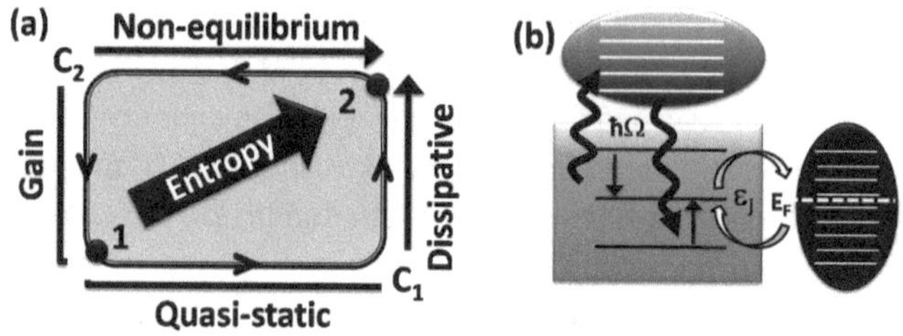

Fig. 5.7: (a) The schematic of a cyclic process that is a combination of quasi-static/dissipative and non-equilibrium with gain allowed by second law of thermodynamics, indicating a process that is dissipative while quasi-static can offer gain when non-equilibrium; (b) a simple illustration representing a single level of a quantum electronic system, that is populated or depopulated by electron-phonon scattering and coupling to a reservoir (R) for electron transport through a non-equilibrium process.

5.4.2 A simple illustration in a single-level quantum system

An illustrative description of such a system may be a ballistic nanowire field-effect transistor (FET) where the transport of electron is a non-equilibrium process, as discussed in Chapter 4, where phonon energy may be harvested. A detail analysis of such results can be studied from [10]; however, here we will provide a simple illustration for developing a basic understanding of the concept. Consider a single level of a quantum electronic system, coupled to a reservoir (R), is undergoing electron-phonon scattering, as depicted in Fig. 5.7(b). Then the power dissipation in such a device due to electron-phonon interaction can be given by (since eq. (5.29)):

$$P_{Dis}^{\mathrm{Sc}} = \frac{1}{h}\int \left(\Sigma^{\mathrm{Sc-In}}(E)i[G^e(E) - G^{e+}(E)] - \Gamma^{\mathrm{Sc}}(E)n(E)\right)EdE \tag{5.33}$$

where the spectral density of states $A(E) = i[G^e(E) - G^{e+}(E)] = G^e(E)[\Gamma^{\mathrm{Sc}}(E) + \Gamma^R(E)]G^{e+}(E)$ and $n(E) = G^e(E)[\Sigma^{Sc-In}(E) + \Sigma^{R-In}(E)]G^{e+}(E) = G^e(E)[\Sigma^{Sc-In}(E) + \Gamma^R(E)f_{FD}^R(E)]G^{e+}(E)$ so that:

$$\begin{aligned}\Sigma^{\mathrm{Sc-In}}(E)i[G^e(E)-G^{e+}(E)]-\Gamma^{\mathrm{Sc}}(E)n(E) &= A(E)\left[\Sigma^{\mathrm{Sc-In}}(E)-\Gamma^{\mathrm{Sc}}(E)\frac{n(E)}{A(E)}\right]\\ &= A(E)\left[\Sigma^{\mathrm{Sc-In}}(E)-\Gamma^{\mathrm{Sc}}(E)\left(\frac{\Sigma^{\mathrm{Sc-In}}(E)+\Gamma^R(E)f_{FD}^R(E)}{\Gamma^{\mathrm{Sc}}(E)+\Gamma^R(E)}\right)\right]\\ &= \frac{A(E)}{1+\Gamma^{\mathrm{Sc}}(E)/\Gamma^R(E)}\left[\Sigma^{\mathrm{Sc-In}}(E) - \Gamma^{\mathrm{Sc}}(E)f_{FD}^R(E)\right]\end{aligned}$$

$$\Rightarrow P_{Dis}^{\mathrm{Sc}} = \frac{1}{h}\int \frac{A(E)}{1+\Gamma^{\mathrm{Sc}}(E)/\Gamma^R(E)}\left[\Sigma^{\mathrm{Sc-In}}(E) - \Gamma^{\mathrm{Sc}}(E)f_{FD}^R(E)\right]EdE \tag{5.33a}$$

The term $[\Sigma^{Sc-In}(E) - \Gamma^{Sc}(E) f^{R}_{FD}(E)]$ in eq. (5.33a) is positive for some phonon modes and negative for others, and thus manipulating the single eigenstate of quantum system depending on the Fermi level of the reservoir at a given temperature, the net sum of such terms over the phonon modes can be made negative as a total, so that the device exhibits a power gain. In such approach phonon energy can be harvested utilizing the non-equilibrium transport of charges in suitably designed nanostructure devices.

5.4.3 A proposed device design using nanowire FET to harvest phonon energy

It is apparent from the above discussion that a non-equilibrium process is necessary for achieving the power 'gain' in a so-called dissipative system. As it has been discussed earlier in the previous chapter, a sub-mean free path nanowire FET (i.e. device length is less than the scattering MFP) exhibits ballistic transport except the unavoidable phonon scattering. Although such phonon scattering broadens the sub-band energy levels thereby driving the process towards quasi-static like in bulk, a mismatch in the effective mass between the quantized channel and source/drain reservoir, on the other hand, may lead to energy level sharpening thereby making the transport highly non-equilibrium. Fortunately, this can be achieved in practice by engineering the source/drain doping. For instance, the electron effective mass in intrinsic Si-channel along [100] is 0.19 m_e while it is 0.291 m_e in Si with a doping level of 10^{20} cm^{-3} (that can be used as source/drain), which further increases up to 0.388 m_e for 10^{21} cm^{-3} doping concentration [11]. Thus, although usually the effective mass mismatch develops a contact resistance resulting in a current degradation in the device as discussed in Chapter 4, a sub-MFP NWFET, in turn, can be engineered by optimizing the doping levels of source/drain to make the electronic transport suffering phonon scattering non-equilibrium enough to achieve a power gain, based on the thermodynamic concept as discussed above.

Now, two such lossy effects, however, being mutually counter-affecting phenomena, i.e. level broadening due to phonon scattering and level sharpening due to effective mass mismatch, result in power dissipation and power gain per energy in the alternating sub-band modes of a Si-NWFET of above specifications but with a higher source/drain doping (8×10^{20} cm^{-3}), as shown in the plots of Fig. 5.8. It is imperative to note at this point that arbitrarily increasing the source/drain doping concentration will not help in this, since accordingly the Fermi level will also get raised that can conversely affect by accommodating a higher dissipative mode than a gaining one. Thus a relevant design optimization is necessary regarding such reservoir engineering. It can be observed from the example of Fig. 5.8 that the reservoir FD distribution, a measure of the occupancy of electrons, incorporates only the modes so that net gain becomes higher than the net loss, in such case, thereby providing with an amount of harvested phonon energy (e.g. ~0.3 μW here).

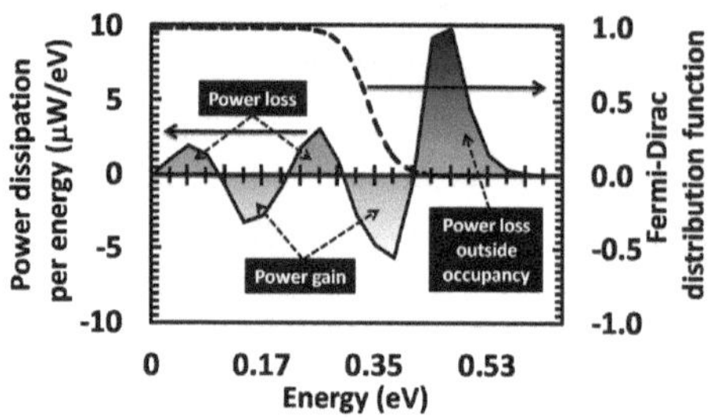

Fig. 5.8: Plot of energy current in a Si-NWFET (specifications as shown in Fig. 5.5) with source/drain doping level 8×10^{20} cm^{-3}, where area under the curve represents power dissipation, if positive (gain if negative), due to phonon scattering. The Fermi-Dirac distribution function of the reservoirs at room temperature (300 K) is also plotted to estimate the occupied modes contributing to power loss or gain. The net power gain in such case is obtained to be 0.3 μW.

Thus, if one varies the reservoir doping, the counter-effect of increasing effective mass mismatch and the FD distribution encompassing higher dissipative mode would result in either a degradation (due to power loss) or a so-called anomalous enhancement (due to power gain) of the device current as a result of electron-phonon scattering. Such a result for the above NWFET device has been plotted in Fig. 5.9 which shows the phonon scattering to degrade the device current (compared to a theoretical ballistic value) for the reservoir doping level up to about 4×10^{20} cm^{-3} (at ON state), above which it exhibits an anomalous enhancement of the current.

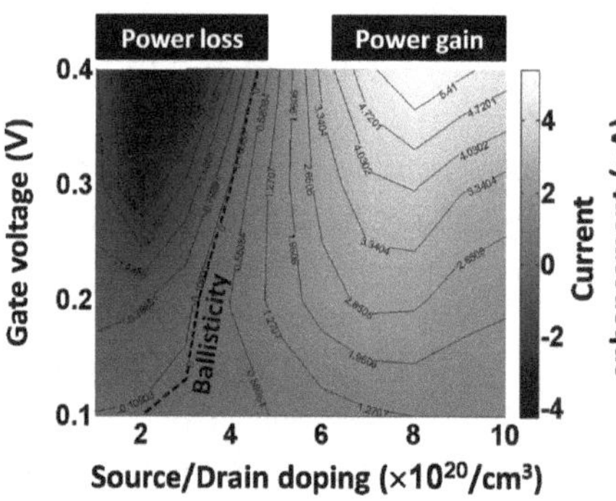

Fig. 5.9: The plot current enhancement due to electron-phonon scattering in Si NWFET (with specifications as shown in Fig. 5.5) with reservoir doping concentration at different gate voltages, indicating power loss and gain regions along with a locus of absolute ballisticity.

Interestingly, such a trend of data clearly indicates that there must be an intermediate value of source/drain doping concentration, which would result in a device current that is neither degraded nor enhanced from the theoretical ballistic current. It upholds the possibility of getting an ideal ballistic transport even at room temperature where phonon scattering phenomena are present without any effect. Thus, in the parameter space of electric field (i.e. gate voltage) and reservoir doping (i.e. doping that determines effective mass mismatch and the occupancy), we can get a locus of absolute ballisticity (as shown in Fig. 5.9), which separates out the regions of power loss and gain, and thus can also be considered as a kind of a transition line between conventional loss and unconventional gain. It also makes us comprehend that such ideal ballisticity depends not only on the physical dimension (i.e. sub-MFP length) of a device but also on a function of other parameters that can introduce the non-equilibrium phenomena in the system dynamics.

Reference

[1] Srivastava, G. P. *The Physics of Phonons*. CRC press, 2022.

[2] Chatterjee, S., Sikdar, S., Chowdhury, B. N., and Chattopadhyay, S. "Investigation of the performance of strain-engineered silicon nanowire field effect transistors (ε-Si-NWFET) on IOS substrates." *Journal of Applied Physics* 125, no. 8 (2019): 082506.

[3] Bienfait, A., Satzinger, K. J., Zhong, Y. P., Chang, H.-S., Chou, M.-H., Conner, C. R., Dumur, É., et al. "Phonon-mediated quantum state transfer and remote qubit entanglement." *Science* 364, no. 6438 (2019): 368–371.

[4] Dumur, É., Satzinger, K. J., Peairs, G. A., Chou, M.-H., Bienfait, A., Chang, H.-S., Conner, C. R., et al. "Quantum communication with itinerant surface acoustic wave phonons." *NPJ Quantum Information* 7, no. 1 (2021): 173.

[5] Qiao, H., Dumur, É., Andersson, G., Yan, H., Chou, M.-H., Grebel, J., Conner, C. R., et al. "Splitting phonons: Building a platform for linear mechanical quantum computing." *Science* 380, no. 6649 (2023): 1030–1033.

[6] Datta, S. *Quantum transport: Atom to transistor*. Cambridge university press, 2005.

[7] Agrawal, A., Chakraborty, W., Li, W., Ryu, H., Markman, B., Hoon, S. H., Paul, R. K., et al. "Silicon RibbonFET CMOS at 6nm gate length." In *2024 IEEE International Electron Devices Meeting (IEDM)*, pp. 1–4. IEEE, 2024.

[8] Anufriev, R., Ramiere, A., Maire, J., and Nomura, M. "Heat guiding and focusing using ballistic phonon transport in phononic nanostructures." *Nature communications* 8, no. 1 (2017): 15505.

[9] Dittman, R. H. and Zemansky, M. W. "Heat and thermodynamics seventh edition." (2021).

[10] Chowdhury, B. N. and Chattopadhyay, S. "Unusual impact of electron-phonon scattering in Si nanowire field-effect-transistors: A possible route for energy harvesting." *Superlattices and Microstructures* 97, (2016): 548–555.

[11] Miyao, M., Motooka, T., Natsuaki, N., and Tokuyama, T. "Change of the electron effective mass in extremely heavily doped n-type Si obtained by ion implantation and laser annealing." *Solid State Communications* 37, no. 7 (1981): 605–608.

6 Light-matter interaction: photodetectors, solar cells, LEDs, lasers, and meta-structures

Numerous literatures are available on optoelectronic devices such as photodetectors, solar cells, light-emitting diodes, lasers, and others, including such devices made up of quantum structures. Thus, instead of further elaborating on such topics, with a very brief description of such devices along with their advancement in the quantum realm, in this chapter, we will focus on developing a general mathematical framework for a comprehensive understanding of the light-matter interaction in such devices. It is imperative to note that all such light-matter, or more precisely, electron-photon, interactions are phase-breaking processes, and thus can be described on the basis of the NEGF model, following the same basic structure of discussion on phonons in Chapter 5. Finally, we will briefly discuss some exotic quantum behaviour of light while interacting with quantum dots and meta-structures.

6.1 Quantum photodetector

The optoelectronic devices, which are capable of converting photons with particular wavelengths (i.e. an objective measure of colour) into electron-hole pairs, can be utilized as photodetectors. Such devices are widely used for high-resolution imaging, night-vision, molecule/biomaterial sensing, and environment monitoring, as well as in communication and surveillance [1]. Based upon the wavelength of photons subject to detection, they are identified as X-ray-, UV-, visible-, IR-, or microwave-photodetectors, which are practically realized in semiconductor devices exhibiting the corresponding band gap values.

While traditionally PMT and InGaAs [2] have been used for detection of visible and IR wavelengths, respectively, through the recent development of quantum dots (QDs) of novel material classes, it is now possible to capture the entire spectrum of light from X-rays to microwave with very high external quantum efficiency (i.e. the ratio of the photogenerated electrons/holes to the incident photons) [1]. Such superior photodetector performance is attributed to the size- and material-dependent tunable energy gaps between the quantum states of the QDs. The QD-based photodetectors are usually of three device categories [1, 2]: 1) photoconductor, where QD-embedded photosensitive layer is placed between two electrodes that drive the photocarriers by a voltage bias; 2) photodiode, where a QD-embedded layer is sandwiched between two different metal electrodes that create the diode operating on the basis of photocarrier transport; and 3) phototransistor, which is usually a back-gated MOSFET with the channel material consisting of the QDs. Recently, the present author, with others, has explored the concept of novel QD-based voltage-controlled photocapacitors that can efficiently detect each wavelength of a full band spectrum [3–5].

 | https://doi.org/10.1515/9783111545950-006

6.1.1 Quantum solar cell

Solar cells are now surpassing the other major global energy resources, such as coal, natural gas, and hydropower. From a physical point of view, the solar cell is a semiconductor junction device where the incident light of wavelengths belonging to the solar spectrum generates electron-hole pairs, which are separated out in opposite directions due to the built-in potential slope at the junction, even when no voltage bias is applied. Thus, it conducts a current under solar illumination (i.e. photocurrent) without any applied voltage, called short-circuit current, and thereby acts as an energy resource.

From the perspective of performance parameters, the high value of short-circuit current (I_{SC}), equivalently a large open-circuit voltage (i.e. V_{OC}, the voltage require to reach zero current), high power conversion efficiency (i.e. η, the ratio of total electrical power generated to total incident solar power), and greater fill-factor (FF, the ratio of maximum power generated to $I_{SC}V_{OC}$) are crucial requirement for a high performing solar cell. Interestingly, the theoretical value of the maximum possible efficiency of a single p-n junction solar cell, known as the Shockley-Queisser limit, is ~33%, while the state-of-the-art commercial Si-based solar cells are able to offer only ~22–24% efficiency. The crystalline-Si-based industrial-level solar cells have recently reached ~25% efficiency or slightly more; a detail on such development and the current challenges can be found in [6, 7]. Among alternative materials, GaAs thin film (although very expensive) on a Si wafer provided maximum efficiency (~29%) as a single junction; however, a multi-junction solar cell made up of InGaP/AlGaAs/InGaAsP/InGaAs has recently been reported to achieve an efficiency of ~48%. Nevertheless, such approaches are still far from industry-level production technology. Several device designs with different materials, including dye-sensitized, organic, QD-sensitized electrolyte-based, solid-state QD-based, and perovskite solar cells, are being thoroughly investigated for further enhancing the power conversion efficiency. A detail of such various materials and designs of solar cells offering state-of-the-art efficiencies is summarized in Green et al. [8].

The quantum well (QW) and QD-based solar cells are currently under investigation due to their promise to provide efficiencies greater than the Shockley-Queisser limit [9]. The photogenerated carriers in the QW devices are transported either by thermoionic emission (if the barriers are low enough) or through tunnelling (if the barriers are adequately thin). Such multi-junction devices have been shown to exhibit >40% efficiency. On the other hand, the solid-state QD-based solar cells are expected to offer even more efficiency due to multiple exciton generation. Usually, such devices are designed either as a metal/QD-interlayer/semiconductor Schottky-junction, or as a heterojunction between semiconductors and colloidal QDs with/without a graphene interlayer. PbS are predominantly used as QDs; however, B-doped Si-QD has provided maximum efficiency amongst other QD-based solar cells, although it is still far from the theoretical expectation. Recently, a voltage-tunable QD-based MOS solar cell design was proposed by the present author with others that is predicted to achieve ~70% efficiency [10].

6.1.2 Quantum LED

A light-emitting diode (LED) is a p-n junction device that emits light due to radiative recombination of electron-hole at the interface under an applied voltage bias; apparently, the colour (i.e. wavelength) of the emitted light is dependent on the semiconductor bandgap. However, a carrier confining layer placed at the junction can increase the efficacy of such LEDs. Usually, the conventional white LEDs are prepared using two approaches: 1) by combining the LEDs generating primary colours, i.e. red, green, and blue; and 2) by coating the blue LEDs with different phosphors for converting into respective colours in the visible spectrum of white light. The modern LEDs are mostly made of $In_xGa_{1-x}N$ QWs sandwiched between two GaN layers, the former exhibiting a tunable bandgap varying from 3.4 (i.e. UV) to 0.69 eV (IR) as x changes from 0 to 1 [11].

The last decade, on the other hand, observed the organic LEDs to be industrially commercialized as a display device owing to their self-emission properties, high contrast ratio values, along with mechanical flexibility. However, a drive to achieve higher colour saturation (i.e. steepness of the peak) and electrical stability has led to the utilization of semiconductor QDs due to their narrow spectrum, size-dependent colour tunability offering high colour gamut, and material reliability along with self-emissive property [12]. A QD-based LED consists of a layered structure of different semiconductors between the electrodes: a hole-injection layer, a hole-transport layer, an emissive layer comprising the QDs, and an electron transport layer in consecutive succession.

The QDs have already been implemented in commercial liquid crystal display panels, and the QD-based LEDs have recently been demonstrated with comparable luminance and external quantum efficiency with respect to organic LEDs. However, the use of toxic heavy metals such as Cd, Hg, Pb, etc. and low emission bandwidth and quantum yield of materials like InP are the present challenges of QD-based LEDs.

6.1.3 Quantum laser

Lasers are optoelectronic devices that emit amplified coherent light as a low-divergence beam exhibiting a particular wavelength with extremely small line width (i.e. almost monochromatic). The basic mechanism of lasing action involves absorption of light that excites an atom from its ground state, and the stimulated emission of light by transition of the excited atom to its ground state mediated by the incident light. Since two such phenomena, i.e. absorption and stimulated emission, are fundamentally two reciprocal but equally probable processes, in order to obtain the amplified emission of light, it is necessary to maintain more excited atoms than the ground state atoms, which is commonly known to be *population inversion*. The probability of stimulated emission can be enhanced by a large number of photons in the medium, which is mostly achieved by confining photons in an optical cavity. Thus, although a laser consists of two primary elements: 1) a gain medium for stimulated emission, and 2) an optical pump to excite the atoms, most lasers further utilize an optical cavity resonator for enhanced amplification.

The quantum lasers are usually made of double-heterostructure QWs, multi-quantum-well diodes, quantum cascading systems, and QDs. The basic principle of QW/cascading lasers is to create a number of quantized levels where the non-radiative (i.e. phonon-assisted) decay-time for $n+1$th to nth level is greater than that for nth to $n-1$th level, and the electron is injected by tunnelling to the $n+1$th level to achieve the population inversion of this level with respect to nth level. On the other hand, the QD lasers operate on the principle of radiative electron-hole recombination between the lowest/highest 0D quantized states in conduction and valence bands, when at least two electrons from the highest level in the valence band are excited to the lowest level of conduction band due to single input stimulator resulting to their radiative recombination leading to enhanced photogeneration [13, 14].

For lasing applications, the most attractive candidate for QD material has been the III-V semiconductors such as $In_xGa_{1-x}As$ epitaxially grown on (001)-GaAs, or InAs grown on InP. Based on the growth conditions and shell material, the QD size/shape, as well as the induced strain, can be manipulated in order to achieve the desired laser wavelength along with its line shape function. The luminescence wavelengths of such QD lasers can range from ~1 to ~ 1.8 μm, covering the entire telecom band [13]. On the other hand, II-VI materials such as CdSe/glass can offer lasers of visible wavelengths, tunable in the range of 520 (green) to 690 nm (red) [14].

6.2 Electron-photon interaction: the NEGF-based mathematical framework

Photons are quantum (i.e. kind of particle) of light energy associated with electromagnetic (EM) waves, which have exchange symmetry leading to exhibit Bose-Einstein commutation relations and the corresponding statistical distribution. Thus, interactions of photons with electrons are quite similar to those of phonons as described in Chapter 5. However, the two most important differences are: 1) unlike phonons (which are related to non-radiative decay), the photons can be emitted or absorbed as a quantum of EM radiation, and 2) photons result in a transition of electrons from bound state to free or quasi-free states, or *vice versa*. For instance, such bound and free/quasi-free states may be a valence band (i.e. compensated hole) and a conduction band state, respectively; or, a state inside a material with an energy lower than its electron affinity and the vacuum state, respectively. Thus, the absorption/emission of photons is directly associated with electron-hole pair generation/recombination processes in the former case, while resulting in photoelectric effect and luminescence in the latter case. Hence, it is evident that the electron-photon coupling term (τ) is completely different from that for phonons. Another interesting difference is the polarization states of photons, which are quite unlike those of the phonons (that are associated with relative quantized vibration of the atoms). The photon polarization indicates the direction of the oscillatory electric field associated with the EM wave and thus is inherently transverse in nature (although for real photons only, which is not the case for virtual photons that appear in high-energy particle scattering).

Nevertheless, let us first summarize all the NEGF-based equations for electron-photon interactions identical to those for phonons, as elaborated in Chapter 5. Note that we have only replaced the angular frequency Ω of phonons with that of the photon to be ω.

The Hamiltonian

$$\hat{H} = \sum_i \varepsilon_i c_i^+ c_i + \sum_a \hbar\omega_a (b_a^+ b_a + 1/2) + \sum_{i,j,a} \left(\tau_{ij}^a c_i^+ c_j b_a + H.C. \right) \tag{6.1}$$

The equations of motion for electrons and photons

$$i\hbar \frac{d}{dt} c_i = \varepsilon_i c_i + \sum_{j,a} \left(\tau_{ij}^a c_j b_a + \tau_{ij}^a * c_j b_a^+ \right) \tag{6.2}$$

$$i\hbar \frac{d}{dt} b_a = \hbar\omega_a b_a + \sum_{i,j} \tau_{ji}^a * c_i c_j^+ \tag{6.3}$$

The non-equilibrium Green's function and photon self-energy

$$[G^e(E)] = \left(E[I] - [\varepsilon] - [\Sigma^{\mathrm{Ph}}(E)] \right)^{-1} \tag{6.4}$$

$$[\Sigma^{\mathrm{Ph}}(E)] = \sum_a \big((N_{\omega a} + 1)[\tau^a][n(E + \hbar\omega_a) + p(E - \hbar\omega_a)][\tau^a]^+ + N_{\omega a}[\tau^a]^+[n(E - \hbar\omega_a) + p(E + \hbar\omega_a)][\tau^a] \big) \tag{6.5}$$

$$[n(E)] = [G^e(E)][\Sigma^{\mathrm{Ph-In}}(E)][G^e(E)]^+ \tag{6.6}$$

$$[\Sigma^{\mathrm{Ph-In}}(E)] = \sum_a (N_{\omega a} + 1)[\tau^a][n(E + \hbar\omega_a)][\tau^a]^+ + N_{\omega a}[\tau^a]^+[n(E - \hbar\omega_a)][\tau^a] \tag{6.7}$$

The photonic impulse response time

$$[\langle t_{\text{Ph}} \rangle] = \frac{\hbar}{2}(-\operatorname{Im}[\Sigma^{\text{Ph}}]^{-1}) \tag{6.8}$$

$$[G^{\text{Rt}}(t, t_0)] = G_0 e^{-\frac{i}{\hbar}[\varepsilon](t-t_0)} + \underset{E\to(t-t_0)}{F.T.}\left([G^e_{\text{No-Sc}}(E)][\Sigma^{\text{Ph}}(E)][G^e(E)]\right) \tag{6.9}$$

Harvested photon power

$$[P_G^{\text{Ph}}] = -\frac{1}{h}\int E\left([\Sigma^{\text{Ph-In}}(E)][i([G^e(E)] - [G^e(E)]^{+})] - [\Gamma^{\text{Ph}}(E)][n(E)]\right)dE \tag{6.10}$$

Photocurrent

$$I^{\text{Ph}} = \frac{e}{h}\int Tr\left([\Sigma^{\text{Ph-In}}(E)][i([G^e(E)] - [G^e(E)]^{+})] - [\Gamma^{\text{Ph}}(E)][n(E)]\right)dE \tag{6.11}$$

6.3 Electron-photon coupling potential

When light, or more precisely photons, incident on a material, particularly with completely occupied (e.g. valence band or HOMO) and completely unoccupied (e.g. conduction band or LUMO) successive energy bands or levels, such as semiconductors or insulators or similar molecules (and also in some cases, metal nanostructures), so that the photon-energy is at least equal to the energy bandgap, an electron, by absorbing the photon, exhibits a transition from its occupied (energetically lower) level to an unoccupied (energetically higher) level leaving a hole behind. Thus, a photon is annihilated, resulting in a corresponding quasi-free electron-hole pair generation. Such a photogenerated electron-hole pair can recombine through the exact opposite transition, leading to the creation of a photon. All kinds of optoelectronic and photonic device operations are fundamentally based upon these two basic mechanisms, which can further be associated with phonons and other scattering phenomena.

Now, in the line of Feynman's approach described in Chapter 2, the creation of such an electron-hole pair by annihilation (i.e. absorption) of a photon can conceptually be thought as a picture, where an electron (which indeed is a quasi-free virtual hole in the valence band), with the effective mass of hole and momentum opposite to it (in order to compensate the opposite sign of charge), is scattered by the incoming photon and thereby gets deviated from its trajectory to move with the effective mass of the electron (in fact the quasi-free real electron in the conduction band) along its momentum. Such electron-hole pair generation and the equivalent picture of scattering are schematically depicted in Fig. 6.1(a). Interestingly, the exact opposite directions of electron and photon as a whole describe the radiative recombination process, as shown in Fig. 6.1(b).

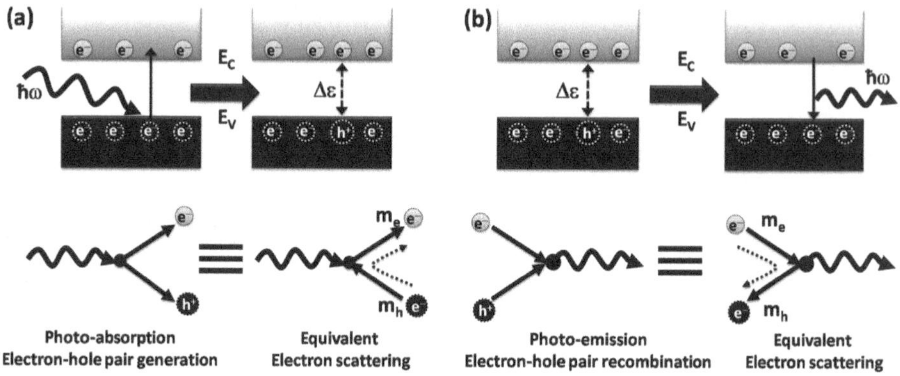

Fig. 6.1: The schematic representation of the physical process of absorption/emission of a photon by a material resulting in/from the generation/recombination of electron-hole pair and the equivalent picture of an analogous electron-photon scattering process: (a) photo-absorption and (b) photo-emission.

Thus, such an electron-hole pair generation can be modelled by considering the annihilation of an electron (c_h), having an effective mass ($m_h{}^*$) of hole and its opposite momentum ($-\vec{k}_h$), and a photon (b), with a momentum $\vec{k}_{ph}$, followed by the subsequent creation of an electron ($c_e{}^+$), outgoing with the effective mass $m_e{}^*$ and momentum $\vec{k}_e$. Thus, the electron-photon interaction Hamiltonian for such a photogeneration process, accompanied by the transition of an electron from the initial state $|i\rangle \equiv 1/\sqrt{V}e^{i(-\vec{k}_h)\cdot\vec{r}}$ to the final state $|f\rangle \equiv 1/\sqrt{V}e^{i\vec{k}_e\cdot\vec{r}}$ can be obtained as

$$H_{e-h}^{Gph} = \langle f|c_e{}^+\left[\left(\frac{\vec{p}}{2m}\right).(-e\vec{A}) + (-e\vec{A})\cdot\left(\frac{\vec{p}}{2m}\right)\right]c_h|i\rangle \tag{6.12}$$

where the photon field is given by $\vec{A} = \vec{A}_0(be^{i\vec{k}_{ph}\cdot\vec{r}} + b^+e^{-i\vec{k}_{ph}\cdot\vec{r}})$ (with $\vec{A}_0 = \vec{\varepsilon}\sqrt{\hbar/(2\omega\varepsilon_M V)}$, ε_M being the dielectric constant, V being the photon absorbing volume, and $\vec{\varepsilon}$ being the polarization vector) and the electron-photon interaction Hamiltonian is obtained from $\left(\frac{[\vec{p}+(-e\vec{A})]^2}{2m}\right) - \left(\frac{[\vec{p}]^2}{2m}\right)$ by neglecting the $\left(\frac{(-e\vec{A})^2}{2m}\right)$ term (note that $\frac{e^2}{2m}\sim 10^{-8}$). At this point, it is worthy to note that $\vec{k}_{ph} + (-\vec{k}_h) = \vec{k}_e$ owing to the momentum conservation principle, even in such phase-breaking processes. Now, in eq. (6.12), applying the first operator on the left and the second operator on the right, we get:

$$H_{e\text{-}h}^{Gph} = (1/V)\int \left[e^{-i\vec{k}_e\cdot\vec{r}} c_e^{+} \left(\left(\frac{-\hbar \vec{k}_e}{2m_e *} \right) .(-e\vec{A}) + (-e\vec{A}) \cdot \left(\frac{\hbar(-\vec{k}_h)}{2m_h *} \right) \right) c_h e^{i(-\vec{k}_h)\cdot\vec{r}} \right] dV$$

$$= (1/V)\int \left[e^{-i\vec{k}_e\cdot\vec{r}} c_e^{+} \frac{e\hbar}{2} \left(\frac{\vec{k}_e}{m_e *} + \frac{\vec{k}_h}{m_h *} \right) .\vec{A}_0 (b e^{i\vec{k}_{ph}\cdot\vec{r}} + b^{+} e^{-i\vec{k}_{ph}\cdot\vec{r}}) c_h e^{i(-\vec{k}_h)\cdot\vec{r}} \right] dV$$

$$\Rightarrow H_{e\text{-}h}^{Gph} = \left[\frac{e\hbar}{2} \left(\frac{\vec{k}_e}{m_e *} + \frac{\vec{k}_h}{m_h *} \right) .\vec{A}_0 \right] (c_e{}^{+} c_h b) \tag{6.13a}$$

Similarly, the electron-hole pair recombination Hamiltonian is obtained to be

$$H_{e\text{-}h}^{Rph} = \left[\frac{e\hbar}{2} \left(\frac{\vec{k}_e}{m_e *} + \frac{\vec{k}_h}{m_h *} \right) .\vec{A}_0 \right] (c_h{}^{+} c_e b^{+}) \tag{6.13b}$$

It is interesting to note from eqs. (6.13a) and (6.13b) that photogeneration Hamiltonian is associated with the term $(c_e{}^{+} c_h b)$ indicating creation of electron in the conduction band $(c_e{}^{+})$ through annihilation of electron in valence band (c_h) and a photon (b), while the photorecombination Hamiltonian involves the term $(c_h{}^{+} c_e b^{+})$ that represents the creation of electron in valence band $(c_h{}^{+})$ and a photon (b^{+}) through annihilation of electron in the conduction band (c_e). This is, indeed, how the second quantization field operators mathematically uphold the physical processes. In such processes, the photonic interaction potential for electron-hole pair generation/recombination is thus given by

$$\tau_{e\text{-}h}^{ph} = \left[\frac{e\hbar}{2} \left(\frac{\vec{k}_e}{m_e *} + \frac{\vec{k}_h}{m_h *} \right) .\vec{A}_0 \right] \tag{6.14}$$

At this point, it is worth noting that the direction of EM vector potential of photon field $\vec{A}_0$ represents its polarization-state $\vec{\varepsilon}$, given by $\vec{\varepsilon}_{R/L} \equiv 1/\sqrt{2}(1, \mp i, 0)$ for a photon propagating along Z-axis (owing to the transverse nature of real photon field $\vec{k}_{ph} \cdot \vec{A}_0 = 0$) where $|R\rangle$ and $|L\rangle$ represent right and left circular polarizations, i.e. RCP and LCP, respectively. It may further be noted that the horizontal ($|H\rangle$) and vertical ($|V\rangle$) polarizations are related linearly to RCP/LCP as $|H\rangle = 1/\sqrt{2}(|R\rangle + |L\rangle)$ and $|V\rangle = 1/\sqrt{2}(|R\rangle - |L\rangle)$. Now, using $\vec{k}_{ph} \cdot \vec{A}_0 = 0$ and $\vec{k}_{ph} + (-\vec{k}_h) = \vec{k}_e$, the electron-photon coupling potential of eq. (6.14) can be written as $\tau_{e\text{-}h}^{ph} = (e\hbar A_0/2\tilde{m}_e *)\vec{k}_e \cdot \vec{\varepsilon}$ where $\tilde{m}_e *$ is an effective coupling mass of the electron-hole pair. Thus, it indicates that if the spin of an electron is correlated to its momentum, then such electron-photon interaction can create quantum entanglement between electron-spin and photon-polarization [15–17].

6.4 A simple illustration of a photon interacting with a quantum dot

6.4.1 Photogenerated electron/hole pair in photodetectors and solar cells

Suppose an electron, initially in a single occupied state, is excited, by absorbing a photon, to a single unoccupied state, leaving behind a hole in the former one. In any photodetector or solar cell device, such electron-hole pairs are collected before they further recombine. In a bulk semiconductor-based device, there are infinitely many occupied and unoccupied states forming the continuum in the valence band and conduction band, respectively.

However, in a quantum device, one can obtain a number of discrete occupied and unoccupied states. For understanding the basic mechanism of such device operations, here we are considering only one discrete occupied (at $E = \varepsilon_h$) and one discrete unoccupied (at $E = \varepsilon_e$) state (such as in a QD), as illustrated by the corresponding band diagram in Fig. 6.2(a). A photon of frequency ω can excite an electron from the bound-state (at $E = \varepsilon_h$) in the valence band to the quasi-free state (at $E = \varepsilon_e$) in the con-

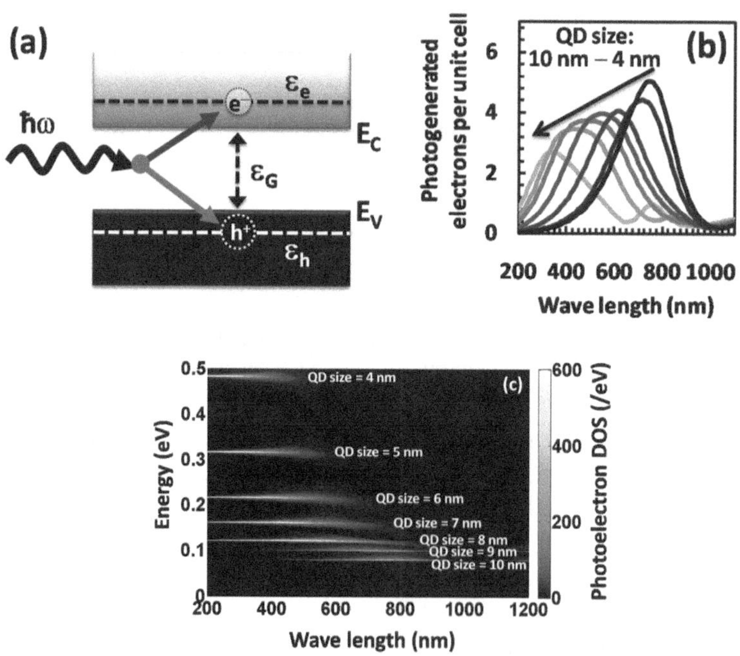

Fig. 6.2: (a) Band diagram of a quantum dot exhibiting photogeneration of electron-hole pair associated with two discrete levels of conduction and valence band; (b) simulated plot of photogenerated electrons in Si-QDs of size varying from 10 to 4 nm; and (c) the corresponding spectral density (i.e. density of state in /eV) with energy of electron and wavelength of photon.

duction band, thereby creating the electron-hole pair, which needs to be separated out from each other (to prevent recombination) by a driving voltage (as usually done in photodetectors) or through in-built material band-bending by utilizing p-n or M-S junctions (as commonly done in solar cells).

Using eqs. (6.4)–(6.7) and (6.14), the number of photogenerated electrons, as well as their spectral density in such devices, can be calculated. The results of such calculations in a Si-QD (considering one discrete state in each of the conduction and valence bands) of different sizes are plotted in Fig. 6.2(b),(c). It is apparent from the plots of Fig. 6.2(b) that, firstly, the photogeneration of electrons shows a maximum (i.e. a peak) at a wavelength of the incident light, and secondly, such peak position exhibits a blue-shift with decreasing size of the QD. Such peaks originate due to maximum possible photogeneration at about the energy gap between quantized electron and hole states, while the blue-shift of such peaks is due to an apparent increasing energy gap (i.e. the increasing energy of eigenstates) with QD-size reduction. Moreover, the plots of Fig. 6.2(c) show that such a blue shift corroborates with the corresponding shift of the most intense region in the spectrum of density of states of photogenerated electrons as the QD-size reduces.

As mentioned, this is owing to the apparent increase in quantized energy of both electron and hole ($\propto 1/\text{size}^2$) and thereby increasing their gap that the photon energy needs to exceed (or at least level up) for exciting the electron. Further, it can be noticed from Fig. 6.2(b) that the peak value of photogeneration decreases with such miniaturization of QD, which is attributed to the reduced probable occupancy of electrons at higher energy levels at a finite temperature (e.g. room temperature in this case) owing to the Fermi-Dirac distribution of contact leads. The results for a real device, however, will be a little (in cases may be greatly) different from such calculated ones due to several physical factors (apart from the technological constraints) such as involvement of multiple electronic states and their mutual interaction and broadening due to contacts (i.e. coupling with collector devices), as well as caused by the inevitable phonon scattering. Such a primitive description, nevertheless, is helpful to grossly visualize how to select the design parameters of the quantum devices to respond to a particular desired wavelength of light.

To get a quantitative estimation of such photogeneration, for instance, in the case of strong confinement of electrons, the total number of photogenerated electrons in the excited state, neglecting its broadening due to photon interaction, is thus given by (since eq. (6.6))

$$N_{\mathrm{e}} = \int n_{\mathrm{e}}(E)dE = \int G^{\mathrm{e}}(E)\Sigma^{\mathrm{Ph-In}}(E)G^{\mathrm{e}+}(E)dE = N_{\omega}\left|\tau_{\mathrm{e-h}}^{\mathrm{ph}}\right|^{2}\int [\delta(E-\varepsilon_{\mathrm{e}})]^{2} n_{\mathrm{h}}(E-\hbar\omega)dE$$

$$= N_{\omega}\left|\tau_{\mathrm{e-h}}^{\mathrm{ph}}\right|^{2}\int \left(\frac{1}{2\pi\hbar}\int e^{-\frac{i}{\hbar}(E-\varepsilon_{\mathrm{e}})t}dt\right)\delta(E-\varepsilon_{\mathrm{e}})n_{\mathrm{h}}(E-\hbar\omega)dE$$

[replacing one delta function by the integral]

$$= N_\omega \left|\tau_{e-h}^{ph}\right|^2 \int \left(\frac{1}{2\pi\hbar}\int e^{-\frac{i}{\hbar}(E-\varepsilon_e)t}\delta(E-\varepsilon_e)n_h(E-\hbar\omega)dE\right)dt$$

$$= N_\omega \left|\tau_{\mathrm{e-h}}^{\mathrm{ph}}\right|^2 n_{\mathrm{h}}(\varepsilon_{\mathrm{e}}-\hbar\omega)\frac{t}{2\pi\hbar}$$

$$\frac{d}{dt}N_{\mathrm{e}} = N_\omega(\left|\tau_{\mathrm{e-h}}^{\mathrm{ph}}\right|^2/2\pi\hbar)n_{\mathrm{h}}(\varepsilon_{\mathrm{e}}-\hbar\omega) \tag{6.15}$$

where $N_\omega = (I/\hbar\omega\bar{c})V$, I being the intensity of incident light propagating with speed $\bar{c}$ within the medium. It is imperative to note that dN_{e}/N_ω is the probability of photogeneration (i.e. the number of generated electrons per incident photon), and thus, eq. (6.15) is identical to the famous Fermi's golden rule, which, in turn, confirms the consistency of the NEGF model with the standard quantum theory. Now, if the hole distribution is peaked at $E = \varepsilon_{\mathrm{h}}$, then eq. (6.15) shows that the photogeneration rate is directly dependent on the difference $\Delta\varepsilon = (\varepsilon_{\mathrm{e}} - \hbar\omega) - \varepsilon_{\mathrm{h}} = \varepsilon_G - \hbar\omega$, where ε_G is the energy gap between occupied and unoccupied states of a molecule, or the bandgap in case of common semiconductors. Thus, in order to detect a particular colour of light (given by $\lambda = c/\omega$), one would need a semiconductor with a bandgap $\sim\hbar\omega$ or a system with that order of energy gap between the occupied and unoccupied states. If such photogenerated electron-hole pairs are separated out before they recombine by appropriately coupling the system with reservoirs, then the device can operate as a solar cell. The corresponding generated power, as well as the solar cell photocurrent, can be obtained from eqs. (6.10) and (6.11). It may be interesting to note that the photogeneration probability is significantly dependent on the electron-photon interaction potential $\tau_{\mathrm{e-h}}^{\mathrm{ph}}$, which, on the other hand, is a function of electron and hole effective masses, and dielectric constant (through $|\vec{A}_0|$, as can be seen from eq. (6.14) of the optoelectronic material that the device is made of.

6.4.2 Photon spectra: absorption/emission in photodetectors and emitters

The process of electron-hole pair generation by absorption of a photon (as in any absorbing material or device) or their recombination to emit a photon (as in LED or other emitters) can be described, since eqs. (6.2) and (6.3), by the equations:

$$i\hbar\frac{d}{dt}c_e = \varepsilon_e c_e + (\tau_{e-h}^{ph}b + \tau_{e-h}^{ph} * b^{+})c_h \tag{6.16a}$$

$$i\hbar\frac{d}{dt}c_{\mathrm{h}} = \varepsilon_{\mathrm{h}}c_{\mathrm{h}} + (\tau_{\mathrm{h-e}}^{\mathrm{ph}}b + \tau_{\mathrm{h-e}}^{\mathrm{ph}} * b^{+})c_{\mathrm{e}} \tag{6.16b}$$

$$i\hbar\frac{d}{dt}b = \hbar\omega b + (\tau_{\mathrm{e-h}}^{\mathrm{ph}} * c_{\mathrm{e}}^{+}c_{\mathrm{h}} + \tau_{\mathrm{h-e}}^{\mathrm{ph}} * c_{\mathrm{h}}^{+}c_{\mathrm{e}}) \tag{6.16c}$$

Equations (6.16a) and (6.16b) lead to

$$\left[i\hbar\frac{d}{dt}+(\varepsilon_e-\varepsilon_h)\right](c_e{}^+c_h)=[\tau^{ph}_{h-e}b+\tau^{ph}_{h-e}{}^{*}b^{+}](c_e{}^+c_e)-[\tau^{ph}_{e-h}b+\tau^{ph}_{e-h}{}^{*}b^{+}](c_h{}^+c_h) \quad (6.17a)$$

$$\left[i\hbar\frac{d}{dt}-(\varepsilon_e-\varepsilon_h)\right](c_h{}^+c_e)=-[\tau^{ph}_{h-e}b+\tau^{ph}_{h-e}{}^{*}b^{+}](c_e{}^+c_e)+[\tau^{ph}_{e-h}b+\tau^{ph}_{e-h}{}^{*}b^{+}](c_h{}^+c_h) \quad (6.17b)$$

which, on adding up, give rise to

$$\left[i\hbar\frac{d}{dt}+(\varepsilon_e-\varepsilon_h)\right](c_e{}^+c_h)+\left[i\hbar\frac{d}{dt}-(\varepsilon_e-\varepsilon_h)\right](c_h{}^+c_e)=0 \quad (6.17c)$$

that has the solution:

$$(c_e{}^+c_h)|_{t=t}=\eta+(c_e{}^+c_h)|_{t=0}e^{\frac{i}{\hbar}(\varepsilon_e-\varepsilon_h)t} \quad (6.18a)$$

$$(c_h{}^+c_e)|_{t=t}=\eta^{+}+(c_h{}^+c_e)|_{t=0}e^{-\frac{i}{\hbar}(\varepsilon_e-\varepsilon_h)t} \quad (6.18b)$$

where $\eta=-\eta^{+}$ is an arbitrary constant. It is imperative to note at this point that eq. (6.18a) represents annihilation of an electron in the occupied state (c_h), i.e. creation of a hole, and creation of an electron in the unoccupied state ($c_e{}^+$), thereby representing the absorption of a photon. Similarly, eq. (6.18b) indicates the emission of a photon. Thus, considering $\tau^{ph}_{e-h}=\tau^{ph}_{h-e}$ since eq. (6.14) for coherent photons, and putting the expressions of eqs. (6.18a) and (6.18b) along with $\eta=-\eta^{+}$ in eq. (6.16c), we get:

$$\left[i\hbar\frac{d}{dt}-\hbar\omega\right]b=\tau^{ph}_{e-h}{}^{*}\left((c_e{}^+c_h)|_{t=0}e^{\frac{i}{\hbar}(\varepsilon_e-\varepsilon_h)t}+(c_h{}^+c_e)|_{t=0}e^{-\frac{i}{\hbar}(\varepsilon_e-\varepsilon_h)t}\right) \quad (6.19)$$

leading to the solution to be:

$$b=-\tau^{ph}_{e-h}{}^{*}\left(\frac{(c_e{}^+c_h)}{\hbar\omega+(\varepsilon_e-\varepsilon_h)}+\frac{(c_h{}^+c_e)}{\hbar\omega-(\varepsilon_e-\varepsilon_h)}\right) \quad (6.19a)$$

Up to this, the natural recombination of electron-hole pairs associated with the intrinsic/minority carrier lifetime of the material has not been considered. If we take this into account, then $(\varepsilon_e-\varepsilon_h)$ would have an extra imaginary part (i.e. $(\varepsilon_e-\varepsilon_h)+i\Delta$) to represent the natural decay of the state occupation. Thus, eq. (6.19a), incorporating the decay, finally results in

$$N^{Abs}_{\omega}=b^{+}b=\left|\tau^{ph}_{e-h}\right|^{2}\left[\frac{n_h p_e}{[\hbar\omega+(\varepsilon_e-\varepsilon_h)]^2+\Delta^2}+\frac{n_e p_h}{[\hbar\omega-(\varepsilon_e-\varepsilon_h)]^2+\Delta^2}\right] \quad (6.20a)$$

$$N_\omega^{\mathrm{Em}} = bb^+ = \left|\tau_{\mathrm{e-h}}^{\mathrm{ph}}\right|^2 \left[\frac{n_\mathrm{e} p_\mathrm{h}}{[\hbar\omega + (\varepsilon_\mathrm{e} - \varepsilon_\mathrm{h})]^2 + \Delta^2} + \frac{n_\mathrm{h} p_\mathrm{e}}{[\hbar\omega - (\varepsilon_\mathrm{e} - \varepsilon_\mathrm{h})]^2 + \Delta^2}\right] \tag{6.20b}$$

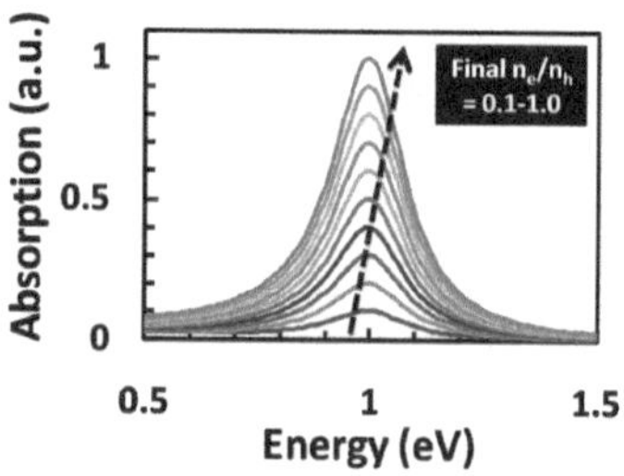

Fig. 6.3: The plots of absorption spectra of photons by a QD with associated single discrete levels in conduction and valence band with an energy gap of 1 eV, revealing the increasing intensity, resulting in the final $n_\mathrm{e}/n_\mathrm{h}$ to increase accordingly.

It may be interesting to note that the first term in each of eqs. (6.20a) and (6.20b) for photon absorption and emission spectra by a material decreases monotonically with frequency and thus of less importance; however, the second term is peaked around the electronic energy gap $(\varepsilon_\mathrm{e} - \varepsilon_\mathrm{h})$ of the material or device with the level broadening given by Δ, and results in the filling up of electron-state by de-occupying the hole-state for photon absorption, and *vice versa* for the emission of photon. For instance, an emitter results in a total emission rate greater than that of absorption, leading to the final state to exhibit $n_h p_e > n_e p_h$ as per eqs. (6.20a) and (6.20b).

It is imperative to note that the initial state for such emission must require the exact opposite, i.e. $n_\mathrm{h} p_\mathrm{e} < n_\mathrm{e} p_\mathrm{h}$, for a net recombination (equivalent to the population inversion in a laser). Similarly, while photons are being absorbed by a material, the final state density will exhibit $n_\mathrm{e} p_\mathrm{h} > n_\mathrm{h} p_\mathrm{e}$, and the intensity of absorption peak will eventually depend on $n_\mathrm{e}/n_\mathrm{h}$ or $p_\mathrm{h}/p_\mathrm{e}$, as illustratively shown in Fig. 6.3. Further, it should be noted that the emission and absorption probability of photons are also dependent on $|\tau_{\mathrm{e-h}}^{\mathrm{ph}}|^2$, and thereby, on the electron and hole effective masses, as well as the dielectric constant of the materials. The LEDs or other emitters for particular desired performances, thus, must be designed on the basis of such material parameters alongside the geometrical dimensions to define the discrete energy gaps.

6.5 Basic scheme of solid-state laser operation

The state-of-the-art solid-state lasers (the most commonly used crystal being the neodymium doped yttrium aluminium garnet ($Y_3Al_5O_{12}$), i.e. Nd:YAG) are four-level quantum systems that operate through the following scheme, as shown in Fig. 6.4: 1) first, the electron in the ground state, i.e. level-1, is excited to level-4 (the highest state) by absorbing a photon from the external pumping source; 2) such electron then experiences a non-radiative decay through interacting with a phonon populating the upper-intermediate metastable state, i.e. level-3; 3) it then de-excites to the lower-intermediate metastable state, i.e. level-2, by radiative recombination emitting a photon; and 4) finally, relaxes to the ground state through a faster non-radiative process to prevent populating level-2. It is apparent from the above discussions on the mathematical description of electron-photon interaction and from Chapter 5 that such an entire loop of processes can be described by the set of equations (eq. 6.21(a)–(h)).

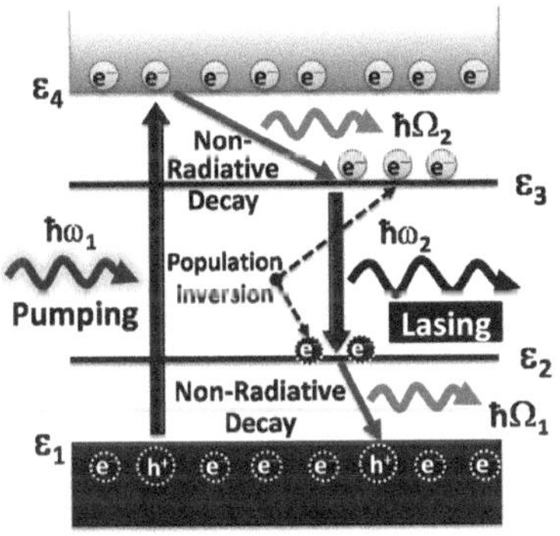

Fig. 6.4: The schematic representation of the physical process of the operating principle of a state-of-the-art four-level quantum solid-state laser.

Non-radiative decay

$$i\hbar\frac{d}{dt}c_4 = \varepsilon_4 c_4 + \tau_{14}^{ph} b_1 c_1 + \tau_{34}^{V} b_{V2} c_3 \quad \text{(6.21c)}$$

$$i\hbar\frac{d}{dt}b_{V2} = \hbar\Omega_2 b_{V2} + \tau_{34}^{V} * c_3^{+} c_4 \quad \text{(6.21d)}$$

$$i\hbar\frac{d}{dt}c_3 = \varepsilon_3 c_3 + \tau_{34}^{V} * b_{V2}^{+} c_4 + \tau_{23}^{ph} b_2 c_2 \quad \text{(6.21e)}$$

Pumping

$$i\hbar\frac{d}{dt}b_1 = \hbar\omega_1 b_1 + \tau_{14}^{ph} * c_1^{+} c_4 \quad \text{(6.21b)}$$

$$i\hbar\frac{d}{dt}b_2 = \hbar\omega_2 b_2 + \tau_{23}^{ph} * c_2^{+} c_3 \quad \text{(6.21f)}$$

Laser emission

$$i\hbar\frac{d}{dt}c_2 = \varepsilon_2 c_2 + \tau_{23}^{ph} * b_2^{+} c_3 + \tau_{12}^{V} b_{V1} c_1 \quad \text{(6.21g)}$$

$$i\hbar\frac{d}{dt}b_{V1} = \hbar\Omega_1 b_{V1} + \tau_{12}^{V} * c_1^{+} c_2 \quad \text{(6.21h)}$$

$$i\hbar\frac{d}{dt}c_1 = \varepsilon_1 c_1 + \tau_{14}^{ph} * b_1^{+} c_4 + \tau_{12}^{V} * b_{V1}^{+} c_2 \quad \text{(6.21a)}$$

Relaxation

where $c_i, i=1,\ldots,4$ are the electron (annihilation) field operators in the corresponding four levels (with energy ε_i) of the system, b_1, b_2 accordingly refer to the photon (annihilation) field operators of the pumping source and emitted laser, while b_{V1}, b_{V2} are the phonon (annihilation) field operators associated with non-radiative electronic transitions between level-2→1 and level-4→3, respectively; τ^{ph}, τ^{V} represent electron-photon and electron-phonon coupling potentials.

Now, eq. (6.21f) for the emitted photon leads to

$$i\hbar\frac{d}{dt}(b_2 b_2{}^+) = \tau_{23}^{ph} * b_2{}^+ c_2{}^+ c_3 - \tau_{23}^{ph} b_2 c_3{}^+ c_2 \tag{6.22}$$

Similarly, level-2 and level-3, which are directly associated with photon emission, give rise to

$$i\hbar\frac{d}{dt}(c_2{}^+ c_2) = (\tau_{23}^{ph} * b_2{}^+ c_2{}^+ c_3 - \tau_{23}^{ph} b_2 c_3{}^+ c_2) + (\tau_{12}^{V} b_{V1} c_2{}^+ c_1 - \tau_{12}^{V} * b_{V1}{}^+ c_1{}^+ c_2)$$

which, since eqs. (6.22) and (6.21h) result in,

$$\frac{d}{dt}(c_2{}^+ c_2) = \frac{d}{dt}(b_2 b_2{}^+) - \frac{d}{dt}(b_{V1} b_{V1}{}^+) \tag{6.23a}$$

In the same way, we get:

$$\frac{d}{dt}(c_3{}^+ c_3) = -\frac{d}{dt}(b_2 b_2{}^+) + \frac{d}{dt}(b_{V2} b_{V2}{}^+) \tag{6.23b}$$

It may be worth noting that eq. (6.23a) physically implies that the level-2 is populated due to lasing out the photons (i.e. electrons transit from level-3 to level-2) and depopulated due to non-radiative relaxation to the ground state through phonon interaction. Similarly, eq. (6.23b) depicts the fact that level-3 is populated by non-radiative decay from level-4 through phonon emission, while it emits lasing photons and gets depopulated by transiting to level-2. Thus, for continuing the lasing action, it is necessary to keep level-3 being populated and level-2 being depopulated at a faster rate than the rate of photons lasing out. This is the so-called 'population inversion' that physically depends on the phonon-coupling strength $|\tau_{34}^{V}|$ between level-3 and level-4, $|\tau_{12}^{V}|$ between level-1 and level-2, and the photonic coupling strength $|\tau_{23}^{ph}|$ between level-2 and level-3. Further, it is also necessary to continuously stimulate the electrons from level-1 to level-4 that, after each lasing cycle, reach back to the ground state (i.e. level-1). The corresponding equations can be obtained similarly by following the above method. Now interestingly, the commutation relation ($[b_2, b_2{}^+] = 1$) of photons (here importantly for the lasing ones) concludes to the fact that the photon emission rate given by $\frac{d}{dt}(b_2 b_2{}^+)$ is identical to its absorption rate $\frac{d}{dt}(b_2{}^+ b_2)$ related to the associated states, i.e. level-2 and level-3, through eqs. (6.23a) and (6.23b) in the same manner. Thus, reflecting back a fraction of the lased out photons and feeding them again into the lasing system can lead to achieve sustained 'population inversion'; the method is commonly known as resonating and usually done by creating optical cavity.

6.6 Meta-optics and exception topology

Meta-optics describes the optical behaviour of so-called metamaterials that are the designed arrays of specific structures (may be of metal or dielectrics), with a subwavelength dimension and repeatability of such structures in the array. Consequently, the conventional laws of light waves are observed to be violated in meta-optics, which rather exhibit exotic phenomena such as the appearance of negative refractive index, inverse Cherenkov radiation, super-lensing (i.e. imaging beyond the diffraction limit of light), invisibility cloaking, and unidirectional transport of light, as well as light-trapping. Further, meta-optical systems offer significant control to achieve desired amplitude, phase, and polarization of light beams. A detailed discussion on this can be found in Chen et al. [18], Chang et al. [19], and Zhao et al. [20]. Here, on the other hand, we will develop a generic mathematical model on the basis of the NEGF framework to describe the light-matter interaction in such metamaterials and the corresponding behaviour of light, usually termed as meta-optics. It may be noted that the 3D structures of this type are conventionally called metamaterials, while their corresponding 2D counterparts are often identified as metasurfaces.

While describing the phenomena of light scattering through such a quantized structure (for simplicity, we neglect the phonons), since eq. (6.2), let's start by calculating:

$$\left(i\hbar\frac{d}{dt}c_i\right)c_j^{+}=\varepsilon_i c_i c_j^{+}+\sum_k\left(\tau_{ik}^{\alpha}b_\alpha+\tau_{ki}^{\alpha}*b_\alpha^{+}\right)c_k c_j^{+}$$

$$c_i\left(-i\hbar\frac{d}{dt}c_j^{+}\right)=\varepsilon_j c_i c_j^{+}+\sum_l\left(\tau_{jl}^{\alpha}b_\alpha+\tau_{lj}^{\alpha}*b_\alpha^{+}\right)c_i c_l^{+}$$

that (on subtracting from each other) lead to

$$i\hbar\frac{d}{dt}\left(c_i c_j^{+}\right)=\sum_k\left(\tau_{ik}^{\alpha}b_\alpha+\tau_{ki}^{\alpha}*b_\alpha^{+}\right)c_k c_j^{+}-\sum_l\left(\tau_{jl}^{\alpha}b_\alpha+\tau_{lj}^{\alpha}*b_\alpha^{+}\right)c_i c_l^{+} \tag{6.24}$$

Multiplying eq. (6.24) with $\tau_{ji}^{\alpha}*$ and summing over the states, we get:

$$\Rightarrow i\hbar\frac{d}{dt}\left(\sum_{i,j}\tau_{ji}^{\alpha}*c_i c_j^{+}\right)=\sum_{i,j,k}\tau_{ji}^{\alpha}*\left(\tau_{ik}^{\alpha}b_\alpha+\tau_{ki}^{\alpha}*b_\alpha^{+}\right)c_k c_j^{+}-\sum_{i,j,l}\tau_{ji}^{\alpha}*\left(\tau_{jl}^{\alpha}b_\alpha+\tau_{lj}^{\alpha}*b_\alpha^{+}\right)c_i c_l^{+}$$

$$\begin{aligned}\Rightarrow i\hbar\frac{d}{dt}\left(\sum_{i,j}\tau_{ji}^{\alpha}*c_i c_j^{+}\right)-\sum_{i,j,k}\tau_{ji}^{\alpha}*\left(\tau_{ik}^{\alpha}c_k c_j^{+}\quad\tau_{jk}^{\alpha}c_l c_k^{+}\right)b_\alpha\\+\sum_{i,j,k}\tau_{ji}^{\alpha}*\left(\tau_{ki}^{\alpha}*c_k c_j^{+}-\tau_{kj}^{\alpha}*c_i c_k^{+}\right)b_\alpha^{+}\end{aligned} \tag{6.25}$$

Now, in the second summation term of eq. (6.25), interchanging the dummy indices $i\leftrightarrow j$ in the first term and in $i\leftrightarrow k$ the second term, we have:

$$\sum_{i,j,k} \tau_{ji}^{a} * \left(\tau_{ki}^{a} * c_k c_j^{+} - \tau_{kj}^{a} * c_i c_k^{+}\right) b_a^{+} = \sum_{i,j,k} \left(\tau_{ij}^{a} * \tau_{kj}^{a} * c_k c_i^{+} - \tau_{jk}^{a} * \tau_{ij}^{a} * c_k c_i^{+}\right) b_a^{+} = 0 \tag{6.26a}$$

while the first summation term of eq. (6.25), using $i \leftrightarrow j$ followed by $j \leftrightarrow k$ in its second term, leads to

$$\begin{aligned}\sum_{i,j,k} \tau_{ji}^{a} * \left(\tau_{ik}^{a} b_a c_k c_j^{+} - \tau_{jk}^{a} b_a c_i c_k^{+}\right) &= \sum_{i,j,k} \left(\tau_{ij}^{a} * \tau_{ik}^{a} c_k c_j^{+} - \tau_{ij}^{a} * \tau_{ik}^{a} c_j c_k^{+}\right) b_a \\ &= \sum_{i,j,k} \left(\tau_{ij}^{a} * \tau_{ik}^{a} c_k c_j^{+} - \tau_{ik}^{a} * \tau_{ij}^{a} c_k c_j^{+}\right) b_a \\ &= \sum_{i,j,k} \left(\tau_{ij}^{a} * \tau_{ik}^{a} - \tau_{ik}^{a} * \tau_{ij}^{a}\right) c_k c_j^{+} b_a \\ &= 2i \sum_{i,j,k} \operatorname{Im}\left(\tau_{ij}^{a} * \tau_{ik}^{a}\right) c_k c_j^{+} b_a\end{aligned}$$

$$i\hbar \frac{d}{dt}\left(\sum_{i,j} \tau_{ji}^{a} * c_i c_j^{+}\right) = 2i \sum_{i,j,k} \operatorname{Im}\left(\tau_{ij}^{a} * \tau_{ik}^{a}\right) c_k c_j^{+} b_a \tag{6.26b}$$

Thus, since eq. (6.3) where we are only writing $\tilde{\omega}_a$ in place of ω_a, eq. (6.26a) is obtained to be,

$$i\hbar \frac{d}{dt} b_a(t) = \hbar \tilde{\omega}_a b_a(t) + (-i/\hbar) \int_0^t dt' [2i \sum_{i,j,k} \operatorname{Im}\left(\tau_{ij}^{a} * \tau_{ik}^{a}\right) c_k(t') c_j^{+}(t)] b_a(t') \tag{6.27}$$

that can be simplified to the expression:

$$i\hbar \frac{d}{dt} b_a = (\hbar \omega_a - i\gamma_a) b_a \tag{6.28}$$

which apparently indicates a photonic resonator with resonant frequency ω_a and loss represented by γ_a. It is worthy to note at this point that the phase and decay part of the term $c_k(t') c_j^{+}(t) b_a(t')$ depend on the repeatability of the structures (like Bloch wave) and thus can be manipulated by designing the metamaterial or metasurface. However, it is imperative to further notice from eq. (6.27) that the phase part of the term $c_k(t') c_j^{+}(t) b_a(t')$ has a contribution in the loss of photons while loss part of such term has a contribution in the resonant frequency of photons, which shows the inherent counter-affecting of resonance and loss by each other in such light-matter interaction that fundamentally limit the design aspect of metasurface/metamaterial to achieve any arbitrary desired resonant frequency with an arbitrary amount of loss. This, in fact, needs appropriate optimization and relevant geometry and material engineering.

The non-equilibrium Greėn's function for such a photonic resonator is given by

$$G_{\text{ph}}(E) = [E - (\hbar\omega_a - i\gamma_a)]^{-1} \tag{6.29}$$

while the photon spectral density eventually be

$$D_{\text{ph}}(E) = i[G_{\text{ph}}(E) - G_{\text{ph}}^{+}(E)] \equiv 2\gamma_a / [(E - \hbar\omega_a)^2 + \gamma_a^{\ 2}] \tag{6.30}$$

which exhibits a peak at $E = \hbar\omega_a$ with the broadening given by γ_a. Now, such resonant photons scattered through a single quantized structure can be considered to represent an equivalent emission line of an artificial pseudo-atom, i.e. often called a meta-atom (MA). Now, if two such MAs with different resonant frequencies and losses ((ω_1, γ_1) and (ω_2, γ_2)) are placed within a sub-wavelength distance, then their coupled Hamiltonian can be expressed as

$$H(2) \equiv \begin{pmatrix} \hbar\omega_1 - i\gamma_1 & \tau_1 \\ \tau_2 & \hbar\omega_2 - i\gamma_2 \end{pmatrix} \tag{6.31}$$

where τ_1, τ_2 are photonic coupling between the two MAs or resonators, so that $\tau_2 = \tau_1{}^*$ for symmetric photon incidence. It is apparent that the Hamiltonian of eq. (6.31) is non-Hermitian and may be considered to represent the photonic Hamiltonian of the corresponding meta-molecule (MM), or what is commonly called a coupled-lossy-resonators. Interestingly, such a system would be parity-time (PT)-symmetric if $\sigma_1 H^{+} \sigma_1 = H$ which poses the conditions: 1) $\omega_2 = \omega_1$, 2) $\gamma_2 = -\gamma_1$, and 3) τ_1, τ_2 are real. In that case, such a Hamiltonian may be written as

$$H_{\text{PT}}(2) \equiv \begin{pmatrix} \hbar\omega_0 - i\gamma_0 & \tau_1 \\ \tau_2 & \hbar\omega_0 + i\gamma_0 \end{pmatrix} \tag{6.32}$$

so that the eigenvalues are given by,

$$E_{\text{PT}} = \hbar\omega_0 \pm \sqrt{\tau_1\tau_2 - \gamma_0{}^2} \tag{6.33}$$

belonging to the eigenstates,

$$|\psi\rangle_{\text{PT}} \equiv \begin{pmatrix} [\gamma_0 \pm \sqrt{\tau_1\tau_2 - \gamma_0{}^2}]/\tau_2 \\ 1 \end{pmatrix} \tag{6.34}$$

It is worthy to note at this point that astonishingly, eq. (6.33) leads to the real eigenvalues of such PT-symmetric non-Hermitian systems for $\tau_1\tau_2 \geq \gamma_0{}^2$ [21]. And even more interestingly, for the particular condition $\tau_1\tau_2 = \gamma_0{}^2$, both the eigenvalues and the eigenstates become identical, indicating coalescence of the two eigenstates, as shown in the plots of Fig. 6.5(a). This is known to be the Exceptional Point (EP) singular degeneracy (see Fig. 6.5(b)) that is, on the other hand, the point of breaking the PT-symmetry in its parameter space, through which the PT -symmetric phase with real eigenspectrum transits to the broken-PT-symmetry phase with complex conjugate pair of eigenspectrum demonstrating the so-called Exceptional Topology (ET) [22–24]. It is imperative to note that ET or EP appears in such PT-symmetric non-Hermitian systems only for $\gamma_2 = -\gamma_1$, i.e. where the loss of one MA of the MM is exactly counterbalanced by the gain of the other. Nevertheless, ET and EP are also observed in non-PT-symmetric systems, whereas with the identical complex eigenvalue.

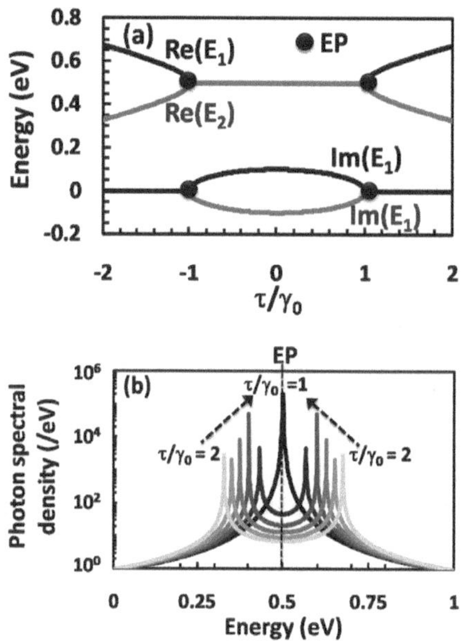

Fig. 6.5: (a) The plots of the real and imaginary parts of eigenvalues of a PT-symmetric CLR-Hamiltonian (given by eq. (6.32)) with inter-resonator coupling per loss, revealing the Exceptional points (EPs); and (b) the corresponding spectral density of emitted photons, showing to be significantly high at the EP singular degeneracy.

It is further worthy to note that the propagating phase (e.g. $e^{i\phi}$) of the photon scattered through the described PT-symmetric system can be given by,

$$\phi = (1/\hbar)[\vec{p}_{\text{ph}} \cdot \vec{r} - E_{\text{PT}}t] = (1/\hbar)[\vec{p}_{\text{ph}} \cdot \vec{r} - (\hbar\omega_0 \pm \sqrt{\tau_1\tau_2 - \gamma_0{}^2})t] \tag{6.35}$$

where the '±' sign in the energy (E_{PT}) term with respect to the momentum ($\vec{p}_{\text{ph}}$) term evidently signifies the forward and reverse motion of the photons, respectively. Thus, at EP, where $\pm\sqrt{\tau_1\tau_2 - \gamma_0{}^2}) = 0$, the photon can move only in one direction [22–24], amazingly violating the classically apparent reversibility principle of light propagation.

Now, a small perturbation ε modifies the Hamiltonian of eq. (6.32) as

$$\tilde{H}_{\text{PT}}(2) \equiv \begin{pmatrix} \hbar\omega_0 - i\gamma_0 & (1+\varepsilon)\tau_1 \\ (1+\varepsilon)\tau_2 & \hbar\omega_0 + i\gamma_0 \end{pmatrix} \tag{6.36}$$

that modifies the eigenvalues as

$$\tilde{E}_{PT} = \hbar\omega_0 \pm \sqrt{\tau_1\tau_2(1+\varepsilon)^2 - \gamma_0{}^2}$$

$$\approx \hbar\omega_0 \pm \sqrt{\tau_1\tau_2(1+2\varepsilon) - \gamma_0{}^2} \text{ [using binomial approximation since } \varepsilon \text{ is small]}$$

Thus, at EP, where $\tau_1\tau_2 = \gamma_0{}^2$, we have,

$$\tilde{E}_{PT} = \hbar\omega_0 \pm \sqrt{\varepsilon(2\tau_1\tau_2)} \tag{6.37}$$

Therefore, the EP singular degeneracy is lifted up through splitting the peak in the photonic spectrum by a factor proportional to $\sqrt{\varepsilon}$. This clearly indicates that the less the perturbing strength ε, the greater is the splitting of EP photonic peak. Thus, it can offer a very high sensitivity for weak signal detection. However, recently, it has been shown that such a scheme cannot provide an enhanced signal-to-noise ratio against the quantum and thermal noises, which also scale as the same square-root dependence [25]. Such a challenge can, nevertheless, be overcome by designing the MM in a manner which does not exhibit EP initially (i.e. $\tau_1\tau_2 \neq \gamma_0{}^2$), but generates EP singularity due to the perturbing Hamiltonian (i.e. $\tau_1\tau_2(1+\varepsilon)^2 = \gamma_0{}^2$). The present author, with others, has recently demonstrated such utilization of EP for detecting pathogens (e.g. SARS-CoV-2) with remarkably enhanced sensitivity [26].

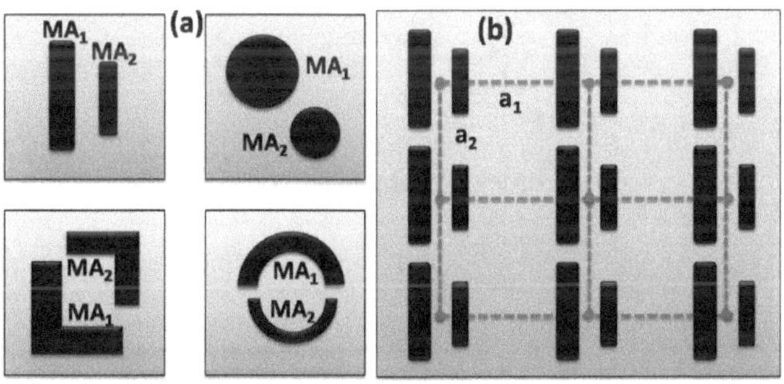

Fig. 6.6: (a) Schematics of various meta-molecules with different geometrical shapes; and (b) an illustration of a metasurface made up of an array of meta-molecules.

What happens now if a 1D-linear chain of such MMs (several conventional MM structures have been illustrated in Fig. 6.6(a)) with repetitive spacing of 'a' are developed? Assuming only the nearest-neighbour MM-coupling (to be $\mu_{12}^{\vec{a}_1}$), the Hamiltonian of a single MM given by eq. (6.31) will then be extended for the chain as

$$H_{1\mathrm{D}}(2) \equiv \begin{array}{c|cccccc} & |MM_1\rangle & |MM_2\rangle & |MM_3\rangle & \cdots & |MM_N\rangle \\ |MM_1\rangle & \begin{pmatrix} \hbar\omega_1 - i\gamma_1 & \tau_1 \\ \tau_2 & \hbar\omega_2 - i\gamma_2 \end{pmatrix} & \begin{pmatrix} 0 & 0 \\ \mu_{12} & 0 \end{pmatrix} & \begin{pmatrix} 0 & 0 \\ 0 & 0 \end{pmatrix} & \cdots & \begin{pmatrix} 0 & \mu_{21} \\ 0 & 0 \end{pmatrix} \\ |MM_2\rangle & \begin{pmatrix} 0 & \mu_{21} \\ 0 & 0 \end{pmatrix} & \begin{pmatrix} \hbar\omega_1 - i\gamma_1 & \tau_1 \\ \tau_2 & \hbar\omega_2 - i\gamma_2 \end{pmatrix} & \begin{pmatrix} 0 & 0 \\ \mu_{12} & 0 \end{pmatrix} & \cdots & \\ |MM_3\rangle & \begin{pmatrix} 0 & 0 \\ 0 & 0 \end{pmatrix} & \begin{pmatrix} 0 & \mu_{21} \\ 0 & 0 \end{pmatrix} & \cdots & \cdots & \\ \cdots & \cdots & \cdots & \cdots & \cdots & \\ |MM_N\rangle & \begin{pmatrix} 0 & 0 \\ \mu_{12} & 0 \end{pmatrix} & & \cdots & \cdots & \end{array} \tag{6.38a}$$

which, assuming the chain to be along $\vec{a}$-vector, can be simply reduced to

$$H_{1\mathrm{D}}(2) \equiv \begin{pmatrix} \hbar\omega_1 - i\gamma_1 & \tau_1 + \mu_{12}^{\vec{a}_1} e^{-i\vec{k}\cdot\vec{a}_1} \\ \tau_2 + \mu_{21}^{\vec{a}_1} e^{i\vec{k}\cdot\vec{a}_1} & \hbar\omega_2 - i\gamma_2 \end{pmatrix} \tag{6.38b}$$

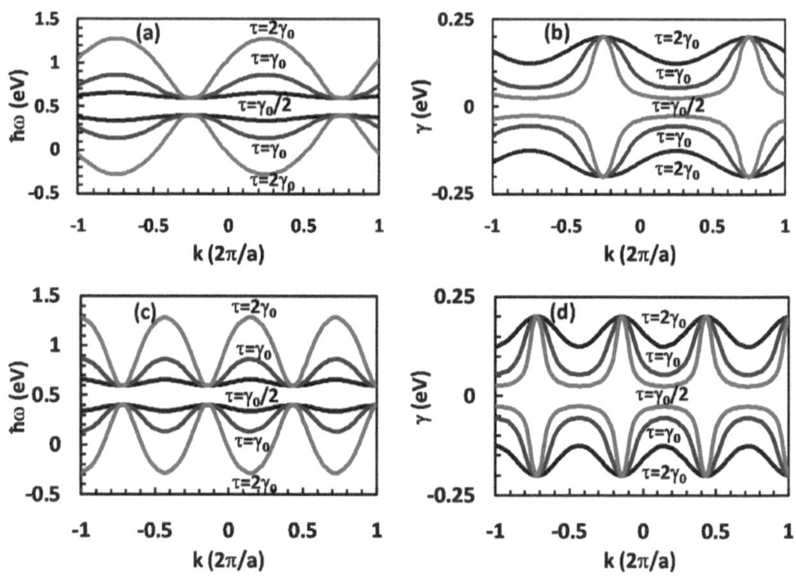

Fig. 6.7: Illustrations of photonic band-structures ($\hbar\omega$ vs. k) and cumulative loss (γ vs. k) in: (a) and (b) 2D-metasurface; and (c) and (d) 3D-metamaterial, for various inter-resonator coupling (τ).

Equation (6.38b) can be extrapolated to 2D-metasurface (see Fig. 6.6(b) as an example) or 3D-metamaterial, respectively, as

$$H_{2D}(2) \equiv \begin{pmatrix} \hbar\omega_1 - i\gamma_1 & \tau_1 + \mu_{12}^{\vec{a}_1} e^{-i\vec{k}\cdot\vec{a}_1} + \mu_{12}^{\vec{a}_2} e^{-i\vec{k}\cdot\vec{a}_2} \\ \tau_2 + \mu_{21}^{\vec{a}_1} e^{i\vec{k}\cdot\vec{a}_1} + \mu_{21}^{\vec{a}_2} e^{i\vec{k}\cdot\vec{a}_2} & \hbar\omega_2 - i\gamma_2 \end{pmatrix} \tag{6.39}$$

$$H_{3D}(2) \equiv \begin{pmatrix} \hbar\omega_1 - i\gamma_1 & \tau_1 + \mu_{12}^{\vec{a}_1} e^{-i\vec{k}\cdot\vec{a}_1} + \mu_{12}^{\vec{a}_2} e^{-i\vec{k}\cdot\vec{a}_2} + \mu_{12}^{\vec{a}_3} e^{-i\vec{k}\cdot\vec{a}_3} \\ \tau_2 + \mu_{21}^{\vec{a}_1} e^{i\vec{k}\cdot\vec{a}_1} + \mu_{21}^{\vec{a}_2} e^{i\vec{k}\cdot\vec{a}_2} + \mu_{21}^{\vec{a}_3} e^{i\vec{k}\cdot\vec{a}_3} & \hbar\omega_2 - i\gamma_2 \end{pmatrix} \tag{6.40}$$

The relation between the real part of eigenvalues of such Hamiltonians with the wave vector ($\vec{k}$) provides the photonic band-structure of the meta-optical system, where the imaginary part represents the cumulative loss of such systems, as illustrated in Fig. 6.7(a)–(d) for 2D-metasurface and 3D-metamaterial. It is interesting to note from the plots of Fig. (6.7) that, firstly, photons in such general class of metamaterials exhibit a photonic bandgap that does not allow any photon to propagate through the material; secondly, band-slope can be zero at non-zero wave vector resulting in zero group-velocity indicating the formation of 2D/3D-quantized standing waves of confined photons; and thirdly, the possible appearance of both positive and negative band-curvatures leading to positive and negative effective mass of photon. Can these be interpreted as particle/anti-particle in the case of distinct photons (instead of a beam) propagating through a metamaterial!

References

[1] Tong, X., Jiang, W., and Wang, Z. M., eds. *Quantum Dot Photodetectors*. 30, Springer Nature, 2021.

[2] Bielecki, Z., Achtenberg, K., Kopytko, M., Mikołajczyk, J., Wojtas, J., and Rogalski, A. "Review of photodetectors characterization methods." *Bulletin of the Polish Academy of Sciences Technical Sciences* 70, no. 2 (2022).

[3] Sikdar, S., Chowdhury, B. N., Ghosh, A., and Chattopadhyay, S. "Analytical modeling to design the vertically aligned Si-nanowire metal-oxide-semiconductor photosensors for direct color sensing with high spectral resolution." *Physica E: Low-dimensional Systems and Nanostructures* 87, (2017): 44–50.

[4] Sikdar, S., Chowdhury, B. N., and Chattopadhyay, S. "Designing InP-Nanowire based vertical metal-oxide-semiconductor capacitors for wavelength selective visible light sensing." In *International Workshop on the Physics of Semiconductor and Devices*, pp. 957–962. Cham: Springer International Publishing, 2017.

[5] Si-Nanowire, O. S. "Analytical Modeling of Vertically." In *Advances in Optical Science and Engineering: Proceedings of the Third International Conference, OPTRONIX 2016*, 194, pp. 173. Springer, 2017.

[6] Battaglia, C., Cuevas, A., and De Wolf, S. "High-efficiency crystalline silicon solar cells: Status and perspectives." *Energy & Environmental Science* 9, no. 5 (2016): 1552–1576.

[7] Sabatino, D., Marisa, R. H., and Garcia, A. S. "Silicon solar cells: Trends, manufacturing challenges, and AI perspectives." *Crystals* 14, no. 2 (2024): 167.

[8] Green, M. A., Dunlop, E. D., Yoshita, M., Kopidakis, N., Bothe, K., Siefer, G., Hao, X., and Jiang, J. Y. "Solar cell efficiency tables (Version 66)." *Progress in photovoltaics: research and applications* 33, (2025): 795–810.

[9] Sayed, I. and Bedair, S. M. "Quantum well solar cells: Principles, recent progress, and potential." *IEEE Journal of Photovoltaics* 9, no. 2 (2019): 402–423.

[10] Sikdar, S., Chowdhury, B. N., and Chattopadhyay, S. "Design and modeling of high-efficiency ga as-nanowire metal-oxide-semiconductor solar cells beyond the Shockley-Queisser limit: An negf approach." *Physical Review Applied* 15, no. 2 (2021): 024055.

[11] Yong-Ho, R., Navamathavan, R., Park, J.-H., and Lee, C.-R. "High-quality uniaxial in x ga1–x n/gan multiple quantum well (mqw) nanowires (nws) on si (111) grown by metal-organic chemical vapor deposition (mocvd) and light-emitting diode (led) fabrication." *ACS Applied Materials & Interfaces* 5, no. 6 (2013): 2111–2117.

[12] Moon, H., Lee, C., Lee, W., Kim, J., and Chae, H. "Stability of quantum dots, quantum dot films, and quantum dot light-emitting diodes for display applications." *Advanced Materials* 31, no. 34 (2019): 1804294.

[13] Norman, J. C., Jung, D., Zhang, Z., Wan, Y., Liu, S., Shang, C., Herrick, R. W., Chow, W. W., Gossard, A. C., and Bowers, J. E. "A review of high-performance quantum dot lasers on silicon." *IEEE Journal of Quantum Electronics* 55, no. 2 (2019): 1–11.

[14] Park, Y.-S., Roh, J., Diroll, B. T., Schaller, R. D., and Klimov, V. I. "Colloidal quantum dot lasers." *Nature Reviews Materials* 6, no. 5 (2021): 382–401.

[15] Greve, D., Kristiaan, L. Y., McMahon, P. L., Pelc, J. S., Natarajan, C. M., Kim, N. Y., Abe, E., et al. "Quantum-dot spin–photon entanglement via frequency downconversion to telecom wavelength." *Nature* 491, no. 7424 (2012): 421–425.

[16] Gao, W. B., Fallahi, P., Togan, E., Miguel-Sánchez, J., and Imamoglu, A. "Observation of entanglement between a quantum dot spin and a single photon." *Nature* 491, no. 7424 (2012): 426–430.

[17] Nag Chowdhury, B. and Chattopadhyay, S. "Entangled electron–photon pair production by channel-exchange in high-energy Compton scattering." *Quantum Information Processing* 20, no. 5 (2021): 179.

[18] Chen, H.-T., Taylor, A. J., and Nanfang, Y. "A review of metasurfaces: Physics and applications." *Reports on Progress in Physics* 79, no. 7 (2016): 076401.

[19] Chang, S., Guo, X., and Xingjie, N. "Optical metasurfaces: Progress and applications." *Annual Review of Materials Research* 48, no. 1 (2018): 279–302.

[20] Zhao, X., Sun, Z., Zhang, L., Wang, Z., Xie, R., Zhao, J., You, R., and You, Z. "Review on metasurfaces: An alternative approach to advanced devices and instruments." *Advanced Devices and Instrumentation* 2022, no. 9765089 (2022): 19. https://doi.org/10.34133/2022/9765089.

[21] Bender, C. M. and Boettcher, S. "Real spectra in non-Hermitian Hamiltonians having P T symmetry." *Physical Review Letters* 80, no. 24 (1998): 5243.

[22] Özdemir, Ş. K., Rotter, S., Nori, F., and Yang, L. "Parity–time symmetry and exceptional points in photonics." *Nature Materials* 18, no. 8 (2019): 783–798.

[23] Miri, M.-A. and Alu, A. "Exceptional points in optics and photonics." *Science* 363, no. 6422 (2019): eaar7709.

[24] Bergholtz, E. J., Budich, J. C., and Kunst, F. K. "Exceptional topology of non-Hermitian systems." *Reviews of Modern Physics* 93, no. 1 (2021): 015005.

[25] Loughlin, H. and Sudhir, V. "Exceptional-point sensors offer no fundamental signal-to-noise ratio enhancement." *Physical Review Letters* 132, no. 24 (2024): 243601.

[26] Chowdhury, N., Basudev, P. L., Johnson, N. P., Rue, R. M. D. L., and Lahiri, B. "Exceptional-point-enhanced superior sensing using asymmetric coupled-lossy-resonator based optical metasurface." *Laser & Photonics Reviews* 19, no. 6 (2025): 2401661.

7 Effects of spin of electron in quantum devices

The spin of an electron (or any elementary particle) is understood as an intrinsic property (like charge or mass) associated with an inherent angular momentum independent of any physical rotation. The effect of such electron-spin was first observed in the Stern-Gerlach experiment in 1921, followed by its physical explanation proposed by Goudsmit and Uhlenbeck in 1925 and elaborated in more detail by Pauli through his famous Exclusion principle based on exchange symmetry (in fact, anti-symmetry) of electrons. However, the consistent theoretical formulation was finally provided by Dirac in his relativistic theory of electrons (or more generally of fermions) [1], which showed that the angular momentum of an electron cannot be conserved (i.e. the rotational symmetric homogeneity cannot be preserved) without the existence of such an additional intrinsic angular momentum exhibiting two possible eigenvalues $\pm\hbar/2$ ($2\pi\hbar$ being universal Planck's constant). It is imperative to note at this point that 'spin' must not be visualized as the spinning of an electron, which is evidenced by its inconsistency in the value of the gyromagnetic ratio (i.e. the ratio of magnetic dipole moment to angular momentum) as shown by Lande.

In the past few decades, electron spin has drawn further attention owing to the Coulomb and Pauli blockades [2, 3], the Kondo effect [4, 5], and, most interestingly, the quantum spin Hall effect (QSHE) [6], leading to the exploration of the topological materials classes [7]. Also, from the technological point of view, the development of qubits based on quantum dots (QDs) and superconductors, as well as topological materials, has offered extensive study on the effects of electron spin. In this chapter, we will show the basic method of how such spin effects can be modelled using NEGF formalism, along with exploring some fundamental ideas on the physical picture of the different types of interactions and the corresponding outcome of multiple interplay.

7.1 Spin-charge dialectics: Coulomb blockade and Kondo effect

In an open system, even if it is a QD coupled to a reservoir, generally at lower energies, the Coulomb interaction between electrons is negligible. However, if the QD is weakly coupled to the reservoir(s) (e.g. tunnel coupling), then a sharp conductance peak is observed at a charging energy equal to the Coulomb potential energy. This clearly indicates that electron-electron interaction creates a Coulomb blockade in such systems, which have been widely studied in the past three decades, along with their utilization in QD-based systems [2]. Interestingly, the Coulomb blockade effect can be rigorously identified in a single-level system, where such a single energy state can split up into two (in the absence of any magnetic field or negligible spin-orbit coupling), due to Coulomb repulsion between two electrons of opposite spins, which otherwise could co-exist in the same energy level according to the Pauli principle. This,

 | https://doi.org/10.1515/9783111545950-007

what we call here, spin-charge dialectics between two electrons at the same energy level, can further lead to the appearance of Kondo effect at low temperatures. The Coulomb blockade effect can be observed as two conductance peaks in a single-level QD (tunnel-coupled with two reservoirs), one of which (i.e. one of the peaks depending on the reservoir Fermi levels) can then be highly sharpened by lowering the temperature, exhibiting Kondo resonance.

In conventional mathematical formalism, the problem is generally dealt with in many-electron theoretical constructs by researchers [8, 9], which leaves the expectation of incorporating the higher orders of perturbation (Kondo himself considered up to second order [10]) to appropriately describe the phenomena. However, such theoretical calculations become more and more challenging as the order of perturbation increases and may require numerical methods. On the other hand, Datta has attempted the problem through the self-consistent field approximation in the NEGF framework [11], which offers a better mathematical approximation method, although it has yet to explain the physical picture unambiguously. Here, we will consider a single-level quantum system with two electrons of opposite spins repelling each other by the Coulomb force, which will then be coupled to a reservoir (that may be extended to any number of reservoirs) to represent a basic physical depiction of the Coulomb blockade and Kondo resonance phenomena.

The equations of two interacting electrons in an isolated single-state quantum system can be given by

$$i\hbar\frac{d}{dt}c_{\uparrow} = \varepsilon_0 c_{\uparrow} + U c_{\uparrow}(c_{\downarrow}{}^{+}c_{\downarrow}) \tag{7.1a}$$

$$i\hbar\frac{d}{dt}c_{\downarrow} = \varepsilon_0 c_{\downarrow} + U(c_{\uparrow}{}^{+}c_{\uparrow})c_{\downarrow} \tag{7.1b}$$

where $(c_{\uparrow}, c_{\downarrow})$ are annihilation operators of electrons with 'up'- and 'down'-spin, respectively, in the same energy level ε_0, while U is the Coulomb interaction potential energy. It may be worth noting that the anti-commutation relations for such electrons in the same energy level with opposite spins are given by

$$c_{\uparrow/\downarrow}(t)c_{\uparrow/\downarrow}(t') = 0 \tag{7.2a}$$

$$\{c_{\uparrow/\downarrow}(t), c_{\downarrow/\uparrow}(t')\} = 0 = \{c_{\uparrow/\downarrow}(t), c_{\downarrow/\uparrow}{}^{+}(t')\} \tag{7.2b}$$

$$\{c_{\uparrow/\downarrow}(t), c_{\uparrow/\downarrow}{}^{+}(t')\} = 1 \tag{7.2c}$$

Now, post-multiplying eq. (7.1a) by $c_{\downarrow}{}^{+}$ and pre-multiplying the Hermitian conjugate of eq. (7.1b) by $c_{\uparrow}$, followed by subtracting the latter from the former, and using the anti-commutation relations of eq. (7.2a)–(7.2c), we get,

$$i\hbar\frac{d}{dt}(c_\uparrow c_\downarrow{}^+) = 0 \tag{7.3}$$

Thus, $(c_\uparrow c_\downarrow{}^+)$, and evidently also its Hermitian conjugate $(c_\downarrow c_\uparrow{}^+)$, are constant. Similarly, if we were to pre-multiply eq. (7.1a) by $c_\downarrow{}^+$ and post-multiply the Hermitian conjugate of eq. (7.1b) by $c_\uparrow$, would eventually lead us to obtain that $(c_\downarrow{}^+ c_\uparrow)$ and $(c_\uparrow{}^+ c_\downarrow)$ are also constant. Thus, physically, it implies that the spin-flipping (i.e. annihilation of one spin and creation of its opposite) is a continuous process in such isolated devices that cannot be identified promptly. Now, when such a system is weakly coupled (e.g. tunnel-coupling) to a reservoir (that itself observes almost no effect of electron spin owing to its large number and closely spaced energy levels), we can approximate (for the electrons in QD) $(c_\uparrow{}^+ c_\downarrow)$ or $(c_\downarrow{}^+ c_\uparrow)$ to be $\langle c^+ c\rangle = f_{\text{FD}}(E - E_{\text{F}})$, whereas $(c_\uparrow c_\downarrow{}^+)$ and $(c_\downarrow c_\uparrow{}^+)$ to be $\langle cc^+\rangle = 1 - f_{\text{FD}}(E - E_{\text{F}})$. Therefore, writing $\langle c^+ c\rangle = f_{\text{FD}}(E - E_{\text{F}})$ in places $(c_\downarrow{}^+ c_\uparrow)$ and $(c_\uparrow{}^+ c_\downarrow)$ in eqs. (7.1a) and (7.1b), for a two-electron QD, we can approximately rewrite the equations as

$$i\hbar\frac{d}{dt}\begin{pmatrix} c_\uparrow \\ c_\downarrow \end{pmatrix} = \begin{bmatrix} \varepsilon_0 & -Uf_{\text{FD}}(E - E_{\text{F}}) \\ -Uf_{\text{FD}}(E - E_{\text{F}}) & \varepsilon_0 \end{bmatrix}\begin{pmatrix} c_\uparrow \\ c_\downarrow \end{pmatrix} \tag{7.4}$$

It may be noted that for a two-hole QD, it would be better to choose $\langle cc^+\rangle = 1 - f_{\text{FD}}(E - E_{\text{F}})$ and rearrange the equations. Now, while such a QD is weakly coupled with a reservoir, considering the reservoir self-energy to be, $\Sigma = \begin{bmatrix} \Sigma_0 & 0 \\ 0 & \Sigma_0 \end{bmatrix}$, for simplicity (it may be calculated in more detail by following the methods described in Chapters 3 and 4), the non-equilibrium Green's function (Fourier transformed into energy space) for the electron transport is obtained to be

$$[G(E)] = \begin{bmatrix} E - \varepsilon_0 - \Sigma_0 & Uf_{\text{FD}}(E - E_{\text{F}}) \\ Uf_{\text{FD}}(E - E_{\text{F}}) & E - \varepsilon_0 - \Sigma_0 \end{bmatrix}^{-1} \tag{7.5}$$

Thus, the density of states and the corresponding electrical conductance ($\frac{dI}{dV}$) of such a QD can be obtained following the methods described in Chapters 3 and 4.

Such a QD occupied with two electrons in a single energy level (at $\varepsilon_0 = 0$) coupled to two reservoirs at various temperatures is simulated following the above model and plotted in Fig. 7.1. It is apparent from the plots that the energy level is split into two, exhibiting two peaks, however, with different broadening depending upon the occupation probability owing to coupling with the reservoir Fermi levels. When the temperature is reduced, the weakly coupled spin-state gets sharpened significantly, indicating Kondo resonance. Physically, after Coulomb splitting, the lower-energy electron couples strongly with the reservoir Fermi level, whereas the higher-energy electron couples weakly along with the repulsion suffered from the lower-energy state. Thus, we argue here that it gets confined more intensely, thereby exhibiting the

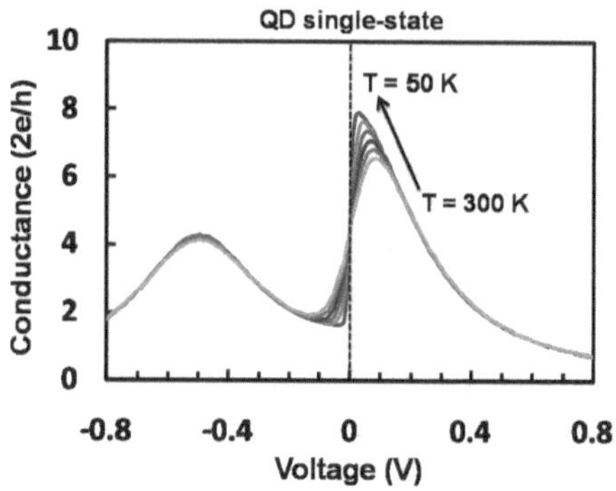

Fig. 7.1: The plot of conductance of a QD coupled to two reservoirs varying the reservoir temperature from 300 to 50 K.

highly sharpened peak, with increasing intensity as thermal broadening is reduced by decreasing the temperature.

7.2 Spin-charge-phonon trialectics: superconductivity

Now, when a third party, namely the phonon, enters into the spin-charge game! In particular cases, we may get the exotic phenomenon, known as superconductivity, so that a material exhibits absolute zero electrical resistance and completely screens out the magnetic fields below a critical temperature.

According to the basic concept of the widely accepted Bardeen-Cooper-Schrieffer (BCS) theory of superconductivity [12], two electrons can interact through a virtual exchange of phonons, resulting in an attractive potential between them, which, when it dominates the Coulomb repulsion between the degenerate-spin electrons, can create Cooper pairs giving rise to superconductivity. The required form of Hamiltonian for such a phenomenon was assumed in the theory that showed a large-scale success in explaining all the essential features of superconductivity in a wide range of material classes, with a few exceptions, like high-pressure hydride superconductors that show superconductivity at much higher temperatures.

Here, based on the fundamental idea of BCS theory, we will try to develop a basic picture of how the spin and charge of two electrons can interact with a phonon that can lead to a superconducting phase in a material. We start by considering two such electrons interacting through a phonon and also exhibiting Coulomb repulsion. Then the equations of the motion of two such electrons and the phonon are given by, as described above and in Chapter 5:

$$i\hbar\frac{d}{dt}c_1 = \varepsilon_1 c_1 + (\tau_{12}b + \tau_{12}{}^{*}b^{+})c_2 + Uc_1(c_2{}^{+}c_2) \tag{7.6a}$$

$$i\hbar\frac{d}{dt}c_2 = \varepsilon_2 c_2 + (\tau_{21}b + \tau_{21}{}^{*}b^{+})c_1 + U(c_1{}^{+}c_1)c_2 \tag{7.6b}$$

$$i\hbar\frac{d}{dt}b = \hbar\Omega b + (\tau_{12}{}^* c_1{}^+ c_2 + \tau_{21}{}^* c_2{}^+ c_1) \tag{7.6c}$$

where the extra term $U = e^2/4\pi\varepsilon_0\varepsilon_r r$ represents the Coulomb repulsion between two electrons. Now the term,

$$\begin{aligned}
i\hbar\frac{d}{dt}(bc_2) &= \left[i\hbar\frac{d}{dt}b\right]c_2 + b\left[i\hbar\frac{d}{dt}c_2\right] \\
&= [\hbar\Omega bc_2 + \tau_{21}{}^* c_2{}^+ c_1 c_2] + [\varepsilon_2 bc_2 + \tau_{21}{}^* bb^+ c_1 + U(c_1{}^+ c_1)bc_2] \\
&= [\varepsilon_2 + \hbar\Omega + U(c_1{}^+ c_1)](bc_2) + [-\tau_{21}{}^*(c_2{}^+ c_2) + \tau_{21}{}^*(bb^+)]c_1
\end{aligned}$$

Thus, using non-equilibrium Green's function for the phonon and second electron coupling, the joint field operator is found to be

$$(bc_2) = (bc_2)|_{t=0} + \int dt' G^{(2)}_{+\hbar\Omega}(t,t')[-\tau_{21}{}^*(c_2{}^+ c_2) + \tau_{21}{}^*(bb^+)]c_1 \tag{7.7a}$$

Similarly, we get,

$$(b^+ c_2) = (b^+ c_2)|_{t=0} + \int dt' G^{(2)}_{-\hbar\Omega}(t,t')[\tau_{12}(c_2{}^+ c_2) + \tau_{21}(b^+ b)]c_1 \tag{7.7b}$$

Applying eqs. (7.7a) and (7.7b) in eq. (7.6a), the equation for the first electron is obtained to be

$$\begin{aligned}
i\hbar\frac{d}{dt}c_1 = {} & \varepsilon_1 c_1 + \tau_{12}(bc_2)|_{t=0} + \int dt' G^{(2)}_{+\hbar\Omega}(t,t')[-\tau_{12}\tau_{21}{}^*(c_2{}^+ c_2) + \tau_{12}\tau_{21}{}^*(bb^+)]c_1 \\
& + \tau_{12}{}^*(b^+ c_2)|_{t=0} + \int dt' G^{(2)}_{-\hbar\Omega}(t,t')[\tau_{12}{}^*\tau_{12}(c_2{}^+ c_2) + \tau_{12}{}^*\tau_{21}(b^+ b)]c_1 \\
& + Uc_1(c_2{}^+ c_2)
\end{aligned} \tag{7.8a}$$

We get the equation for the second electron in a similar method to be

$$\begin{aligned}
i\hbar\frac{d}{dt}c_2 = {} & \varepsilon_2 c_2 + \tau_{21}(bc_1)|_{t=0} + \int dt' G^{(1)}_{+\hbar\Omega}(t,t')[-\tau_{21}\tau_{12}{}^*(c_1{}^+ c_1) + \tau_{21}\tau_{12}{}^*(bb^+)]c_2 \\
& + \tau_{21}{}^*(b^+ c_1)|_{t=0} + \int dt' G^{(1)}_{-\hbar\Omega}(t,t')[\tau_{21}{}^*\tau_{21}(c_1{}^+ c_1) + \tau_{21}{}^*\tau_{12}(b^+ b)]c_2 \\
& + U(c_1{}^+ c_1)c_2
\end{aligned} \tag{7.8b}$$

Now post-multiplying eq. (7.8a) by c_2 and pre-multiplying eq. (7.8b) by c_1, and using the anti-commutation rules of (c_1, c_2), we respectively obtain:

$$\left[i\hbar\frac{d}{dt}c_1\right]c_2=\left[\varepsilon_1+\int dt'G^{(2)}_{+\hbar\Omega}(t,t')\tau_{12}\tau_{21}*(bb^+)+\int dt'G^{(2)}_{-\hbar\Omega}(t,t')\tau_{12}*\tau_{21}(b^+b)\right](c_1c_2)$$

and

$$c_1\left[i\hbar\frac{d}{dt}c_2\right]=\varepsilon_2(c_1c_2)+\int dt'G^{(1)}_{+\hbar\Omega}(t,t')[-\tau_{21}\tau_{12}*c_1(c_1{}^+c_1)c_2+\tau_{21}\tau_{12}*(bb^+)(c_1c_2)]$$

$$+\int dt'G^{(1)}_{-\hbar\Omega}(t,t')[\tau_{21}*\tau_{21}c_1(c_1{}^+c_1)c_2+\tau_{21}*\tau_{12}(b^+b)(c_1c_2)]+Uc_1(c_1{}^+c_1)c_2$$

$$=\Bigg(\varepsilon_2+\int dt'G^{(1)}_{+\hbar\Omega}(t,t')[-\tau_{21}\tau_{12}*+\tau_{21}\tau_{12}*(bb^+)]$$

$$+\int dt'G^{(1)}_{-\hbar\Omega}(t,t')[\tau_{21}*\tau_{21}+\tau_{21}*\tau_{12}(b^+b)]+U\Bigg)(c_1c_2)$$

$$[\text{since } c_1(c_1{}^+c_1)c_2=(c_1c_1{}^+)(c_1c_2)=(1-c_1{}^+c_1)(c_1c_2)=(c_1c_2)]$$

Thus, by adding, we get,

$$i\hbar\frac{d}{dt}(c_1c_2)=(\varepsilon_1+\varepsilon_2)(c_1c_2)+\int dt'[\tau_{12}\left((b^+b)G^{(1)}_{-\hbar\Omega}(t,t')+(bb^+)G^{(2)}_{+\hbar\Omega}(t,t')\right)\tau_{21}*$$

$$+\tau_{12}*\left((bb^+)G^{(1)}_{+\hbar\Omega}(t,t')+(b^+b)G^{(2)}_{-\hbar\Omega}(t,t')\right)\tau_{21}](c_1c_2)$$

$$+\left[U-\int dt'\tau_{21}\left(G^{(1)}_{+\hbar\Omega}(t,t')\tau_{12}*-G^{(1)}_{-\hbar\Omega}(t,t')\tau_{21}*\right)\right](c_1c_2) \tag{7.9}$$

Considering $\tau_{12}=\tau_{21}=\tau$ (since both the electrons have equal footings on the phonon-interaction) and then applying the Fourier transform to the energy domain, the effective Hamiltonian for the electron pair is obtained to be

$$H_{12}\equiv(\varepsilon_1+\varepsilon_2)+\tau[N_{\text{Ab}}\left(G^{(1)}_{-\hbar\Omega}(E)+G^{(2)}_{-\hbar\Omega}(E)\right)+N_{\text{Em}}\left(G^{(1)}_{+\hbar\Omega}(E)+G^{(2)}_{+\hbar\Omega}(E)\right)]\tau*$$

$$+[U-\tau\left(G^{(1)}_{+\hbar\Omega}(E)-G^{(1)}_{-\hbar\Omega}(E)\right)\tau*] \tag{7.10a}$$

By exchanging 1 and 2 in eq. (7.10a), we find:

$$H_{21}\equiv(\varepsilon_2+\varepsilon_1)+\tau[N_{\text{Ab}}\left(G^{(2)}_{-\hbar\Omega}(E)+G^{(1)}_{-\hbar\Omega}(E)\right)+N_{\text{Em}}\left(G^{(2)}_{+\hbar\Omega}(E)+G^{(1)}_{+\hbar\Omega}(E)\right)]\tau*$$

$$+[U-\tau\left(G^{(2)}_{+\hbar\Omega}(E)-G^{(2)}_{-\hbar\Omega}(E)\right)\tau*] \tag{7.10b}$$

It is imperative to note at this point that the third term $[U-\tau(G^{(1)}_{+\hbar\Omega}(E)-G^{(1)}_{-\hbar\Omega}(E))\tau*]$ of eqs. (7.10a) or the corresponding term of (7.10b) represents the screened Coulomb potential, where $(G^{(1)}_{+\hbar\Omega}(E)-G^{(1)}_{-\hbar\Omega}(E))$ is positive due to the phonon energy gap $2\hbar\Omega$.

Interestingly, since $|\tau|^2 = \frac{(D\hbar\beta)^2/2\rho V}{\hbar\Omega}$ (see Chapter 5), the screening potential is found to be inversely related to $\hbar\Omega$ to the first order, which is consistent with BCS theory. Further, owing to the exchange anti-symmetric property of the electrons with opposite spins, we get comparing eqs. (7.10a) and (7.10b):

$$G^{(2)}_{+\hbar\Omega}(E) - G^{(2)}_{-\hbar\Omega}(E) = G^{(1)}_{+\hbar\Omega}(E) - G^{(1)}_{-\hbar\Omega}(E)$$

$$\Rightarrow G^{(1)}_{-\hbar\Omega}(E) + G^{(2)}_{+\hbar\Omega}(E) = G^{(1)}_{+\hbar\Omega}(E) + G^{(2)}_{-\hbar\Omega}(E) \tag{7.11}$$

Thus, eq. (7.11) clearly suggests that the momentum changes of the two electrons of same energy level and opposite spins due to phonon-interaction are equal and opposite, leading to an inherently entangled Cooper pair state $|-\vec{k},\uparrow\rangle_1|+\vec{k},\downarrow\rangle_2 + |+\vec{k},\uparrow\rangle_1|-\vec{k},\downarrow\rangle_2$, which is also in consonance with the BCS theory.

Now our arguments here are:

Firstly, two electrons can attract each other, surpassing the Coulomb repulsion in a material if $\tau\left(G^{(1)}_{+\hbar\Omega}(E) - G^{(1)}_{-\hbar\Omega}(E)\right)\tau^* > U$, however, provided the self-energy term (i.e. the second term of eq. (7.10a) or eq. (7.10b)) due to phonon-interaction is extremely small (at least less than the third term). Such a maximum value of phonon self-energy (in comparative strength of the third term) determines the critical temperature of superconductivity in a material. Evidently, with a considerable decrease in temperature, $N_{\text{Ab}}(\hbar\Omega/K_BT)$ and $N_{\text{Em}}(\hbar\Omega/K_BT)$ may become negligible, resulting in the superconducting phase.

Secondly, the phonon-interaction potential is directly proportional to the lattice strain (Ξ) as $\tau = D(\vec{\nabla}\cdot\vec{u}+\Xi)$, and thus may increase the screening potential $\tau\left(G^{(1)}_{+\hbar\Omega}(E) - G^{(1)}_{-\hbar\Omega}(E)\right)\tau^*$ in a strained material, so that, a superconducting phase can be achieved at a higher temperature by surpassing phonon self-energy. This can, at least in principle, explain the high-pressure phase of hydrides to show superconductivity at comparatively higher temperatures. However, with increasing phonon-interaction potential (τ) due to strain, the phonon self-energy also increases, and thus constrains the critical temperature at a finite limit.

And thirdly, in a negative dielectric constant material (e.g. negative capacitance oxides or meta-materials), the Coulomb potential itself can be attractive and thus can show zero resistance under certain conditions. However, that is fundamentally different from superconductivity since they cannot form entangled Cooper pairs (as discussed above), originating from, as we argue here, the phonon-assisted screening potential creating the spin-momentum correlation. It is imperative to note that the repulsive Coulomb potential and attractive spin-phonon interaction potential create a potential well surpassing the phonon self-energy at a relevant temperature to form the bound Cooper pair state that is fundamental to superconducting properties. Thus, it is an outcome of the spin-charge-phonon trialectics.

7.3 Quantum Hall effect and topology

Quantum Hall effect is one of the most intriguing observations of the last century, first explored experimentally by Klitzing in 1980 [13], that has led to the revelation of a fundamental aspect of physics (or more precisely, the physical properties of matter), i.e. the topological symmetry. In the conventional Hall effect, the electrons (or holes), while transporting through the driving electric field within a material, bend towards its perpendicular direction by a Lorentz magnetic field (applied along the other perpendicular direction), and thereby develop a voltage (called Hall voltage) across such perpendicular direction. The same phenomenon was studied by Klitzing in the 'inversion'-channel of a Si MOSFET (prepared by Dorda and Pepper) applying a strong magnetic field (~15 T) at liquid helium temperatures. Interestingly, in this case, it showed a peculiar feature that the Hall conductance exhibits an integral multiple of a universal constant (i.e. the quantum conductance e^2/h) that decreases at integral steps with increasing strength of the magnetic field owing to Landau magnetic quantization. This is known as the integral quantum Hall effect (iQHE). Astonishingly, such features are robust against the presence of any impurity scattering centres. Two years later, Tsui and Störmer discovered the fractional quantum Hall effect (fQHE) that exhibits newly emerging steps at several particular fractional multiples of quantum conductance [14]. Laughlin attributed the phenomenon to the electromagnetic gauge invariance and the existence of a mobility-gap [15], whereas Thouless with others (TKNN) described the iQHE on the basis of a finite Fermi level gap and a non-degenerate ground state, while fQHE owing to a degenerate ground-state instead, with a discrete symmetry breaking [16].

Here we argue that the quantum Hall effect, be it iQHE or be it the fQHE, can be theoretically modelled quite in a same manner that of a nanostructure FET or a gate-defined QD in 2D electrons gas, where, instead of applying electrical voltage as in the latter case, the transverse quantization is achieved (in the former case) by magnetic field (i.e. Landau quantization). And the Hall conductance here is numerically equal to the gate leakage conductance. In order to model it theoretically in the NEGF framework, it can be assumed that the magnetically quantized spectral density of states are occupied from the longitudinal current source (see Fig. 7.2), and the Hall current is a result of carrier escape through Hall probes (H_1 and H_2) at the transverse direction (i.e. no inflow from the Hall probes). Thus, the Hall current due to a small collector voltage V at the Hall probe H_2 is obtained to be, following Chapter 4, as

$$I = \frac{e}{h}\int T_{H_1 \to H_2}(E) f_{\mathrm{FD}}(E - E_{\mathrm{F}})dE \tag{7.12a}$$

where the transmission coefficient is

$$T_{H_1 \to H_2}(E) = \delta(E - \varepsilon_n)(\Gamma_{H_2} - \Gamma_{H_1}) \tag{7.12b}$$

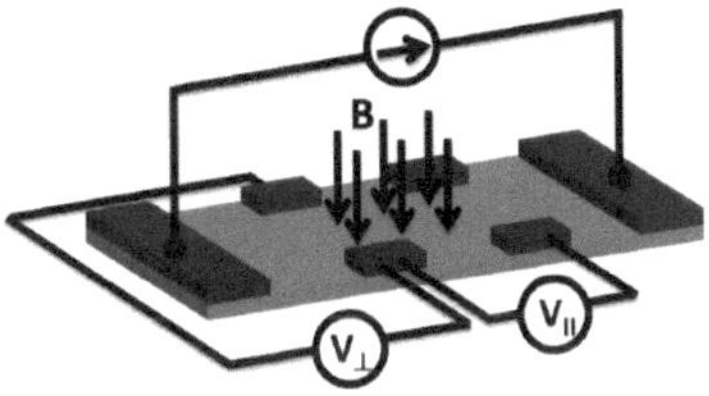

Fig. 7.2: The schematic of the quantum Hall experiment applying a magnetic field on a 2D sheet under a constant longitudinal current source.

It is apparent that the corresponding Hamiltonian due to the constant magnetic field $\vec{B} = B\hat{z}$ along the perpendicular direction to the transport plane is given by

$$\hat{H} = (\hat{p}_y - eBy)^2/2m* \tag{7.13}$$

that leads to the quantized states in eq. (7.13) to be $\varepsilon_n = (n + 1/2)\hbar\omega_B$, where $\omega_B = \left(\frac{e}{m*}\right)B$. Interestingly, from the discussions of Chapters 3 and 4, the broadening due to a Hall probe here is given by

$$\begin{aligned}\Gamma_{H_2} &= i(\Sigma - \Sigma^+) = -2\tau \sin ka = -2\tau(1 - \cos^2 ka)^{1/2} \approx -2\tau(1 - 1/2\cos^2 ka)\\ &\approx -2\tau[1 - 1/2(1 + (E - \varepsilon_n + eV)/2\tau)^2]\\ &\approx -2\tau[1 - 1/-2(1 + 2(E - \varepsilon_n + eV)/-2\tau)]\\ \Rightarrow \Gamma_{H_2} &\approx (E - \varepsilon_n + eV) - \tau \end{aligned} \tag{7.14a}$$

Similarly, we get,

$$\Gamma_{H_1} \approx (E - \varepsilon_n) - \tau \tag{7.14b}$$

Therefore, eq. (7.12)–(7.14) leads us to

$$I = \frac{e}{h}\sum_n \int \delta(E - \varepsilon_n)(\Gamma_{H_2} - \Gamma_{H_1})f_{FD}(E - E_F)dE = V\frac{e^2}{h}\sum_n \int \delta(E - \varepsilon_n)f_{FD}(E - E_F)dE$$

Thus, the Hall conductance is obtained to be

$$\sigma_H = \frac{e^2}{h}\sum_n \int \delta(E - \varepsilon_n)f_{\mathrm{FD}}(E - E_{\mathrm{F}})dE \tag{7.15a}$$

$$\Rightarrow \sigma_H = \frac{e^2}{h}\sum_n f_{\mathrm{FD}}(\varepsilon_n - E_{\mathrm{F}}) \tag{7.15b}$$

which exhibits decreasing steps with increasing magnetic field as shown in Fig. 7.3(a) at the values of integral multiples of the quantum conductance. It is imperative to note that it is only possible if $E_F \neq \varepsilon_n$ that is consistent with Thouless' proposition [16].

It is worth noting from the above derivation that no periodic potential is needed to consider for theoretically describing the QHE, rather, an effective mass approxima-

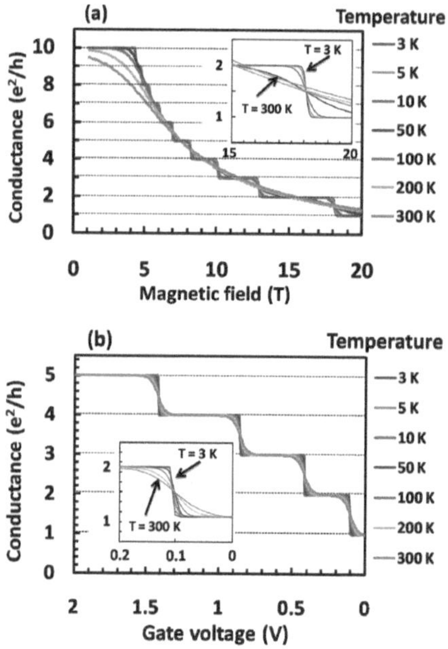

Fig. 7.3: (a) The plots of Hall conductance with increasing magnetic field obtained from the NEGF simulation; and (b) the electrical conductance of a nanostructure FET with gate voltage. Temperature is considered to vary from 300 to 3 K in both cases.

tion also works well to explain the phenomenon. Even if the local impurity levels are considered that are magnetically non-responsive, then they will only shift the step positions along the magnetic field axis. We argue on the basis of this model that only if there are defect levels that lift up degeneracy due to the magnetic field, they may get manifested as intermediate plateaus in the Hall conductance exhibiting the fQHE.

Now, interestingly, the same conductance steps (i.e. integral multiples of the quantum conductance) can be observed in a nanostructure FET or a gate-defined QD in 2D electron gas with decreasing gate voltage (see Fig. 7.3(b)), just as the increasing magnetic field in QHE. However, it is imperative to note at this point that the nanostructure FET characteristics are not equivalent to the quantum Hall effect, and these two phenomena are completely different from the point of view of physics, since they belong to two different types of topological classes.

It is worthy to make a note of the fact that the increasing (or decreasing) gate-voltage shifts the quantized states into (or out of) the Fermi level thereby creating (or destructing) an occupied state (i.e. changes the energy eigenvalues additively), while magnetic field strengthen the quantization without creating or destructing new occupied states (i.e. changes the energy eigenvalues multiplicatively), thereby representing the two different kinds of 'numbers' (i.e. *C-number* and *V-number*, respectively, as proposed in Chapter 1), although numerically identical. In order to understand this difference, it is better to see the picture in the topological parametric space, i.e. by transforming the Hall conductance (of eq. (7.15a)) from energy space (E) into the 2D-wave vector space ($\vec{k}$). Let us start with writing:

$$dE\delta(E-\varepsilon_n)=d[\vartheta(E-\varepsilon_n)]=dk\frac{d}{dk}[\vartheta(k-\kappa_n)]=\frac{dk_x dk_y}{2\pi k}\delta(k-\kappa_n) \tag{7.16}$$

and at a low temperature,

$$\frac{1}{2\pi}\delta(k-\kappa_n)f_{\mathrm{FD}}(\hbar^2k^2/2m*-E_{\mathrm{F}})=\frac{\partial}{\partial k}\left[\frac{1}{2\pi}\vartheta(k-\kappa_n)f_{\mathrm{FD}}(\hbar^2k^2/2m*-E_{\mathrm{F}})\right] \tag{7.17}$$

where the Fermi-Dirac distribution is a unit step function leading to $f_{\mathrm{FD}}[1-f_{\mathrm{FD}}]=0$, so that, the quantized conductance (due to gate-voltage tuned or magnetically achieved quantization), in general, can be written as

$$\begin{aligned}
\sigma_Q &= \frac{e^2}{h}\sum_n\int\frac{dk_x dk_y}{2\pi k}\delta(k-\kappa_n)f_{\mathrm{FD}}(\hbar^2k^2/2m^*-E_{\mathrm{F}})\\
&= \frac{e^2}{h}\sum_n\frac{1}{2\pi}\int dk_x dk_y\frac{1}{k}\frac{\partial}{\partial k}[\vartheta(k-\kappa_n)f_{\mathrm{FD}}(\hbar^2k^2/2m^*-E_{\mathrm{F}})]\\
&= \frac{e^2}{h}\sum_n\frac{1}{2\pi}\int dk_x dk_y\frac{1}{k}\frac{\partial}{\partial k}[kA_\phi]\\
&\qquad\left[\text{where} A_\phi(k)=\frac{1}{k}\vartheta(k-\kappa_n)f_{\mathrm{FD}}(\hbar^2k^2/2m*-E_{\mathrm{F}})\right]\\
&= \frac{e^2}{h}\sum_n\frac{1}{2\pi}\int dk_x dk_y\hat{z}.\left(\vec{\nabla}_k\times\vec{A}\right)=\frac{e^2}{h}\sum_n\frac{1}{2\pi}\oint kd\phi A_\phi\\
&\qquad[\text{since Stokes' Theorem}]\\
&= \frac{e^2}{h}\sum_n\frac{1}{2\pi}\oint d\phi[\vartheta(k-\kappa_n)f_{\mathrm{FD}}(\hbar^2k^2/2m*-E_{\mathrm{F}})]\\
\Rightarrow \sigma_Q &= \frac{e^2}{h}C
\end{aligned} \tag{7.18}$$

where

$$C=\frac{1}{2\pi}\oint d\phi\sum_n[\vartheta(k-\kappa_n)f_{\mathrm{FD}}(\hbar^2k^2/2m^*-E_{\mathrm{F}})] \tag{7.19}$$

is a positive integer for any quantized conductance (i.e. for both the gate-voltage tuned, as well as the magnetic quantization) at low temperature, and known to be the Chern number. However, the corresponding Berry connections:

$$A_\phi(k)=\frac{1}{k}\vartheta(k-\kappa_n)f_{\mathrm{FD}}(\hbar^2k^2/2m^*-E_{\mathrm{F}}) \tag{7.20}$$

are quite different for the two. For instance, in the parametric space of 2D-wave vector, it is a torus for magnetic quantization, whereas just close to a Gabriel's Horn for

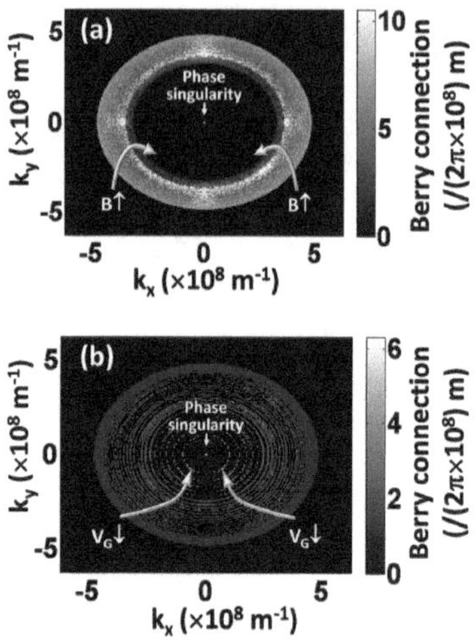

Fig. 7.4: The plots of simulated Berry connections on the topological parametric space for: (a) magnetic quantization (QHE), where the magnetic field is varied from 1 to 5 T; and (b) electrical quantization by varying gate voltage from 2 to 0.5 V.

gate-voltage tuned quantization; however, both have a singularity at the centre, as shown in Fig. 7.4(a) and (b), respectively.

7.4 Spin-orbit coupling: quantum spin Hall effect and topological materials

In some materials of two-dimensional structure (i.e. electrons are confined along one of the dimensions), if the spin-orbit coupling is strong enough, they can exhibit a quantum Hall effect even in the absence of any external magnetic field. This is known as the quantum spin Hall effect (QSHE) [17-20]. A few recent experiments have confirmed such phenomena, such as in HgTe-based quantum well, antiferromagnet-proximitized grapheme, InAs-based tri-layer quantum well, etc. [21-23]. Thus, quite like QHE described above, such materials intrinsically exhibit the topologically symmetric properties and may lead to exotic material behaviours, in cases, such as semi-metallic gapless edge-states, Fermi arcs, as well as Majorana-bound-states in topological superconductors [7].

Not going into the details of the theoretical models and analyses of this phenomenon, let us briefly present the core concept here, illustrating a 2D material case. It may be worthy to mention that the quantum relativistic behaviour of the electron leads to, as described in Dirac theory, the conservation of total angular momentum due to its orbital motion and the intrinsic spin. Thus, spin and orbital motion affect each other, thereby generating a potential known as spin-orbit coupling that can be obtained directly from the Dirac equation as

$$\hat{H}_{SO} = \lambda \underline{\underline{\vec{\sigma}}} \cdot (\vec{E} \times \vec{p}) \tag{7.21}$$

where $\vec{E}$ is the electric field, $\vec{p}$ is the momentum operator, the components of $\underline{\underline{\vec{\sigma}}}$ are Pauli matrices representing the spin of the electron, and $\lambda = q\hbar/(4m^{*2}c^2)$. Now, if a linear electric field ($\vec{E} = E_0(x\hat{i} + y\hat{j})$) is applied on a 2D material (let in the XY-plane) with strong spin-orbit interaction, its total Hamiltonian may be expressed as

$$\hat{H} = p^2/2m^* + \hat{H}_{SO} + V(x,y) = p^2/2m^* + \lambda \underline{\underline{\sigma}}_3 E_0(xp_y - yp_x) + (E_0/2)(x^2 + y^2) \tag{7.22}$$

where $V(x,y) = (E_0/2)(x^2 + y^2)$ is the potential associated with the electric field. It is to be noted that since $\vec{E} \times \vec{p} = E_0(xp_y - yp_x)\hat{k}$, only the Z-component of electron spin, i.e. along perpendicular to the surface of the 2D material, contributes to $\hat{H}_{SO}$. Now, owing to the eigenvalues of $\underline{\underline{\sigma}}_3$ being ± 1, suppose, for 'up' and 'down' spin along the Z-axis, respectively, eq. (7.22) can be algebraically manipulated by a coordinate transformation $(x,y) \to (m^* E_0)^{-1/4}(x,y)$ to express as

$$\hat{H}_{\uparrow\downarrow} = a[p_{\bar{x}}{}^2 + p_{\bar{y}}{}^2 + \bar{x}^2 + \bar{y}^2 \pm b(\bar{x}p_{\bar{y}} - \bar{y}p_{\bar{x}})] \tag{7,23}$$

Interestingly, for $b = 2$ (b being dependent on the strength of spin-orbit interaction), eq. (7.23) turns into:

$$\hat{H}_{\uparrow\downarrow} = a[(p_{\bar{x}} \pm \bar{x})^2 + (p_{\bar{y}} \mp \bar{y})^2] \tag{7.24}$$

leading to Landau quantization as shown above for the quantum Hall effect, although here in the absence of any external magnetic field. This is why it is known as the QSHE, where now the material has an intrinsically similar topological character as for the quantum Hall effect created by a magnetic field. Further, it can be observed from eq. (7.23) that the effective angular momentum component $\bar{L} = \bar{x}p_{\bar{y}} - \bar{y}p_{\bar{x}}$ has now positive and negative eigenvalues for 'up' and 'down' spin, respectively, indicating up-spin to be *chiral* and down-spin to be *anti-chiral* so that spin and momentum of the electrons are locked to each other. Thus, accordingly, opposite spin-currents will be obtained for the applied electric field. It is apparent that such spin-momentum locking and opposite spin-currents are robust against any disorder, just as in the case of the quantum Hall effect, the only difference being the spin Hall conductivity that is quantized as $(\hbar/2)[(e^2/h)/e] = e/4\pi$.

References

[1] Dirac, P. A. M. *The principles of quantum mechanics*. p. 27. Oxford university press, 1981.

[2] Aleiner, I. L., Brouwer, P. W., and Glazman, L. I. "Quantum effects in Coulomb blockade." *Physics Reports* 358, no. 5–6 (2002): 309–440.

[3] Ruoyu, L., Hudson, F. E., Dzurak, A. S., and Hamilton, A. R. "Pauli spin blockade of heavy holes in a silicon double quantum dot." *Nano letters* 15, no. 11 (2015): 7314–7318.

[4] Checkelsky, J. G., Bernevig, B. A., Coleman, P., Qimiao, S., and Paschen, S. "Flat bands, strange metals and the Kondo effect." *Nature Reviews Materials* 9, no. 7 (2024): 509–526.

[5] Pustilnik, M. and Glazman, L. "Kondo effect in quantum dots." *Journal of Physics: Condensed Matter* 16, no. 16 (2004): R513.

[6] Bernevig, B. A. and Zhang, S.-C. "Quantum spin Hall effect." *Physical review letters* 96, no. 10 (2006): 106802.

[7] Hasan, M. Z. and Kane, C. L. "Colloquium: Topological insulators." *Reviews of modern physics* 82, no. 4 (2010): 3045–3067.

[8] Meir, Y., Wingreen, N. S., and Lee, P. A. "Transport through a strongly interacting electron system: Theory of periodic conductance oscillations." *Physical review letters* 66, no. 23 (1991): 3048.

[9] Kouwenhoven, L. and Glazman, L. "Revival of the Kondo effect." *Physics world* 14, no. 1 (2001): 33.

[10] Kondo, J. "Resistance minimum in dilute magnetic alloys." *Progress of theoretical physics* 32, no. 1 (1964): 37–49.

[11] Datta, S. *Quantum transport: Atom to transistor*. Cambridge university press, 2005.

[12] Bardeen, J., Cooper, L. N., and Schrieffer, J. R. "Theory of superconductivity." *Physical review* 108, no. 5 (1957): 1175.

[13] Klitzing, K. V., Dorda, G., and Pepper, M. "New method for high-accuracy determination of the fine-structure constant based on quantized Hall resistance." *Physical review letters* 45, no. 6 (1980): 494.

[14] Tsui, D. C., Stormer, H. L., and Gossard, A. C. "Two-dimensional magnetotransport in the extreme quantum limit." *Physical Review Letters* 48, no. 22 (1982): 1559.

[15] Laughlin, R. B. "Quantized Hall conductivity in two dimensions." *Physical Review B* 23, no. 10 (1981): 5632.

[16] Thouless, D. J., Mahito Kohmoto, M. P. N., and Den Nijs, M. "Quantized Hall conductance in a two-dimensional periodic potential." *Physical review letters* 49, no. 6 (1982): 405.

[17] Murakami, S., Nagaosa, N., and Zhang, S.-C. "Dissipationless quantum spin current at room temperature." *Science* 301, no. 5638 (2003): 1348–1351.

[18] Sinova, J., Culcer, D., Qian Niu, N. A. S., Jungwirth, T., and MacDonald, A. H. "Universal intrinsic spin Hall effect." *Physical review letters* 92, no. 12 (2004): 126603.

[19] Kane, C. L. and Mele, E. J. "Quantum spin Hall effect in graphene." *Physical review letters* 95, no. 22 (2005): 226801.

[20] Kane, C. L. and Mele, E. J. "Z 2 topological order and the quantum spin Hall effect." *Physical review letters* 95, no. 14 (2005): 146802.

[21] Konig, M., Wiedmann, S., Brune, C., Roth, A., Buhmann, H., Molenkamp, L. W., Xiao-Liang, Q., and Zhang, S.-C. "Quantum spin Hall insulator state in HgTe quantum wells." *Science* 318, no. 5851 (2007): 766–770.

[22] Ghiasi, T. S., Petrosyan, D., Ingla-Aynés, J., Bras, T., Watanabe, K., Taniguchi, T., Mañas-Valero, S., et al "Quantum spin Hall effect in magnetic graphene." *Nature Communications* 16, no. 1 (2025): 5336.

[23] Meyer, M., Baumbach, J., Krishtopenko, S., Wolf, A., Emmerling, M., Schmid, S., Kamp, M., et al "Quantum spin Hall effect in III-V semiconductors at elevated temperatures: Advancing topological electronics." *Science Advances* 11, no. 43 (2025): eadz2408.

8 NEGF modelling of a single-qubit operation: initialization-manipulation-measurement

The concept of 'information' has been portrayed in Chapter 1 in something like HSV (hue-saturation-value) colour model, in the sense: (1) information is inherently probabilistic in nature, (2) information may offer objective property of a system (for maximum entropy), and (3) information can be quantitatively measured (in terms of *bits*). It has also been mentioned that information can be deterministically stored in *bits* (i.e. the units of classical computer) that are practically realized utilizing MOSFETs, while may be stored probabilistically in exponentially higher order of amounts (e.g. $\sim 2^n$) in qubits (i.e. the units of quantum computer), which, in reality, can be achieved in superconductors, photonic systems, quantum dots (QDs), trapped atoms/ions, as well as in topological materials. The qubits, especially when appropriately entangled, not only can reduce the required space for information to be stored but also can offer a polynomial runtime in comparison to *bits* (that exhibit superpolynomial runtime) for computing several demanding problems, thereby exponentially speeding up such computations. However, developing reliable qubits is still a challenge, particularly, due to the fragility of their state of coherence being affected by environmental noises. For a detail comprehensive understanding on this, the readers may study [1, 2].

A quantum computer generally comprises a number of its subsystems, such as control hardware that may include voltage sources, waveform generators, lasers and microwaves, and noise shielding systems like vacuum chambers and cryogenic setups; however, its key ingredient is a quantum register, i.e. made up of an array of qubits, which can be subjected to perform physical operations to realize the relevant quantum algorithms. In this context, a five-point minimum requirement, from technological perspective, in order to developing a quantum computer has been grossly summarized by DiVincenzo [3]:

1) An exactly countable numbers of physically distinguishable qubits (without fudging) that can be described by precisely enumerable Hilbert space.
2) All the qubits can be prepared to exhibit an initialized state (e.g. $|0\rangle$), which will be one of the computational basis states ($|0\rangle$ and $|1\rangle$), such as all spin-down electrons, or all horizontally polarized photons, or all trapped atoms in their motional-ground-state.
3) The qubits should preserve their coherent state as long as possible (i.e. until measured) by keeping them isolated from the environmental noise or decoherence. A small decay of the state coherence is tolerable depending on the quantum error correction protocol.
4) It should allow 'a controlled sequence of unitary transformations' of qubit states, i.e. the possible logic operations on each of the qubits as well as on a pair or a collection of selected qubits (to create inter-qubit entanglement).

 | https://doi.org/10.1515/9783111545950-008

5) The measurement techniques of qubit states must be 'projective', not 'protective', ensuring the irreversible state-collapse in order to obtain information about the system.

Since a qubit operates on two computational basis states ($|0\rangle$ and $|1\rangle$), it is apparent that any two-state quantum systems can be used as qubits as mentioned in Chapter 1. Here, we will briefly discuss the basic operational principles of a few among such qubit devices, which have already drawn significant attention from the perspective of large-scale production of quantum computers.

8.1 Photonic qubits

Photons are evidently the most attractive candidate for single-qubit operations due to their two intrinsic quantum states (i.e. polarizations), minimal decoherence, and operability as *flying qubits* for quantum communication. The orthogonal polarizations (e.g. horizontal/vertical or LCP/RCP) of single photons or their non-overlapping propagation paths are most frequently used as the two states of qubits (see Fig. 8.1, where the fundamental scheme of photonic qubit operation is illustrated). However, photon losses during generation and detection as well as by scattering and absorption while propagating through a medium are the key challenges for practical photonic computing. They also suffer from entangling difficulty and scalability issues along with the storing problems. It is only after the proposal by Knill, Laflamme, and Milburn [4] that the scalable scheme (KLM scheme) of photonic qubits have emerged, which comprises linear optics, single photon generation/detection, and classical feed-forward operability (i.e. to perform measurement on the photonic qubits in the middle of a circuit, execute relevant quantum operations, and carry forward in the same circuit direction). Technically, in such method, only selected heralding modes of photons from the source are teleported for gate operations at the cost of a large numbers of *ancilla* photons, leading to an interference phase shift with nonlinear dependence on the photon number.

At this point, it is worthy to note that the basic operation of a qubit comprises three consecutive steps: (1) *initialization* (or *preparation*), (2) *manipulation* (or *logic gate operations*), and (3) *measurement* (or *read-out*), as discussed in Chapter 1. For *initialization* of photonic qubit state, the most popular technology to achieve indistinguishable single photon source is the spontaneous parametric down-conversion (SPDC) mechanism [5], where a photon from a laser source (called *pump photon*) is converted into an entangled pair of photons (termed as *signal* and *idler*) in a nonlinear dielectric medium with *rank-3 susceptibility tensor* (where each component of electric polarization depends on all the components of electric field and their products). Such SPDC is a nondeterministic process with very low probability of conversion (efficiency ~10^{-6}), although can lead to provide a single photon, i.e. the *signal*, heralded by detecting the other, i.e.

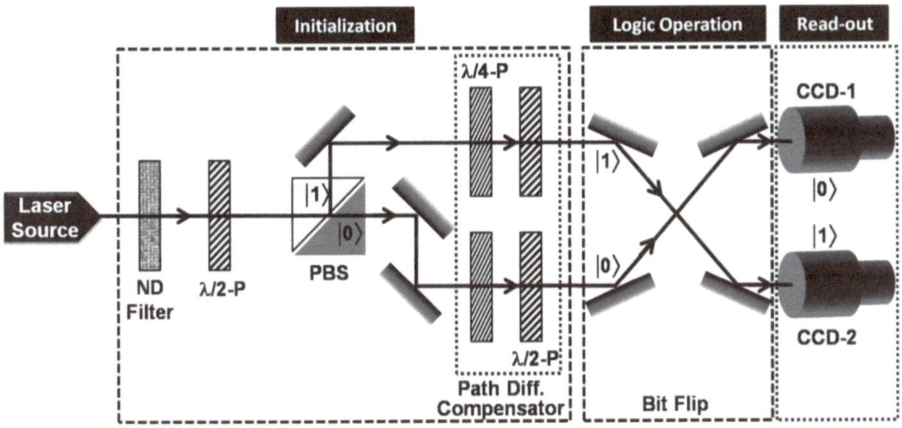

Fig. 8.1: Schematic of the basic operational modes of a photonic qubit, i.e. *initialization*, *manipulation*, and *read-out*, illustrating a bit-flip logic operation. The same scheme can be implemented on a scalable (i.e. wave-guided material) platform.

the *idler*. The photon-pair entanglement may be obtained in their polarization states or in time-bin encoding as well as in their transverse spatial modes or even in frequencies. In any case, such process is a non-equilibrium one, and non-equilibrium Green's function (NEGF) can be the most appropriate framework for their theoretical modelling for describing the photon emission, on the basis of discussion in Chapter 6. It is interesting to note that such probabilistic single photon source can be transformed into a deterministic one through either parallel or sequential approaches. The parallel approach is to de-multiplex the pulsed laser photons through an array of SPDC sources, followed by heralding and rerouting of the photons via relevant optical network, thereby introducing appropriate delay to provide a number of simultaneous single photons at the output. On the other hand, in the sequential approach, a single SPDC source is used to time-multiplexing of the photons by utilizing pertinent temporal delay network to provide single heralded photons sequentially at a fixed repetition rate. A detail discussion on this can be obtained in [6].

The *manipulation* of photonic qubit states are generally performed by using integrated electro-optic modulators. However, detection of such photons with high efficiency, low reset time, low dark count rate, and good photon-number-resolving (PNR) capacity is not trivial. For such purpose, single photon avalanche photodiodes are used that offer fast reset (~100 ns) and low dark counts (~100 cps). However, to achieve high detection efficiency and PNR, the superconducting nanowire single-photon detectors (SNSPDs) are employed, that exhibit faster reset times (<100 ns) and provide an efficiency >90% at the telecom wavelength bands (around 1,310 and 1,550 nm) [6]. Nevertheless, the cryogenic temperature (~1 K) requirement for superconductor to operate is the key limitation for large-scale practical implementation.

8.1.1 Trapped-ion qubits

The scheme of trapped-ion qubit exploits the long-coherence electronic levels of particular ions with hydrogen-like spectrum (e.g. group II), confined in radiofrequency (RF) traps (e.g. Paul trap) with long lifetime, as the qubit states. Depending on the energy levels and frequencies, such qubits are generally categorized into four types: (1) Zeeman qubits based on Zeeman-split magnetic sublevels having tens of MHz frequencies; (2) hyperfine qubits that utilize hyperfine levels of the ions with frequencies ~GHz order; (3) fine-structure qubits with tens of THz frequencies; and (4) optical qubits that exhibit optical transition between the levels with frequencies of few hundreds of THz [7]. The state-of-the-art ion-qubit state *manipulation* processes are performed at few hundred kHz, which are much faster in regard to their lifetimes in the order of seconds. Further, another remarkable advantage of trapped-ion qubits is, alongside such internal electronic levels, the existence of their vibrational states associated with phonon modes (~MHz) in the ion-trapping potential. Such vibrational states are shared among multiple ions (called *motional* states) in the trap with normal modes of coupled oscillations, which, by laser manipulation, can entangle the internal electronic states of different ions [7].

The *initialization* of a trapped-ion qubit is performed by optical pumping, where a laser pumps it to an electronic excited state of the ion (in its *motional ground* state with zero phonon, achieved by *laser-cooling* technique, i.e. Doppler cooling followed

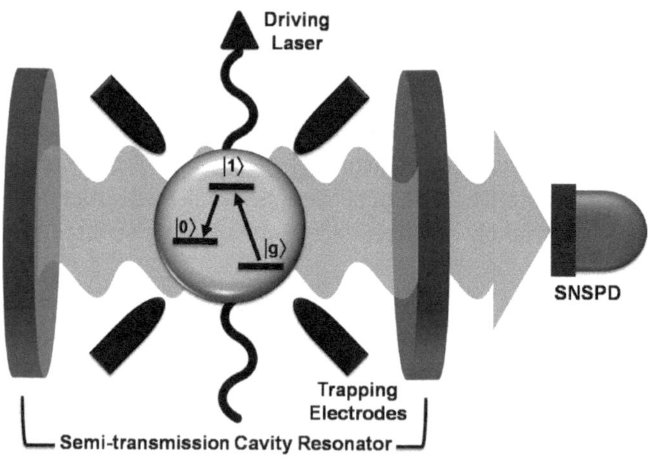

Fig. 8.2: The schematic of a trapped-ion qubit: the ion is confined by trapping potential of electrodes where a driving laser excites it to $|0\rangle$ or $|1\rangle$from the *motional ground state* for *initialization*, where *manipulation* is performed by semi-transparent cavity-resonators, and finally the *read-out* photons (after all losses) are detected by an SNSPD.

by sideband cooling [8]) that readily decays to its electronic ground state ($|0\rangle$), which does not couple to the laser. On the other hand, their *manipulation* for several logic operations are performed by using near-field microwaves, Raman transitions, or resonant lasers with very high fidelity (>0.99). And the qubit-state read-out is accomplished through illuminating the trapped-ion by a detection-laser, which projects its superposed state either to the bright state (e.g.$|1\rangle$) that scatters significantly large numbers of photons, or to the dark state (e.g.$|0\rangle$) that rarely scatters any photon [7] (illustrated schematically in Fig. 8.2).

While the principal advantages of trapped-ion qubits are their natural reproducibility, significantly high fidelities in each qubit operation, long coherence time, as well as very high coherence time to gate-operation time ratio, the key challenges include slower 'absolute' gate speed (even compared to classical computer), off-resonant photon scattering, undesired state trapping resulting in leakage error, scalability issues due to lack of control on specific ions in a large array, and the requirement for numerous lasers.

8.1.2 Superconducting qubits

Superconductors are materials that exhibit zero resistance (!) and expel bulk magnetic field (known as Meissner effect) below a critical temperature (~K, mostly <30 K). Such property is attributed to (due to the seminal BCS theory [9]) the formation of Cooper pairs, where two electrons, conventionally having opposite spins and momenta, together carry a net zero momentum in absence of a finite magnetization. Interestingly, the magnetic flux (Φ) and charge (Q) in superconducting circuits show the same uncertainty relation like position and momentum, i.e. $[\Phi, Q] = i\hbar$, and the energy associated with them (i.e. inductive and capacitive, respectively) are exactly analogous to that of a quantum harmonic oscillator resulting in a number of equidistance quantized energy levels. Thus, for the purpose of developing superconductor-based qubits, in order to define the qubit sub-space (i.e. two selective states out of many), the Hilbert space needs to be engineered through an anharmonicity in the Hamiltonian by introducing a Josephson junction (i.e. a thin tunnelling barrier between two superconductors) in the circuit (for a detailed study, see [10-12]). The superconducting qubits, as schematically shown in Fig. 8.3, can operate based on three physical quantities: (1) *charge*, where a single potential well is created by omitting the inductor; (2) *flux*, where a double-well potential is created (making the Josephson junction energy much greater than the capacitive energy), so that two ground states are in quantum superposition of the clockwise and counter-clockwise persistent currents once the applied flux is almost a half integral of flux quanta; and (3) *phase*, where an asymmetric double well is biased to generate a metastable anharmonic well providing two lowest energy levels as the qubit states.

Usually, the superconducting qubit frequencies are of the order of ~10 GHz, which are subjected to *manipulation* by resonant microwave pulses of duration ~1–10 ns. For such logic gate operations, the temperature must be kept at ~10 mK (i.e. thermal energy corresponding to ~0.1 GHz) using dilution refrigerators to minimize the thermal noise. In case of multi-qubit gate operations, the adjacent qubits get naturally correlated through capacitive or inductive interactions; however, for long-distance (~mm) correlation, microwave photons through transmission lines are utilized to couple the qubits as well as for non-local quantum correlation measurements [10, 11]. Superconducting qubit *read-outs* are performed through a threshold of Josephson junction critical current to distinguish the two qubit states. Also the quantum non-demolition measurements [13] are carried out giving rise to state-dependent phase shifts in the transmission line EM waves.

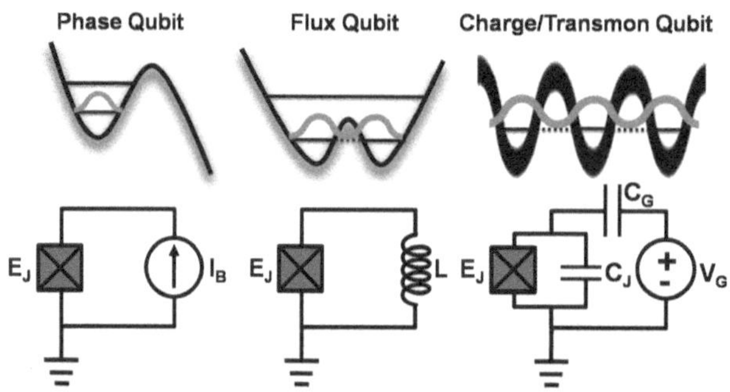

Fig. 8.3: The confining potentials, engineered computational eigenstates, and the equivalent circuit diagrams of superconducting qubits illustrating the phase qubit, flux qubit, and charge/transmon qubit.

The most potential advantage of superconducting qubits is the observance of quantum superposition at macroscopic scale that reduces the fabrication processing cost significantly, although their performances are mostly challenged by decoherence effects, especially the charge noise. Interestingly, a scheme of charge qubit with two capacitively shunted superconductors, termed as *transmon*, has been exploited to potentially decrease the sensitivity to charge noise [10, 11]. Nevertheless, the requirement of cryogenic temperature is a serious issue for large-scale production of superconductor-based quantum computers.

8.1.3 Quantum dot (QD) qubits

QD-based qubits are predominantly of two kinds: (1) single quantum dot (SQD) [14], where the two spins of a single electron define the two qubit states and (2) double quantum dot (DQD) [15], where possible locations (in either of the two QDs) of the electron (correlated or non-correlated with the spin state) represent the corresponding qubit states. Loss and DiVincenzo first proposed the scheme of SQD [14], where voltage gates can create the QD by confining electrons as well as can be utilized for logic operations by coherent *manipulation* of its superposed states. Here, a strong magnetic field is generally required to lift up the spin degeneracy so as to energetically separate the spin-up and spin-down states. Such QDs are then coupled to a reservoir for *initialization* and *read-out* through controlled tunnelling to a lower energy state and from a higher energy state of the QD, respectively. However, in practice, the *read-outs* are often performed by using auxiliary QDs or quantum point contact [16].

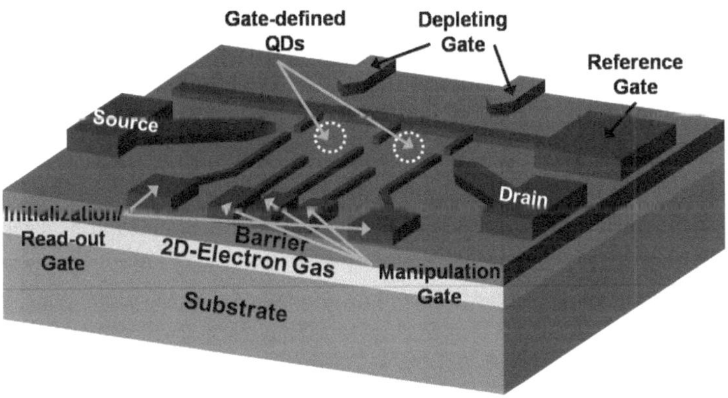

Fig. 8.4: The schematic of a double quantum dot (DQD) qubit: the QDs are formed in the 2D-electron gas material (sandwiched between the substrate and barrier material) by applying appropriate gate voltages (using depleting and manipulating gates). *Initialization* and *read-out* are controlled by corresponding gate voltages to vary the source/drain barrier with QDs, and the *manipulation* is carried out by relevant voltages at the respective gates to control the inter-dot tunnelling.

In DQD structures (see Fig. 8.4), the *manipulation* is carried out by voltage gates as well as by applying magnetic field to localize one or more electron(s) in either left or right QDs. For instance, a two-electron (±1/2-spin each) DQD system comprises one spin-singlet and three spin-triplet states, thereby forming a 0 or 1 spin together. Either magnetic field can be applied to split the triplet, or a suitable control voltage (by lowering the tunnel barrier) can increase the inter-QD virtual tunnelling to lower the singlet energy level (which does not occur for the triplet due to Pauli principle). Further, gate voltages on two QDs can be set asymmetrically to allow unidirectional virtual tunnelling for desired *manipulation* of qubit states. However, nuclear spin induces significant decoherence effects into such spin qubit systems. Also electron-phonon interaction is another unavoidable source of decoherence in the QD-based qubits. For a comprehensive understanding on QD spin-based qubits, the reader can see [17]. It is imperative to note at this point that the QD-based spin/charge qubit scheme is evidently simpler than other qubit schemes, but mostly attractive due to its compatibility with the mature CMOS technology. In this context, recently a high-volume cryogenic testing method to probe single electron across

300 mm wafer comprising Si/SiGe DQD [18] and ^{29}Si-based DQD [19] spin qubits have been developed, the latter being further advanced to achieve a two-qubit entangled gate with a lifetime of ~100 μs [20].

8.1.4 Topological qubits

Amidst all the different kinds of qubits suffering from decoherence and thus requiring error-correction models and fault-tolerant computing schemes, an extraordinary proposition came out to develop topological qubits, which are inherently robust, from their hardware level, against the decoherence inducing possible noises. The concept of such topologically protected robustness against any perturbation has been discussed in Chapter 7. Physically, a topological material, in general, implies that some of its fundamental properties are unresponsive to non-transitional (i.e. unlike quantum phase transition) changes in material parameters due to some kind of protected symmetry. Such topological equivalence/differences of the materials are mostly observed in their reciprocal lattice spaces. It can be understood by the analogy of comparison between a glass of water, a cup, and a bottle opener. Conventionally, the glass and the cup seem to be similar and different from the opener; however, from another perspective, the case may be different. For instance, the glass cannot be continuously deformed to transform it into a structure of a cup (without creating a hole in it since the cup has a handle), and thus they are topologically different; however, the cup and the opener can be inter-transformed into one another just by continuous deformation of their shapes thereby making them topologically equivalent. Thus, against a continuous deformation, the topology of a structure is protected.

With the advantage of such coherence protection from environmental decoherence effects, the topological qubit schemes are generally based upon a kind of 2D-quasiparticle excitations, termed as *anyons*, for information encoding as well as manipulation [21]. The exchange properties of *anyons* can be thought of something in-between, or a more general class of, the symmetric and anti-symmetric behaviour of *bosons* and *fermions*, respectively; and accordingly, instead of Bose-Einstein or Fermi-Dirac distributions, the *anyons* follow non-Abelian braiding statistics (e.g. exchanging twice does not bring back to the same state) [21]. Intriguingly, the degenerate subspace of such non-Abelian *anyons* exhibit vanishing matrix elements corresponding to any local perturbation, and only braiding operations can lead to their unitary transformation, which is the inherent resource of fault-tolerant quantum computation with such topological qubits. Majorana bound states (MBSs) are most attractive among *anyonic* systems. Such qubits are considered as yet to achieve. However, very recently, with interesting but controversial features, MBS has been claimed to be realized by developing a two-site Kitaev chain proximitized by a superconductor [22].

8.2 Basics of a single-qubit operation: initialization-manipulation-measurement

Owing to the superposition principle of a quantum system, any pure state of a qubit (i.e. undisturbed by environment) as discussed in Chapter 1, can be expressed in general by

$$|\psi\rangle = \cos(\theta/2)|0\rangle + e^{i\varphi}\sin(\theta/2)|1\rangle \tag{8.1}$$

where $\cos^2(\theta/2)$ and $\sin^2(\theta/2)$ measure the probability of its basis states $|0\rangle$ and $|1\rangle$, respectively, while φ represents the phase difference between such basis states. Interestingly, any such possible pure state, i.e. any of the corresponding information, can be encoded in a qubit, which, as described in Chapter 1, can be diagrammatically represented by the respective coordinate-(θ,φ) on a sphere of unit radius (known to be the Bloch sphere), its north pole ($\theta \equiv 0^\circ$) indicating $|0\rangle$ and the south pole ($\theta \equiv 180^\circ$) referring to $|1\rangle$. A qubit is believed to remain in a superposed state, in general, of $|0\rangle$ and $|1\rangle$ with such probabilities, whereas, on a single measurement it reads either $|0\rangle$ or $|1\rangle$ (i.e. collapses to either of these states), with a probability of $\cos^2(\theta/2)$ and $\sin^2(\theta/2)$, respectively. It can be verified by performing a large number of such single measurement runs under identical conditions.

At this point, we argue that not only the single measurement run of a qubit state, but any quantum measurement, in general, can be described by the three consecutive steps: (1) *initialization*, (2) *manipulation*, and (3) *read-out*. For instance, as shown in Fig. 8.5(a), the famous Stern-Gerlach experiment, where spin of the single electron in 5s subshell of silver atom is measured by (silver) beam splitting under magnetic field, can also be thought of to be sequenced into three such steps. First, the collimator *initializes* the silver atom electron in a particular spin-state, suppose $|\uparrow\rangle_X$; if the magnetic field (with a little non-uniformity) is applied in *Z*-direction, the state now can be represented as $1/\sqrt{2}(|\uparrow\rangle_Z + |\downarrow\rangle_Z)$, while it can be *manipulated* to $a|\uparrow\rangle_Z + b|\downarrow\rangle_Z)$, in general, by selecting the relevant field direction (this step sometimes is considered also as a part of *initialization*); and finally, the spin states are measured (i.e. *read-out*) on the absorber screen (in original experiment, it was a glass plate, cooled with liquid nitrogen) at different positions for either in $|\uparrow\rangle_Z$ or in $|\downarrow\rangle_Z$ state. This entire scheme can be translated, for instance, into a QD with an excess electron in a discrete state for developing a single physical qubit.

Another example is shown in Fig. 8.5(b) to illustrate an excess electron in a tunnel-coupled DQD structure that can realize a qubit state in the positional basis (i.e. $|L\rangle$ for left and $|R\rangle$ for right QD). The corresponding band alignments of the entire device and the sequential steps of qubit operation, i.e. *initialization-manipulation-measurement*, are schematically illustrated in Fig. 8.5(c). In such a scheme, first, a voltage (V_{G1}) is applied to gate 1 over the left QD to make a single electron enter to its lowest empty state ($|L\rangle$) from left lead; second, a voltage (V_{G2}) is applied on gate 2, so

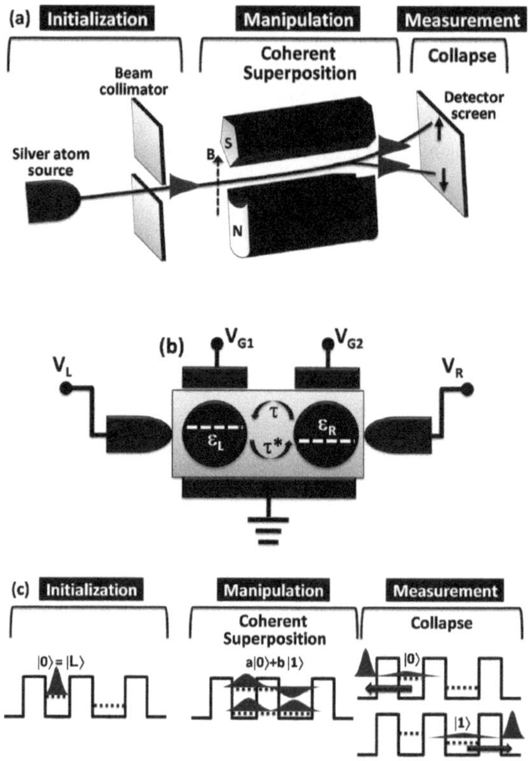

Fig. 8.5: (a) The scheme of Stern-Gerlach experiment to measure the spin of electron represented in qubit-measurement sequences, i.e. *initialization*, *manipulation*, and *read-out*; (b) the basic design of structure of voltage-controlled tunnel-coupled double quantum dot to operate as a qubit; and (c) the schematic of its operational principle in the positional basis of left ($|L\rangle$) and right ($|R\rangle$) QD depicting the device band alignments, eigenstates, and wavefunctions.

that the electron in QD-1 can tunnel to QD-2 back-and-forth, but no electron can enter to the latter from right-lead; such tunnel-coupling between the QDs create the 'bonding' ($|\psi\rangle_B = 1/\sqrt{2}(|L\rangle + |R\rangle)$) and 'anti-bonding' ($|\psi\rangle_{AB} = 1/\sqrt{2}(|L\rangle - |R\rangle)$) states that are shared by both the QDs, and now a small variation of V_{G1} and $V_{G2,}$ as well as their frequencies can manipulate the probability and phase (i.e. (θ, φ)) of the superposition between QD states, in general, as $|\psi\rangle = \alpha(V_{G1}, V_{G2})|L\rangle + \beta(V_{G1}, V_{G2})|R\rangle$; finally, the application of a small voltage bias to both the leads result in a collapse of the superposed state through collection of the electron at either left or right lead, and accordingly a positive or negative device current. Such a device operation for specific geometry and material has been described in detail in [23]. Further, it may be worthy to note that, if one wants to measure in the bonding/anti-bonding basis states, then such device scheme needs to be a little more modified in order to achieve a controlled tunnelling at different energy levels to the measuring lead, as mentioned in Section 8.1.4.

8.3 How to model the physical qubit operation?

Conventionally, a physical qubit, when its interaction with environment is not considered, is assumed to comprise a closed two-level quantum system described by a two-component vector in Hilbert space, the logic gate operations by relevant unitary matrices, and the read-out being represented by a projection-valued measure. On the other hand, the effects of environmental noise are incorporated mostly by Markovian open system formulation (under the approximation that environment forgets past interactions rapidly) while representing the qubit by its density matrix (instead of vector) over the Hilbert space, the logic operations by completely positive trace-preserving map, and the read-out by the corresponding positive operator-valued measure (for detail, see [24]). The non-Markovian errors, however, need to model in case-specific manners.

It is imperative to note that such modelling schemes are abstract algebraic formulation of the qubit operations and do not focus on the real nature of physical interactions between the quantum system and the gates or the measuring instrument or its sub-systems, which, on the other hand, is necessary from the perspective of experimental realization of such qubit operations in practice. Thus, here we are illustrating the physical aspects of qubit operation on the basis of NEGF framework with an example of the abovementioned coupled QD architecture.

8.4 NEGF model to describe multi-level coupled quantum dots

Suppose two semiconductor QDs (QD-1 and QD-2), i.e. with confined many (theoretically infinite) electronic-state systems, are embedded within an insulator that, by its barrier potential, prevents the thermodynamic quasi-static transport of electrons from one another. Also consider the sizes of QDs to be smaller than the electron mean free path in the materials, so that they hardly face any scattering, and assume the semiconductor/insulator interfaces to be smooth enough to minimize the surface roughness scattering phenomena. For now, also suppose that the QDs are not coupled to any probing lead or reservoir that will be included later. Furthermore, we are not considering the impact of phonon scattering here that can be easily incorporated into the model following the methods described in Chapter 5.

In such condition, the Hamiltonian for the two interacting QDs can be given by

$$H = \sum_{i} \varepsilon_1^i c_1^{i+} c_1^i + \sum_{j} \varepsilon_2^j c_2^{j+} c_2^j + \sum_{i,j} \tau_{ij} c_1^{i+} c_2^j + \sum_{i,j} \tau_{ij}{}^* c_2^{j+} c_1^i \tag{8.2}$$

where c_1^i, c_2^j are the second-quantized field (annihilation) operators for electron in ith state of QD-1 and jth state of QD-2, respectively, $\varepsilon_1^i, \varepsilon_2^j$ are the corresponding energy eigenvalues when the QDs are isolated from each other, while τ_{ij} represents the coupling between such states when they interact. It may be interesting to note that the QDs are made isolated from, or interact to, each other in a controlled manner usually by implementing another gate between them. However, it will be shown here how manipulating the eigenstates of QDs by only two gate voltages can play the same role.

Nevertheless, if the QDs start interacting from time $t = 0$, then the coupled equations for such interacting QDs are obtained from Heisenberg equation as:

$$i\hbar \frac{d}{dt} c_1^j = \varepsilon_1^j c_1^j + \sum_k \tau_{jk} c_2^k \tag{8.3a}$$

$$i\hbar \frac{d}{dt} c_2^k = \varepsilon_2^k c_2^k + \sum_j \tau_{kj} * c_1^j \tag{8.3b}$$

Thus, considering the Green's functions for eqs. (8.3a) and (8.3b) (as described in Chapter 3), the solutions of such equations are given by

$$c_1^j(t) = c_1^j(0) + \sum_{m,k} \int dt_1 G_1^{jm}(t, t_1) \tau_{mk} c_2^k(t_1) \tag{8.4a}$$

$$c_2^k(t) = c_2^k(0) + \sum_{n,j} \int dt_1 G_2^{kn}(t, t_1) \tau_{nj} * c_1^j(t_1) \tag{8.4b}$$

Putting eqs. (8.4b) and (8.4a) into eq. (8.3a) and (8.3b), respectively, we get:

$$i\hbar \frac{d}{dt} c_1^j = \varepsilon_1^j c_1^j + \sum_{k,n,j'} \int \tau_{jk} dt_1 G_2^{kn}(t, t_1) \tau_{nj'} * c_1^{j'}(t_1) + \sum_k \tau_{jk} c_2^k(0) \tag{8.5a}$$

$$i\hbar \frac{d}{dt} c_2^k = \varepsilon_2^k c_2^k + \sum_{j,m,k'} \int dt_1 \tau_{kj} * G_1^{jm}(t, t_1) \tau_{mk'} c_2^{k'}(t_1) + \sum_j \tau_{kj} * c_1^j(0) \tag{8.5b}$$

Thus, the Fourier transform of eqs. (8.5a) and (8.5b) over the difference-time coordinate into the energy (E) domain, the NEGFs for the two such interacting QDs are obtained as:

$$\underline{\underline{G}}_1 = \left[(E\underline{\underline{I}} - \underline{\underline{\varepsilon}}_1) - \underline{\underline{\tau}}\,\underline{\underline{G}}_2 \underline{\underline{\tau}}^+\right]^{-1} = \left[(E\underline{\underline{I}} - \underline{\underline{\varepsilon}}_1) - \underline{\underline{\Sigma}}_2\right]^{-1} \tag{8.6a}$$

$$\underline{\underline{G}}_2 = \left[(E\underline{\underline{I}} - \underline{\underline{\varepsilon}}_2) - \underline{\underline{\tau}}^+ \underline{\underline{G}}_1 \underline{\underline{\tau}}\right]^{-1} = \left[(E\underline{\underline{I}} - \underline{\underline{\varepsilon}}_2) - \underline{\underline{\Sigma}}_1\right]^{-1} \tag{8.6b}$$

where $\underline{\underline{\Sigma}}_2 = \underline{\underline{\tau}}\,\underline{\underline{G}}_2 \underline{\underline{\tau}}^+$ and $\underline{\underline{\Sigma}}_1 = \underline{\underline{\tau}}^+ \underline{\underline{G}}_1 \underline{\underline{\tau}}$ are self-energies originated due to coupling with the respective QDs. Now, the corresponding spectral density functions (i.e. DOS in per energy unit) of the QDs are given by

$$\underline{\underline{A}}_1 = i(\underline{\underline{G}}_1 - \underline{\underline{G}}_1^+) = i\underline{\underline{G}}_1 \underline{\underline{\tau}}(\underline{\underline{G}}_2 - \underline{\underline{G}}_2^+)\underline{\underline{\tau}}^+ \underline{\underline{G}}_1^+ = \underline{\underline{G}}_1 \underline{\underline{\tau}}\,\underline{\underline{A}}_2 \underline{\underline{\tau}}^+ \underline{\underline{G}}_1^+ \tag{8.7a}$$

$$\underline{\underline{A}}_2 = i(\underline{\underline{G}}_2 - \underline{\underline{G}}_2^{+}) = i\underline{\underline{G}}_2\underline{\underline{\tau}}^{+}(\underline{\underline{G}}_1 - \underline{\underline{G}}_1^{+})\underline{\underline{\tau}}\underline{\underline{G}}_2^{+} = \underline{\underline{G}}_2\underline{\underline{\tau}}^{+}\underline{\underline{A}}_1\underline{\underline{\tau}}\underline{\underline{G}}_2^{+} \tag{8.7b}$$

which can further be written as:

$$\underline{\underline{\tau}}\underline{\underline{A}}_2\underline{\underline{\tau}}^{+} = (\underline{\underline{G}}_1)^{-1} i(\underline{\underline{G}}_1 - \underline{\underline{G}}_1^{+})(\underline{\underline{G}}_1^{+})^{-1} \tag{8.8a}$$

$$\underline{\underline{\tau}}^{+}\underline{\underline{A}}_1\underline{\underline{\tau}} = (\underline{\underline{G}}_2)^{-1} i(\underline{\underline{G}}_2 - \underline{\underline{G}}_2^{+})(\underline{\underline{G}}_2^{+})^{-1} \tag{8.8b}$$

so that the level broadenings are obtained to be:

$$\underline{\underline{\Gamma}}_1 = i(\underline{\underline{\Sigma}}_1 - \underline{\underline{\Sigma}}_1^{+}) = i\underline{\underline{\tau}}^{+}(\underline{\underline{G}}_1 - \underline{\underline{G}}_1^{+})\underline{\underline{\tau}} = \underline{\underline{\tau}}^{+}\underline{\underline{A}}_1\underline{\underline{\tau}} = (\underline{\underline{G}}_2)^{-1} i(\underline{\underline{G}}_2 - \underline{\underline{G}}_2^{+})(\underline{\underline{G}}_2^{+})^{-1} \tag{8.9a}$$

$$\underline{\underline{\Gamma}}_2 = i(\underline{\underline{\Sigma}}_2 - \underline{\underline{\Sigma}}_2^{+}) = i\underline{\underline{\tau}}(\underline{\underline{G}}_2 - \underline{\underline{G}}_2^{+})\underline{\underline{\tau}}^{+} = \underline{\underline{\tau}}\underline{\underline{A}}_2\underline{\underline{\tau}}^{+} = (\underline{\underline{G}}_1)^{-1} i(\underline{\underline{G}}_1 - \underline{\underline{G}}_1^{+})(\underline{\underline{G}}_1^{+})^{-1} \tag{8.9b}$$

Thus, the inter-QD transmission coefficient $T = \text{trace}\left[\underline{\underline{\Gamma}}_2\underline{\underline{A}}_1\right] = \text{trace}\left[\underline{\underline{\Gamma}}_1\underline{\underline{A}}_2\right]$ can be written, using eq. (8.8b) and (8.9b), as

$$\begin{aligned} T = \text{trace}\left[\underline{\underline{\Gamma}}_2\underline{\underline{A}}_1\right] &= \text{trace}\left[\left((\underline{\underline{G}}_1)^{-1} i(\underline{\underline{G}}_1 - \underline{\underline{G}}_1^{+})(\underline{\underline{G}}_1^{+})^{-1}\right) i(\underline{\underline{G}}_1 - \underline{\underline{G}}_1^{+})\right] \\ &= -\,\text{trace}\left[\left((\underline{\underline{G}}_1^{+})^{-1} - (\underline{\underline{G}}_1)^{-1}\right)(\underline{\underline{G}}_1 - \underline{\underline{G}}_1^{+})\right] \\ &= -\,\text{trace}\left[((\underline{\underline{G}}_1^{+})^{-1}\underline{\underline{G}}_1 - \underline{\underline{I}}) - (\underline{\underline{I}} - (\underline{\underline{G}}_1)^{-1}\underline{\underline{G}}_1^{+})\right] \end{aligned}$$

$$\Rightarrow T = \text{trace}\left[2\underline{\underline{I}} - \left((\underline{\underline{G}}_1^{+})^{-1}\underline{\underline{G}}_1 + (\underline{\underline{G}}_1)^{-1}\underline{\underline{G}}_1^{+}\right)\right] \tag{8.10}$$

It is apparent that eq. (8.10) can be similarly expressed in terms of $\underline{\underline{G}}_2$ and such expressions are completely symmetric under an unbiased condition until the reservoir leads are connected. It apparently indicates the inter-QD transmission of electrons to be back-and-forth that thereby creates standing waves spanning over the two QDs resulting in their superposition, in general, unless the Green's functions are real leading to $T = 0$. However, the latter condition represents no possible transmission of electron between the QDs indicating these to be isolated.

It is imperative to note that the single states of the two QDs, when tunnel-coupled, create two states: (1) symmetric (called 'bonding') and (2) anti-symmetric (called 'anti-bonding'), which are distributed over, or common to, both the QDs, and differing by an amount of energy called *anti-crossing*. Although usually the energy difference between the QD states ($\Delta\varepsilon = \varepsilon_1 - \varepsilon_2$) is often called *detuning* of the two QDs and their coupling strength to be *anti-crossing*, in the practical devices, they are a little more complex and inter-dependent. Owing to the inter-QD back-and-forth tunnelling or the feedback of QD-2 on QD-1, and *vice versa*, through their coupling, as shown in eqs. (8.6.a) and (8.6.b), the *anti-crossing* and *detuning* continuously impact each other to reach to a stability. As a result, the *anti-crossing* and *detuning* axes suffer an angular shift (i.e. a rotation) from the energy/QD-state-gap axes. This is shown in Fig. 8.6(a) and (b), which plot the variation of normalized spectral densities (DOS in /energy unit) of electron in two single-state coupled QDs, i.e. QD-1 and QD-2, respectively, with energy (E) and QD-state-gap ($\Delta\varepsilon$). The complementary probability of the position of electron to be in QD-1 and QD-2 (which is not always either/or but in both with some fractional values) is also apparent from Fig. 8.6.

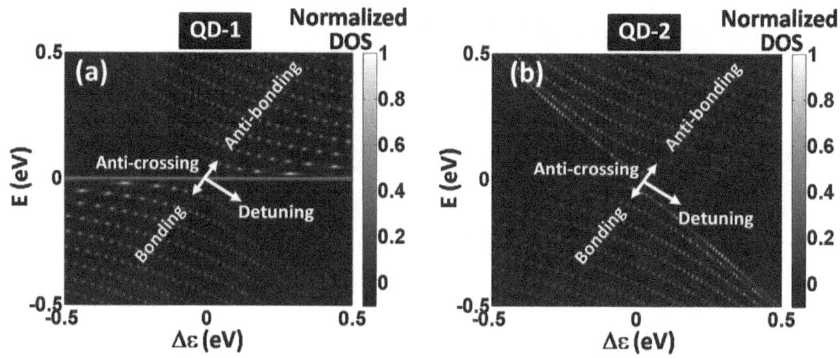

Fig. 8.6: The plots of the variation of normalized spectral densities (DOS in /energy unit) of electron with energy (E) and QD-state-gap ($\Delta\varepsilon$) for two single-state coupled QDs: (a) in QD-1 and (b) in QD-2.

8.4.1 Solution of coupled equations: workable condition for inter-QD coupling

The expressions of coupled Green's functions for the QDs given by eqs. (8.6a) and (8.6b) lead to:

$$\underline{\underline{I}} = (E\underline{\underline{I}} - \underline{\underline{\varepsilon}}_1)\underline{\underline{G}}_1 - \underline{\underline{\tau}}\,\underline{\underline{G}}_2\underline{\underline{\tau}}^+\underline{\underline{G}}_1 \tag{8.11a}$$

$$\underline{\underline{I}} = \underline{\underline{G}}_2(E\underline{\underline{I}} - \underline{\underline{\varepsilon}}_2) - \underline{\underline{G}}_2\underline{\underline{\tau}}^+\underline{\underline{G}}_1\underline{\underline{\tau}} \tag{8.11b}$$

Now, by right-multiplying eq. (8.11a) and left-multiplying eq. (8.11b) with $\underline{\underline{\tau}}$, we get:

$$\underline{\underline{\tau}} = (E\underline{\underline{I}} - \underline{\underline{\varepsilon}}_1)\underline{\underline{G}}_1\underline{\underline{\tau}} - \underline{\underline{\tau}}\,\underline{\underline{G}}_2\underline{\underline{\tau}}^+\underline{\underline{G}}_1\underline{\underline{\tau}} \tag{8.12a}$$

$$\underline{\underline{\tau}} = \underline{\underline{\tau}}\,\underline{\underline{G}}_2(E\underline{\underline{I}} - \underline{\underline{\varepsilon}}_2) - \underline{\underline{\tau}}\,\underline{\underline{G}}_2\underline{\underline{\tau}}^+\underline{\underline{G}}_1\underline{\underline{\tau}} \tag{8.12b}$$

which eventually give rise to:

$$\underline{\underline{\tau}}\,\underline{\underline{G}}_2(E\underline{\underline{I}} - \underline{\underline{\varepsilon}}_2) = (E\underline{\underline{I}} - \underline{\underline{\varepsilon}}_1)\underline{\underline{G}}_1\underline{\underline{\tau}}$$

$$\Rightarrow \underline{\underline{\tau}}\,\underline{\underline{G}}_2\underline{\underline{\tau}}^+ = (E\underline{\underline{I}} - \underline{\underline{\varepsilon}}_1)\underline{\underline{G}}_1\underline{\underline{\tau}}(E\underline{\underline{I}} - \underline{\underline{\varepsilon}}_2)^{-1}\underline{\underline{\tau}}^+$$

$$\Rightarrow \underline{\underline{\tau}}\,\underline{\underline{G}}_2\underline{\underline{\tau}}^+\underline{\underline{G}}_1 = (E\underline{\underline{I}} - \underline{\underline{\varepsilon}}_1)\underline{\underline{G}}_1\underline{\underline{\tau}}(E\underline{\underline{I}} - \underline{\underline{\varepsilon}}_2)^{-1}\underline{\underline{\tau}}^+\underline{\underline{G}}_1$$

$$\Rightarrow (E\underline{\underline{I}} - \underline{\underline{\varepsilon}}_1)\underline{\underline{G}}_1 - \underline{\underline{I}} = (E\underline{\underline{I}} - \underline{\underline{\varepsilon}}_1)\underline{\underline{G}}_1\underline{\underline{\tau}}(E\underline{\underline{I}} - \underline{\underline{\varepsilon}}_2)^{-1}\underline{\underline{\tau}}^{+}\underline{\underline{G}}_1 \qquad \text{(since eq. (8.11a))}$$

$$\Rightarrow \underline{\underline{G}}_1\underline{\underline{\tau}}(E\underline{\underline{I}} - \underline{\underline{\varepsilon}}_2)^{-1}\underline{\underline{\tau}}^{+}\underline{\underline{G}}_1 - \underline{\underline{G}}_1 + (E\underline{\underline{I}} - \underline{\underline{\varepsilon}}_1)^{-1} = \underline{\underline{0}} \qquad (8.13)$$

which can be written in the form of a general equation as:

$$\underline{\underline{G}}_1\underline{\underline{\beta}}\,\underline{\underline{G}}_1 - \underline{\underline{G}}_1 + \underline{\underline{\alpha}} = \underline{\underline{0}} \qquad (8.14)$$

where

$$\begin{cases} \underline{\underline{\alpha}} = (E\underline{\underline{I}} - \underline{\underline{\varepsilon}}_1)^{-1} \\ \underline{\underline{\beta}} = \underline{\underline{\tau}}(E\underline{\underline{I}} - \underline{\underline{\varepsilon}}_2)^{-1}\underline{\underline{\tau}}^{+} \end{cases} \qquad (8.15)$$

The quadratic matrix eq. (8.14) can be factorized, for a non-singular matrix $\underline{\underline{a}}$, as:

$$(\underline{\underline{G}}_1\underline{\underline{a}} + \underline{\underline{c}})(\underline{\underline{b}}\,\underline{\underline{G}}_1 + \underline{\underline{a}}^{-1}) - \underline{\underline{0}}$$

$$\Rightarrow \underline{\underline{G}}(\underline{\underline{a}}\,\underline{\underline{b}})\underline{\underline{G}} + (\underline{\underline{I}} + \underline{\underline{c}}\,\underline{\underline{b}})\underline{\underline{G}} + (\underline{\underline{c}}\,\underline{\underline{a}}^{-1}) = \underline{\underline{0}} \qquad (8.16)$$

Comparing eqs. (8.14) and (8.16), we get the following relations:

$$\begin{cases} \underline{\underline{a}}\,\underline{\underline{b}} = \underline{\underline{\beta}} \\ \underline{\underline{c}}\,\underline{\underline{b}} = -2\underline{\underline{I}} \\ \underline{\underline{c}}\,\underline{\underline{a}}^{-1} = \underline{\underline{\alpha}} \end{cases} \qquad (8.17)$$

which leads to the condition for eq. (8.14) to have finite solution as:

$$\underline{\underline{\alpha\beta}} = -2\underline{\underline{I}}$$

$$\Rightarrow (E\underline{\underline{I}} - \underline{\underline{\varepsilon}}_1)^{-1}\underline{\underline{\tau}}(E\underline{\underline{I}} - \underline{\underline{\varepsilon}}_2)^{-1}\underline{\underline{\tau}}^{+} = -2\underline{\underline{I}} \qquad (8.18)$$

since eq. (8.15). Thus, considering $\Delta\underline{\underline{\varepsilon}} = (\underline{\underline{\varepsilon}}_2 - \underline{\underline{\varepsilon}}_1)$, eq. (18) can be rewritten as:

$$\underline{\underline{\tau}}(E\underline{\underline{I}} - \underline{\underline{\varepsilon}}_2)^{-1}\underline{\underline{\tau}}^{+} = -2(E\underline{\underline{I}} - \underline{\underline{\varepsilon}}_2) - 2\Delta\underline{\underline{\varepsilon}}$$

$$\Rightarrow \underline{\underline{\tau}}(E\underline{\underline{I}} - \underline{\underline{\varepsilon}}_2)^{-1}\underline{\underline{\tau}}^{+}(E\underline{\underline{I}} - \underline{\underline{\varepsilon}}_2)^{-1} + 2\Delta\underline{\underline{\varepsilon}}(E\underline{\underline{I}} - \underline{\underline{\varepsilon}}_2)^{-1} + 2\underline{\underline{I}} = \underline{\underline{0}} \qquad (8.18a)$$

In order to solve eq. (8.18a), a simple method is to suppose (assuming two real numbers (r, φ)):

$$[\underline{\underline{\tau}}(E\underline{\underline{I}} - \underline{\underline{\varepsilon}}_2)^{-1} + \sqrt{2}re^{-i\varphi}\underline{\underline{I}}][\underline{\underline{\tau}}^{+}(E\underline{\underline{I}} - \underline{\underline{\varepsilon}}_2)^{-1} + \sqrt{2}r^{-1}e^{i\varphi}\underline{\underline{I}}] = \underline{\underline{0}} \qquad (8.18b)$$

so that:

$$\Delta\underline{\underline{\varepsilon}} = \frac{1}{\sqrt{2}}(\underline{\underline{\tau}} r^{-1} e^{i\varphi} + \underline{\underline{\tau}}^{+} r e^{-i\varphi}) \tag{8.19}$$

Now since eq. (8.18b) and eq. (8.19), the energy solutions are obtained as:

$$\underline{\underline{\mathrm{EI}}} = \begin{cases} \underline{\underline{\varepsilon}}_2 - \frac{1}{\sqrt{2}}\underline{\underline{\tau}} r^{-1} e^{i\varphi} \\ \underline{\underline{\varepsilon}}_2 - \frac{1}{\sqrt{2}}\underline{\underline{\tau}}^{+} r e^{-i\varphi} \end{cases} = \begin{cases} \underline{\underline{\varepsilon}}_1 + \frac{1}{\sqrt{2}}\underline{\underline{\tau}}^{+} r e^{-i\varphi} \\ \underline{\underline{\varepsilon}}_1 + \frac{1}{\sqrt{2}}\underline{\underline{\tau}} r^{-1} e^{i\varphi} \end{cases} \tag{8.20}$$

Averaging the equal expressions we get the solutions as:

$$\underline{\underline{\mathrm{EI}}} = \begin{cases} \frac{1}{2}(\underline{\underline{\varepsilon}}_1 + \underline{\underline{\varepsilon}}_2) - \frac{1}{\sqrt{2}}(\underline{\underline{\tau}} r^{-1} e^{i\varphi} - \underline{\underline{\tau}}^{+} r e^{-i\varphi}) \\ \frac{1}{2}(\underline{\underline{\varepsilon}}_1 + \underline{\underline{\varepsilon}}_2) + \frac{1}{\sqrt{2}}(\underline{\underline{\tau}} r^{-1} e^{i\varphi} - \underline{\underline{\tau}}^{+} r e^{-i\varphi}) \end{cases} \tag{8.20a}$$

Now the real part of $\underline{\underline{\tau}}$ along with (r, r^{-1}) creates the 'bonding' and 'anti-bonding' states, whereas φ and the phase of $\underline{\underline{\tau}}$ together contribute to the level broadening to such states. It is worthy to note that such 'bonding' and 'anti-bonding' states meet each other at a diabolic point for $r = 1$ where the anti-crossing is minimum (note that it still has a phase degeneracy).

Now, for qubit operations, only the excess single electron of the coupled QDs are subjected to tunnel back-and-forth between each other for *manipulation* (i.e. the relevant logic operations), so that near diabolic point (i.e. $r \approx 1$) eq. (8.19) gives rise to:

$$\Delta\varepsilon = \frac{1}{\sqrt{2}}(\tau e^{i\varphi} + \tau^{*} e^{-i\varphi}) = \frac{1}{\sqrt{2}}[2\mathrm{Re}(\tau e^{i\varphi})]$$

$$\Rightarrow |\varepsilon_2 - \varepsilon_1| \le \sqrt{2}|\tau| \tag{8.21}$$

It is worthy to note that eq. (8.21) is the workable condition for appropriate *manipulation* of the qubit states. Therefore, the materials and geometry of the devices as well as the gate biases for gate-defined QDs are suggested to be engineered on the basis of this fundamental equation.

8.5 Manipulation of DQD-based single qubit: physical action for logic gate operation

It may be important to recall that the only non-trivial logic operation applicable on a single classical *bit* is a NOT gate that transforms the state-0 to state-1 and vice versa. Similarly, a quantum NOT gate applied on a qubit results in the transformations $|0\rangle \overset{NOT}{\rightarrow} |1\rangle$

and $|1\rangle \xrightarrow{NOT} |0\rangle$. If such states are represented as vectors $|0\rangle \equiv \begin{pmatrix} 1 \\ 0 \end{pmatrix}$ and $|1\rangle \equiv \begin{pmatrix} 0 \\ 1 \end{pmatrix}$, then the quantum NOT gate can be expressed as the Pauli matrix $\sigma_1 \equiv \begin{pmatrix} 0 & 1 \\ 1 & 0 \end{pmatrix}$ (this is why it is often called Pauli X-gate). It is apparent that the application of such a NOT gate on a general qubit superposed state, as given by eq. (8.1), i.e. $|\psi\rangle = \cos(\theta/2)|0\rangle + e^{i\varphi} \sin(\theta/2)|1\rangle$, changes it to $\left|\tilde{\psi}\right\rangle = \cos(\theta/2)|1\rangle + e^{i\varphi}\sin(\theta/2)|0\rangle$, thereby eventually altering the probabilities of the states $|0\rangle$ and $|1\rangle$. The question is how such a logic operation can be performed on a physical qubit. Considering the example of coupled QD system (see Fig. 8.5(b) and (c)), this can be done simply by appropriately varying the gate voltages V_{G1} and V_{G2} as shown on the Bloch sphere plotted in Fig. 8.7(a) and (b).

It may be noted that the probability of an electron to be in state $|0\rangle$ (i.e. distributed over QD-1) or $|1\rangle$ (i.e. distributed over QD-2) can be obtained in the NEGF framework from the energy integral of its local spectral densities given by eqs. (8.7a) and (8.7b) while the corresponding phases are included in the local Green's functions. Now, as shown in the plotted Bloch sphere representations of probability and phase in Fig. 3(a) and (b), a positive V_{G1} and a negative V_{G2} suggest a greater probability of $|0\rangle$ (i.e. QD-1), while a negative V_{G1} with a positive V_{G2} results in a greater probability of $|1\rangle$ (i.e. QD-2), so that such a simultaneous change from the former to the latter range of values of V_{G1} and V_{G2} effectively operates as a NOT gate.

It is imperative to note at this point that while changing the qubit state from $|\psi\rangle = \cos(\theta/2)|0\rangle + e^{i\varphi}\sin(\theta/2)|1\rangle$ to $\left|\tilde{\psi}\right\rangle = \cos(\theta/2)|1\rangle + e^{i\varphi}\sin(\theta/2)|0\rangle$ through a NOT gate, what is often overlooked is that the phase difference between states $|0\rangle$ and $|1\rangle$ becomes opposite (i.e. a 2φ-shift) as well. Such change in phase between the so-called OFF and ON states is completely out of the scope of classical *bit* operations, and considered as an important ingredient of *qubit* logic operations. Pauli Z-gate (i.e.

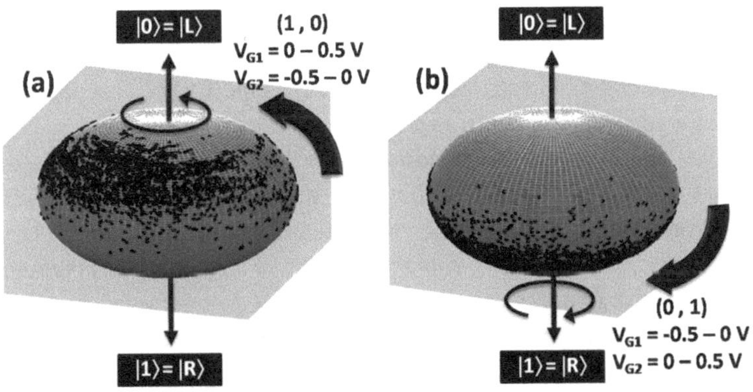

Fig. 8.7: The Bloch sphere of the DQD system plotting the simulated qubit states by varying the gate voltages (V_{G1} and V_{G2}) to make the electron more probabilistic (with different phases) to remain in (a) QD-1 and (b) QD-2.

$\sigma_3 \equiv \begin{pmatrix} 1 & 0 \\ 0 & -1 \end{pmatrix}$), S-gate (i.e. $S \equiv \begin{pmatrix} 1 & 0 \\ 0 & i \end{pmatrix}$), and T-gate (i.e. $T \equiv \begin{pmatrix} 1 & 0 \\ 0 & e^{i\pi/4} \end{pmatrix}$) are examples of such phase changing gates and can be generalized as $P(\varphi) \equiv \begin{pmatrix} 1 & 0 \\ 0 & e^{i\varphi} \end{pmatrix}$. From the perspective of physical operation on practical qubits such as the above DQD system, it can be observed from Fig. 8.7(a) and (b) that, either in the range of $V_{G1} > 0, V_{G2} < 0$ for northern Bloch hemisphere or $V_{G1} < 0, V_{G2} > 0$ for southern Bloch hemisphere, one can vary V_{G1}, V_{G2} to get any desired phase difference between $|0\rangle$ and $|1\rangle$ states of the qubit, and thereby can design any generalized phase-gate. Similarly, the Hadamard gate defined by $H \equiv \frac{1}{\sqrt{2}} \begin{pmatrix} 1 & 1 \\ 1 & -1 \end{pmatrix}$ that can create a superposed state of $|0\rangle$ and $|1\rangle$ while operating on any one of them can be achieved by changing the range of gate voltages from $V_{G1} > 0, V_{G2} < 0$ or $V_{G1} < 0, V_{G2} > 0$ to $V_{G1} > 0, V_{G2} > 0$, which is further analysed in the following section.

8.6 Coupling of DQD with contact leads: read-out

It is imperative to note that although the *manipulation* step of single-qubit operation is usually assumed to be an adiabatic evolution that can be described by linear Schrödinger equation and unitary operators, we have shown above why it is indeed not so trivial to model it in physical devices, where two basis states are continuously *feedbacking* each other while interacting. On the other hand, it is obvious that the *read-out* process, and in cases also the *initialization*, are inherently non-equilibrium in nature, and are, at least conventionally, not believed to be explicable on the basis of Schrödinger evolution. Not going into such debate, rather to say that in this context, NEGF framework is possibly the most efficient technique to theoretically describe and simulate such phenomena. Although it does not (and in fact cannot) describe the result of each single measurement run which is completely random collapse to either of the states, but can model the expectation values of such results (i.e. of a large number of measurement runs) to quite good degree of accuracy accommodating all the physical phenomena that turn out in the physical device during such *measurement* processes.

Physically to execute the processes of *initialization* and *read-out* operations on such qubit devices, for instance, the abovementioned tunnel-coupled QDs need to be connected with two reservoirs, say source (i.e. connected to QD-1) and drain (i.e. connected to QD-2), through corresponding contact leads. A voltage $V_{G1}(0)$ applied on gate-1 can make an electron (from the source) enter the lowest unoccupied quantum state in the discretized conduction band of QD-1 that can be considered as *initialization* (i.e. $|0\rangle$). Similarly after manipulating the electronic superposed state by appropriately varying V_{G1}, V_{G2} as described in the previous section, if a negligibly small bias (V_D) is applied at the drain, the electron, which is positionally distributed over both the QDs, will collapse to either QD-1 or QD-2 and accordingly tunnel into either source or drain leading to probabilistic quantum *read-outs*. Such transport of electron is a non-equilibrium process in general that is modelled in NEGF formalism by incorporating the relevant self-energies of the contact leads corresponding to source and drain. Mathematically this can be done, after getting the self-consistent coupled-$\underline{\underline{G}}_1(E), \underline{\underline{G}}_2(E)$ (by running iterative program loops from eqs. (8.6a) and (8.6b)), by modifying the net Green's function as:

$$\underline{\underline{G}}(E) = \begin{pmatrix} \underline{\underline{G}}_1^{-1}(E) - \underline{\underline{\Sigma}}_{R1}(E) & 0 \\ 0 & \underline{\underline{G}}_2^{-1}(E) - \underline{\underline{\Sigma}}_{R2}(E) \end{pmatrix}^{-1} \tag{8.22}$$

where $\underline{\underline{\Sigma}}_{R1}$ and $\underline{\underline{\Sigma}}_{R2}$ are the self-energies of source and drain, respectively. The off-diagonal block elements of the qubit Green's function are zero to ensure no effect of source on 'QD-2 → drain' tunnelling probability and that of drain on 'QD-1 → source' tunnelling probability. Now, the qubit level broadening due to source is $\underline{\underline{\Gamma}}_{R1} = i(\underline{\underline{\Sigma}}_{R1} - \underline{\underline{\Sigma}}_{R1}^{+})$, which is also a measure of electron escape/enter rate to(/from) source from(/to) QD-1, that leads to the probability of such electron transmission to be $\underline{\underline{\Gamma}}_{R1}\underline{\underline{G}}$. Similarly the electron transmission probability from QD-2 to drain is given by $\underline{\underline{\Gamma}}_{R2}\underline{\underline{G}}^{+}$. Thus, using Landauer formula, the net *read-out* current to drain can be expressed by:

$$I = \frac{e}{h}\int \mathrm{trace}(\underline{\underline{\Gamma}}_{R1}\underline{\underline{G}} - \underline{\underline{\Gamma}}_{R2}\underline{\underline{G}}^{+})(f_S(E) - f_D(E))dE \tag{8.23}$$

$f_S(E)$ and $f_D(E)$ being the Fermi-Dirac distribution functions of source and drain, respectively. It is evident that a positive or negative drain current accordingly indicates collapse of the qubit state to QD-2 (i.e. $|1\rangle$) or QD-1 (i.e. $|0\rangle$).

In a physical qubit made up of two tunnel-coupled QDs, the energy variation of 'bonding' and 'anti-bonding' states as a function of their state-energy gap, as shown in Fig. 8.6(a) and (b), is manifested in practice as a pulse *read-out* current mapping on the plane of two gate voltages. This is commonly known to be the stability diagram of a physical qubit, which is illustrated for the DQD system considered in Fig. 8.8. It is interesting to note that numerous experiments on DQD qubits have indeed demonstrated such stability diagrams [25–28].

It may further be noted that the electronic quantum state-energy values of two QDs and eventually their gap as well as the back-and-forth tunnelling between such QDs can be manipulated by appropriately varying the two gate-voltages (V_{G1}, V_{G2}). That is why the *read-out* current that measures the same electron occupying such states maps onto the gate-voltage plane in the same manner revealing the 'bonding' and 'anti-bonding' states in the stability diagram. It can be observed from the plot of Fig. 8.8 that the regions defined by $V_{G1} > 0, V_{G2} < 0$ or $V_{G1} < 0, V_{G2} > 0$, representing the electron to be most probably confined to QD-1 or QD-2, respectively, exhibit negligibly small negative or positive current. This is attributed to the large state-energy gap be-

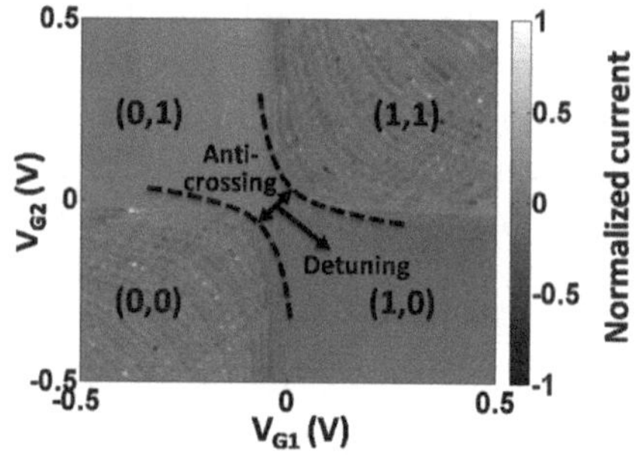

Fig. 8.8: The stability diagram of the DQD qubit simulated using NEGF formalism showing how the anti-crossing and detuning as well as the charge state occupancy are manifested in the *read-out* current map on the parameter space of gate voltages.

tween QDs for such values of gate-voltages resulting in a negligible inter-dot tunnelling (see eq. (8.21)). The exact zero value of current in such regions accordingly indicates $|0\rangle$ and $|1\rangle$, any of which can always be considered as the *initialization* state of the qubit. Varying the gate-voltages within such domains and measuring the current represent the logic operations and their *read-outs* in the basis of $|0\rangle$ and $|1\rangle$.

On the other hand, the other two regions, i.e. $V_{G1} < 0, V_{G2} < 0$ and $V_{G1} > 0, V_{G2} > 0$, physically realize the 'bonding' and 'anti-bonding' states (see the dashed lines in Fig. 8.8) owing to strong tunnel-coupling between the QDs. Apparently, the stability diagram mapping can be considered to depict the equivalent band-structure of a DQD system and the 'bonding' and 'anti-bonding' states to be the band maxima and minima, respectively, that can be represented by $1/\sqrt{2}(|0\rangle + |1\rangle)$ and $1/\sqrt{2}(|0\rangle - |1\rangle)$. Thus, by varying gate voltages in these domains, one can implement logic operations in such basis states. Furthermore, if the initial state is $|0\rangle$ or , just making V_{G1} positive → negative or negative → positive, accordingly, will transform the state to $1/\sqrt{2}(|0\rangle + |1\rangle)$ or $1/\sqrt{2}(|0\rangle - |1\rangle)$, representing the Hadamard gate operation. On the other hand, keeping any of such regions intact, varying the gate-voltages results in a corresponding change in probability or phase or both. Thus, such DQD system can realize, in practice, *initialization* of a qubit state, all the relevant logic operations, and appropriate *read-outs* in an all-electrical means, and such physical operations can be mathematically described by NEGF formalism.

References

[1] Nielsen, M. A. and Isaac, L. Chuang. *Quantum computation and quantum information*, Cambridge university press, 2010.

[2] Hidary, J. D. *Quantum computing: An applied approach*. Vol. 1, Cham, Switzerland: Springer, 2021.

[3] DiVincenzo, D. P. "Topics in quantum computers." *Mesoscopic electron transport*, 657–677. Dordrecht: Springer Netherlands, 1997.

[4] Knill, E., Laflamme, R., and Milburn, G. J. "A scheme for efficient quantum computation with linear optics" *Nature* 409, no. 6816 (2001): 46–52.

[5] Takeuchi, S. "Recent progress in single-photon and entangled-photon generation and applications." *Japanese Journal of Applied Physics* 53, no. 3 (2014): 030101.

[6] Slussarenko, S. and Pryde, G. J. "Photonic quantum information processing: A concise review." *Applied physics reviews* 6, no. 4 (2019).

[7] Bruzewicz, C. D., Chiaverini, J., McConnell, R., and Sage, J. M. "Trapped-ion quantum computing: Progress and challenges." *Applied physics reviews* 6, no. 2 (2019).

[8] Vuletić, V., Chan, H. W., and Black, A. T. "Three-dimensional cavity Doppler cooling and cavity sideband cooling by coherent scattering." *Physical Review A* 64, no. 3 (2001): 033405.

[9] Bardeen, J., Cooper, L. N., and Schrieffer, J. R. "Theory of superconductivity." *Physical review* 108, no. 5 (1957): 1175.

[10] Krantz, P., Kjaergaard, M., Yan, F., Orlando, T. P., Gustavsson, S., and Oliver, W. D. "A quantum engineer's guide to superconducting qubits." *Applied Physics Reviews* 6, no. 2 (2019).

[11] Kjaergaard, M., Schwartz, M. E., Braumüller, J., Krantz, P., Wang, J. I. J., Gustavsson, S., and Oliver, W. D. "Superconducting qubits: Current state of play." *Annual Review of Condensed Matter Physics* 11, no. 1 (2020): 369–395.

[12] Gao, Y. Y., Rol, M. A., Touzard, S., and Wang, C. "Practical guide for building superconducting quantum devices." *PRX quantum* 2, no. 4 (2021): 040202.

[13] Braginsky, V. B., Vorontsov, Y. I., and Thorne, K. S. "Quantum nondemolition measurements." *Science* 209, no. 4456 (1980): 547–557.

[14] Loss, D. and DiVincenzo, D. P. "Quantum computation with quantum dots." *Physical Review A* 57, no. 1 (1998): 120.

[15] DiVincenzo, D. P. "Double quantum dot as a quantum bit." *Science* 309, no. 5744 (2005): 2173–2174.

[16] Van Houten, H. and Beenakker, C. "Quantum point contacts." *Physics today* 49, no. 7 (1996): 22–27.

[17] Burkard, G., Ladd, T. D., Pan, A., Nichol, J. M., and Petta, J. R. "Semiconductor spin qubits". *Reviews of Modern Physics* 95, no. 2 (2023): 025003.

[18] Neyens, S., Zietz, O. K., Watson, T. F., Luthi, F., Nethwewala, A., George, H. C., Henry, E., et al "Probing single electrons across 300-mm spin qubit wafers". *Nature* 629, no. 8010 (2024): 80–85.

[19] Steinacker, P., Stuyck, N. D., Lim, W. H., Tanttu, T., Feng, M., Serrano, S., Nickl, A., et al "Industry-compatible silicon spin-qubit unit cells exceeding 99% fidelity." *Nature* (2025): 1–7.

[20] Steinacker, P., Tanttu, T., Lim, W. H., Stuyck, N. D., Feng, M., Serrano, S., Vahapoglu, E., et al "Bell inequality violation in gate-defined quantum dots." *Nature communications* 16, no. 1 (2025): 3606.

[21] Sarma, S. D., Freedman, M., and Nayak, C. "Majorana zero modes and topological quantum computation." *npj Quantum Information* 1, no. 1 (2015): 1–13.

[22] Haaf, T., Sebastiaan, L. D., Qingzhen Wang, A. M. B., Liu, C.-X., Kulesh, I., Kim, P., Xiao, D., et al "A two-site Kitaev chain in a two-dimensional electron gas." *Nature* 630, no. 8016 (2024): 329–334.

[23] Nag Chowdhury, B. and Chattopadhyay, S. "Dual-Gate GaAs-Nanowire FET for Room Temperature Charge-Qubit Operation: A NEGF Approach." *Advanced Quantum Technologies* 6, no. 4 (2023): 2200072.

[24] Hashim, A., Nguyen, L. B., Goss, N., Marinelli, B., Naik, R. K., Chistolini, T., Hines, J., et al "A practical introduction to benchmarking and characterization of quantum computers." *PRX quantum* 6, (2025): 030202.

[25] Lai, N. S., Lim, W. H., Yang, C. H., Zwanenburg, F. A., Coish, W. A., Qassemi, F., Morello, A., and Dzurak, A. S. "Pauli spin blockade in a highly tunable silicon double quantum dot." *Scientific reports* 1, no. 1 (2011): 110.

[26] MacQuarrie, E. R., Neyens, S. F., Dodson, J. P., Corrigan, J., Thorgrimsson, B., Nathan Holman, M. P., et al "Progress toward a capacitively mediated CNOT between two charge qubits in Si/SiGe." *npj Quantum Information* 6, no. 1 (2020): 81.

[27] Connors, E. J., Nelson, J., Edge, L. F., and Nichol, J. M. "Charge-noise spectroscopy of Si/SiGe quantum dots via dynamically-decoupled exchange oscillations." *Nature communications* 13, no. 1 (2022): 940.

[28] Zwerver, A. M. J., Krähenmann, T., Watson, T. F., Lampert, L., George, H. C., Ravi Pillarisetty, S. A. B., et al "Qubits made by advanced semiconductor manufacturing". *Nature Electronics* 5, no. 3 (2022): 184–190.

9 Functional connectivity in brain: an NEGF framework for neural signal generation/transmission

Brain, more specifically human brain, is hitherto the most powerful and efficient information-processing system available to mankind. The brain is a complex device (or can be regarded as an assembly of devices) that can acquire information, analyse it, solve various problems based on and associated with it, and draw a set of possible conclusions in order to finally make informed decisions. Most astonishingly, we can understand the functionality of our brain by using our brains only. This suggests that the human brain is a peculiar information-processing device that can also self-incarnate its own methods of actions.

Physically, the average weight of a human brain is about 1.3 kg confined within a volume of approximately 1,200 cc, comprising of around 80 billion nerve cells, i.e. neurons, each with about 10,000 synaptic connections [1, 2]. The entire brain functionality for information processing is attributed to the signal generation and transmission through such neurons and synapses. Not going into the debate on the precise values of these components, we can at least imagine from the orders of these numbers how much complex interconnected networking structure it possesses. And even more interestingly, such gigantic network is managed to work extremely efficiently only at the expense of about 20 W [2] of power (which is less than even that a bedside bulb consumes)!

Eventually, the schemes of information processing tasks performed by neurons and inter-neuron synaptic junctions have been abstracted to mathematical logic system, known as artificial neural networks (ANNss). Such ANN models are computationally applicable to solve numerous problems, e.g. now-a-days ranging up to highly critical problems, without interventions of human brains that usually a human brain is needed for, and thus famously known as artificial intelligence (AI). Such a field of research greatly emerged after the remarkable contribution by Hopfield [3] in 1982, where he proposed that, biological organisms and computers, more precisely the neurobiology and integrated circuits, can be collectivized using a large number of simple neuron-equivalent components, thereby offering an efficient content-addressable memory, which is weakly sensitive to any failure of individual sub-part. Such model has since been advanced to the recent ANNs, which can be appropriately trained with enormous information (from many individual human beings) to provide solutions to the much complex problems that an individual human cannot even do in finite time. However, feeding the computer with large datasets poses the challenge of energy limitations, where human brains are still much ahead of any conventional computers (as mentioned earlier) [2].

 | https://doi.org/10.1515/9783111545950-009

Therefore, if we aim to achieve such performance artificially by means of any electronic or optoelectronic device by mimicking human brain, we need to start from understanding the basic mechanism(s) of information processing and storage in brain. And the core of such information tasks are performed within the neurons, and their synaptic interconnects, through generation and transmission of electrodynamic signals. In this short chapter, we will try to develop a basic mathematical framework, based on non-equilibrium Green's function formalism, to model such signal generation and transmission through neurons and synapses. Then, we will explore how such mechanisms can be subjected to abstraction to develop a Green's function-based neural network model applicable to AI.

9.1 Basic structure of human brain

From our textbook knowledge, it is well-known that the human brain is sub-parted into left and right cerebral hemispheres, interconnected through a network of nerve fibres called corpus callosum. Empirical observations indicate that mostly the left one deals with logical/analytical approach, language and speech, and the right one operates for creativity, imagination, adventure, and holistic perception, while both, being linked with the opposite sides of the body, complement each other for sensory activities [4]. The differences between hemispheres in such activities are owing to the asymmetries in structure and composition of neuronal network.

Further, each of the hemispheres is constructed with three segments, i.e. cerebrum, cerebellum, and brainstem, as shown in Fig. 9.1. The cerebrum (i.e. the largest part of human brain making up ~ 85% of its weight and volume) contains four lobes: frontal (for planning and movement, etc.), parietal (for sensing touch and temperature, etc.), temporal (for hearing and memorizing, etc.), and occipital (for vision). The outer layer of cerebrum, known as the cerebral cortex, is believed to be mostly responsible for superior cognitive ability of human beings compared to other animals; and it has been shown to be just a scaled-up version of that of the primates [1]. Such ability is thus attributed to the extremely large (absolute) numbers of neurons, not their relatively larger numbers, in the enlarged cerebral cortex. On the other hand, the inner content of the cerebral cortex is known to be the limbic system that communicates predominantly with the brainstem, which, in turn, connects the cerebrum to cerebellum and spinal cord, and is considered to be responsible for breathing, heart rate, blood pressure, consciousness and sleep [5]. The cerebellum, whereas, fine-tunes motor movement and controls muscle tone that helps in activities like walking or writing [6]. Further, thalamus and hypothalamus, located above the brainstem, are two separate components of human brain, the former modulating arousal, pain, etc., while the latter regulating smooth muscle tone, heart rate, bladder contraction, etc..

However, hippocampus (belonging to the temporal lobe of cerebral cortex) that plays a significant role in learning, memorizing as-well-as spatial decision-making is

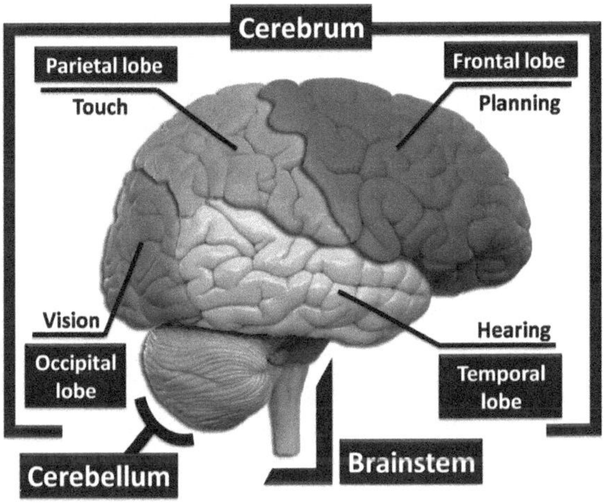

Fig. 9.1: A simple schematic diagram of human brain showing the principal segments along with their key functions.

one of the most interesting part of human brain from the perspective of studying neuropsychiatric disorders like Alzheimer's disease, epilepsy, etc. [7] as well as developing memory-based learning algorithms of ANNs for AI [8]. The parallels in functions of hippocampus and AI agents suggest that the study of these two can complement the comprehensive understanding on both the subjects.

Currently, the 3D neuronal architectures are visualized by reconstruction of the tissues based on light-sheet imaging coupled to brain atlas [9, 10]. Very recently, such method has been scaled up beyond the limit of atlases for unbiased mapping of local neuronal activity as well as connectivity using 3D deep learning model [11]. Conventionally, the neural dynamics in space-time is considered to originate from the array of short- or long-range axonal connections that can be represented as a topologically complex connectivity matrix known to be the connectome. However, recently, the extension of discrete connectome model to neural field theory, constrained by the intrinsic geometry of the brain, has been conceptualized to offer an alternative model that only needs the shape of the brain (and not its detail axonal connectivity matrix) to provide its spatially patterned activity [12]. Nevertheless, it is still essential to model the neuronal activity as well as connectivity from physical point of view in order to understand the mechanism of signal generation and transmission in the brain system.

9.2 Physical mechanism of signal generation and transmission through neurons and synapses

A nerve cell, i.e. the neuron, as shown in Fig. 9.2, consists of three basic components: (1) *soma*, the cell body (containing its nucleus), which acts as the central processing unit, (2) *dendrites*, the branches that receive information from other neurons, and (3) *axon*, which transmits the information, i.e. the signal generated at the soma, to its terminal branches. Such terminal branches of axon of a neuron are connected to the dendrites of other neurons through small junction-gaps called *synapses*. For a comprehensive understanding, a detail on this can be found in [13].

After receiving the signals by dendrites, soma integrates them with relevant weights, and if the integrated signal is sufficiently strong to excite its trigger zone, i.e. axon hillock, an electrical signal, called action potential (kind of a threshold voltage), is fired. Such signal is then transported by the axon to its terminal and communicated to dendrites of other neurons via synapses, either electrically or chemically. Generally, the electrical synaptic information transmission is much rapid compared to the chemical synaptic communication, where in the latter case the signal arrived at axon terminal leads to the release of several chemical messengers, known as the neurotransmitters (that have localized and short-term effect) or neuromodulators (that have long range and long-lasting effect).

The neurotransmitters secreted from the axon terminal of one neuron (i.e. presynaptic) bind with the receptors of the dendrite of another neuron (i.e. post-synaptic) and thereby open up the ion channels to flow through and create small synaptic potentials (often called synaptic strength). This, in turn, results in either an 'excitatory' or an 'inhibitory' effect to, respectively, enhance or reduce the possibility of corresponding action potential generation in the post-synaptic neuron. All the cognitive processes, e.g. sensation, perception, memory, and spatial decision-making, are attributed to the balance between such excitatory and inhibitory neuronal activities. At this point, it is worthy to note that inside and outside the neuron, there are plenty of ions (e.g. K^+, Na^+, Ca^{2+}, and Cl^-) in the aqueous medium that balances the entire electrodynamic environment in and around the neurons.

It is widely considered that learning and memorizing mechanisms of brain are associated with the long-term alteration in the synaptic strength, commonly known to be the synaptic plasticity. It is a capacity to change the synaptic potential between two neurons through their repetitive and consecutive communications. Usually, the synaptic potentials are short term (i.e. lasting a few seconds or minutes); however, repetitive activations of such synaptic strengths can lead to the release of a second or multiple neurotransmitters that alter the very structure of the synapse itself through protein synthesis in the neurons, and thereby result in the long-term potentiation (LTP) for excitatory activation while long-term depression (LTD) for the inhibitory one [13]. Usually, it is believed that the excitatory and inhibitory neurons are of different types of nerve cells (e.g. glutamatergic and GABA-ergic, respectively), so that the excitatory neurons encode information while the inhibitory neurons modulate their processing; and this can be regarded as the cellular origin of learning and memorizing functionality of brain. However, very recently, it has been claimed that the instantaneous gating along with regulatory power of local inhibitory neural circuits can lead to plasticity thereby encoding contextually different concepts [14].

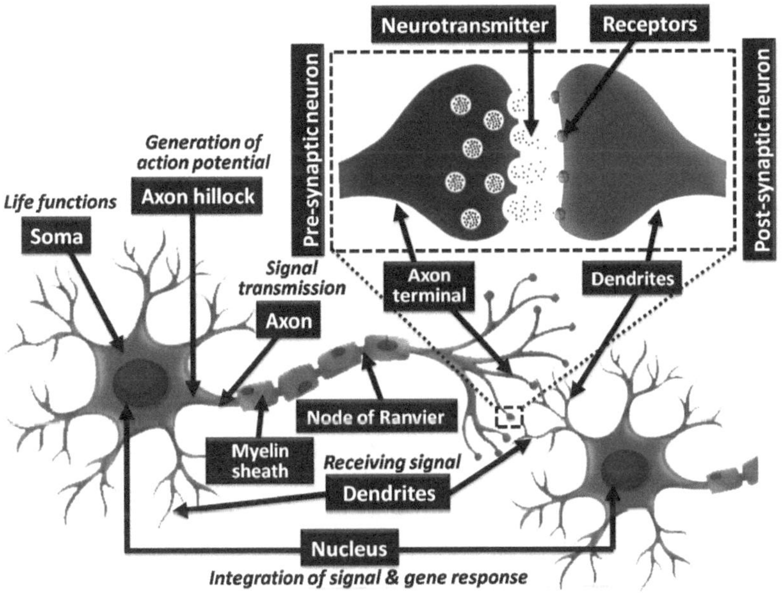

Fig. 9.2: A brief schematic of neuron showing its major constituent parts; the inset depicts the synaptic junction.

9.3 Neuron membrane potential: a basic model for equilibrium

As mentioned, electrical signals are generated at the axon hillocks of neurons and transmitted to the axon terminals, despite the fact that neurons are not good electrical conductors. Such conduction of signals are carried out in the electric potential of ions (e.g. K^+, Na^+, Ca^{2+}, and Cl^-) in the aqueous medium inside and outside the nerve cells, separated from each other by their plasma membranes (see Fig. 9.3(a)). Some intermediate gates in the membrane become ON only if a localized voltage is applied temporarily, e.g. may be due to the transmitting signal, to let the ions exchange through them. It has been observed that, otherwise, the equilibrium potential difference across the membrane (i.e. potential inside in reference to the outside region of neuron walls) varies typically in the range of –40 to –90 mV (e.g. for human neurons it is ~–70 mV), called the resting membrane potential [13]. Such resting potential is usually calculated from Goldman equation which is, in fact, an extrapolation of Nernst equation for electrochemistry (a detail on this can be found in [13]). However, since ion movements through the membrane are generally restricted unless the gates are turned open, an electrostatic model should be more pertinent to comprehend the physical origin of the membrane potential. Here, we are describing such a basic level of formalism and also explain why the order of membrane potential is several tens of mV.

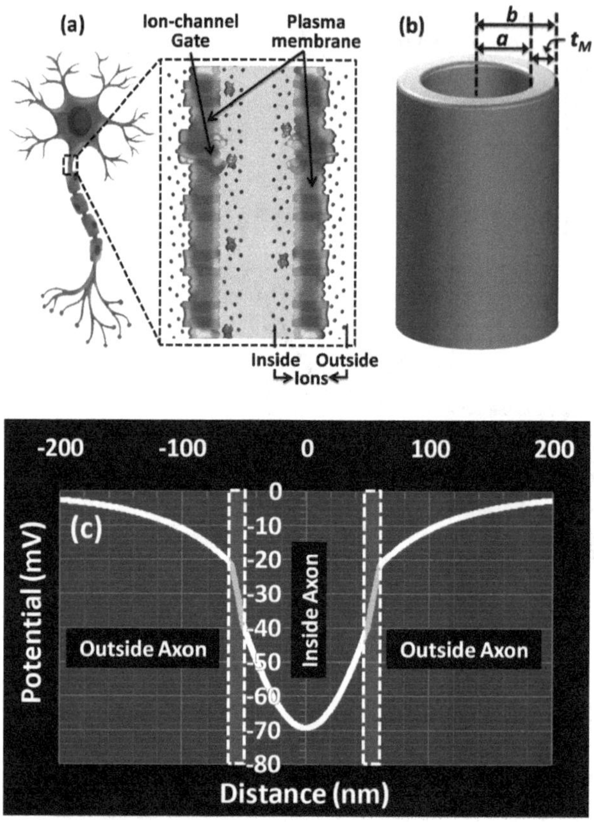

Fig. 9.3: (a) Schematic representation of human neuronal axon surrounded by the plasma membrane containing ion-channel gates; (b) simplified geometrical equivalent structure of such axon for modelling; and (c) simulated result based on the theory developed here showing the plot of potential across the axon.

For basic formulation of the problem, as shown in Fig. 9.3(b), let us consider the neuronal axon to be a long straight cylinder (of radius a) containing a variety of mobile charges, surrounded by a cylindrical shell of plasma membrane (of thickness $t_M = b - a$, b being the outer radius) that isolates it from the outer aqueous medium also filled up with several free charges. Then, since Poisson's equation, the electric potential ($\varphi(\vec{r})$) can be given by

$$\vec{\nabla} \cdot \overleftrightarrow{\varepsilon}_r(\vec{r}) \cdot \vec{\nabla}\varphi(\vec{r}) = -\frac{1}{\varepsilon_0}\sum_j q_j n_j(\vec{r}) \tag{9.1}$$

where $\overleftrightarrow{\varepsilon}_r(\vec{r})$ represents the relative permittivity of the region (assuming separate values for interior of axon, plasma membrane and outside medium), ε_0 is the permittivity of vacuum, and $q_j, n_j(\vec{r})$ are the charge and concentration distribution of jth mobile ion in such regions. Thus, assuming cylindrical symmetry, the radial distribution

(r varying from 0 at the centre of axon to ∞ at a distant point outside neuron) of such potential at any particular point in and around the axon can be expressed as

$$\frac{1}{r}\frac{\partial}{\partial r}\varepsilon_r(r)\left(r\frac{\partial}{\partial r}\varphi(r)\right) = -\frac{1}{\varepsilon_0}\sum_j q_j n_j(r) \tag{9.1a}$$

At a finite temperature T, the equilibrium concentration of ions can be obtained from conventional Boltzmann distribution (K_B is the Boltzmann constant) as

$$n_j(r) = \begin{cases} n^{\mathrm{In}}{}_j(0)e^{-\frac{\langle q\rangle_{\mathrm{In}}(\varphi_{\mathrm{In}}(0)-\varphi(r))}{K_BT}}, & r<a \\ 0, & a\le r\le b \\ n^{\mathrm{Out}}{}_j(\infty)e^{-\frac{\langle q\rangle_{\mathrm{Out}}(\varphi(r)-\varphi_{\mathrm{Out}}(\infty))}{K_BT}}, & r>b \end{cases} \tag{9.2}$$

since the potential is negative inside neuronal axon if it is assumed to be zero (as reference) outside it at a considerably large distance. Here the average charge is given by $\langle q\rangle = \frac{\sum_j q_j \int n_j(\vec{r})dV}{\sum_j \int n_j(\vec{r})dV}$. While the solution of eq. (9.2) within the membrane material (i.e. $a\le r\le b$, which does not contain any free charges) is simply given by

$$\varphi(r) = \varphi_{M0}\ln(k_M r) + \varphi_{M1} \tag{9.3}$$

so that:

$$\frac{\partial}{\partial r}\varphi(r) = \varphi_{M0}\left(\frac{1}{r}\right) \tag{9.3a}$$

$$\frac{1}{r}\frac{\partial}{\partial r}\left(r\frac{\partial}{\partial r}\varphi(r)\right) = 0 \tag{9.3b}$$

interestingly, we found the exact solution inside the axon to be of the form:

$$\varphi(r) = \varphi_0 \ln\left(\frac{2}{(kr)^2+1}\right) \tag{9.4}$$

where $\varphi_0 = \varphi_{\mathrm{In}}(0)/\ln(2)$ so that:

$$\frac{\partial}{\partial r}\varphi(r) = -k^2\varphi_0\left(\frac{2r}{(kr)^2+1}\right) \tag{9.4a}$$

$$\frac{1}{r}\frac{\partial}{\partial r}\left(r\frac{\partial}{\partial r}\varphi(r)\right) = -k^2\varphi_0\left(\frac{2}{(kr)^2+1}\right)^2 \tag{9.4b}$$

which, since eqs. (9.1a) and (9.2), leads to

$$-k^2\varphi_0\left(\frac{2}{(kr)^2+1}\right)^2=-\frac{1}{\varepsilon_0\varepsilon_r^{\text{In}}}\sum_j q_j n^{\text{In}}{}_j(0)e^{-\frac{\langle q\rangle_{\text{In}}(\varphi_{\text{In}}(0)-\varphi(r))}{K_BT}}$$

$$\Rightarrow k^2(\varphi_{\text{In}}(0)/\ln(2))\left(\frac{2}{(kr)^2+1}\right)^2=\left(\sum_j\frac{q_j n^{\text{In}}{}_j(0)}{\varepsilon_0\varepsilon_r^{\text{In}}}\right)e^{-\frac{\langle q\rangle_{\text{In}}\varphi_{\text{In}}(0)}{K_BT}}\left(\frac{2}{(kr)^2+1}\right)^{\frac{\langle q\rangle_{\text{In}}\varphi_{\text{In}}(0)/\ln(2)}{K_BT}}$$

Thus, equating the constant power of the function of radial distance and the constant factor multiplied with it, we, respectively, have

$$\varphi_{\text{In}}(0)=2\ln(2)K_BT/\langle q\rangle_{\text{In}} \tag{9.5}$$

and

$$k^2(\varphi_{\text{In}}(0)/\ln(2))=\left(\sum_j\frac{q_j n^{\text{In}}{}_j(0)}{\varepsilon_0\varepsilon_r^{\text{In}}}\right)e^{-\frac{\langle q\rangle_{\text{In}}\varphi_{\text{In}}(0)}{K_BT}}$$

which, using eq. (9.5), further leads to:

$$k^2=\frac{1}{8}\left(\langle q\rangle_{\text{In}}\sum_j\frac{q_j n^{\text{In}}{}_j(0)}{\varepsilon_0\varepsilon_r^{\text{In}}K_BT}\right)\approx\frac{1}{8}\left(\frac{\langle q\rangle_{\text{In}}{}^2}{\varepsilon_0\varepsilon_r^{\text{In}}K_BT}\sum_j n^{\text{In}}{}_j(0)\right)=\frac{1}{8}\left(\frac{\langle q\rangle_{\text{In}}{}^2 n_{\text{In}}}{\varepsilon_0\varepsilon_r^{\text{In}}K_BT}\right)$$

$$\Rightarrow k^2=1\Big/\left(2\sqrt{2}\lambda_D^{\text{Neu}}\right)^2 \tag{9.6}$$

where $\lambda_D^{\text{Neu}}=\sqrt{1/\frac{\langle q\rangle^2 n_{\text{In}}}{\varepsilon_0\varepsilon_r^{\text{In}}K_BT}}$ is the conventional Debye length of the medium inside the neuronal axon, typically of the order of 1 nm for human neurons. Thus, since eq. (9.4), the final solution inside the axon is obtained as:

$$\varphi(r)=2K_BT/\langle q\rangle_{\text{In}}\ln\left(\frac{2}{1+\left[r/(2\sqrt{2}\lambda_D^{\text{Neu}})\right]^2}\right) \tag{9.7}$$

Now, since we have considered the reference of the potential to be outside the neuron at a considerable distance, i.e. $\varphi_{\text{Out}}^{Ref}(\infty)=0$, the resting membrane potential gets equal to $\varphi_{\text{In}}(0)$ given by eq. (9.5). It may be interesting to note that $2\ln(2)\approx 1.4$, the thermal voltage at room temperature is $K_BT\approx 25$ mV, and the average charge inside neuron is generally negative, which is an weighted average of (+1) for K^+ and Na^+, (+2) for Ca^{2+}, and (–1) for Cl^- as well as protein and organic anions, the concentration of the negative ones being higher. Thus, if such weighted average of charge is ~(–0.5), then the resting membrane potential would be ~–70 mV.

On the other hand, outside the membrane, the net charge for positive and negative ion concentrations do not differ much (although positive as resultant), and thus eq. (9.2) can be approximated to write the Poisson's eq. (9.1a) as:

$$\frac{1}{r}\frac{\partial}{\partial r}\left(r\frac{\partial}{\partial r}\varphi(r)\right) = -\sum_j \frac{q_j n^{\mathrm{Out}}{}_j(\infty)}{\varepsilon_0 \varepsilon_r^{\mathrm{Out}}}\left(1 - \frac{\langle q\rangle_{\mathrm{Out}}(\varphi(r) - \varphi_{\mathrm{Out}}(\infty))}{K_B T}\right)$$

$$\Rightarrow \left[\frac{\partial^2}{\partial r^2} + \frac{1}{r}\frac{\partial}{\partial r} - \frac{1}{(\lambda_D^{\mathrm{Out}})^2}\right](\varphi(r) - \varphi_{\mathrm{Out}}(\infty) - K_B T/\langle q\rangle_{\mathrm{Out}}) = 0 \tag{9.8}$$

which is modified Bessel equation of second kind of zero order, the solution being K_0, since it must be well behaved at infinity. Thus, the solution outside the neuron membrane can be written as:

$$\varphi(r) = K_B T/\langle q\rangle_{\mathrm{Out}} + \varphi_{\mathrm{Out}}(\infty) - C_1 K_0(r/\lambda_D^{\mathrm{Out}}) \tag{9.9}$$

where C_1 is a positive constant. Since the modified Bessel function of second kind of zero order $K_0 \to 0$ as $r \to \infty$, in order to consider $\varphi_{\mathrm{Out}}^{\mathrm{Ref}}(\infty) = 0$, one needs to set $\varphi_{\mathrm{Out}}(\infty) = -K_B T/\langle q\rangle_{\mathrm{Out}}$.

Now, the continuity of electric displacement ($\varepsilon \frac{\partial}{\partial r}\varphi(r)$) at the interface of the different media, i.e. the interior of axon and the inner surface of membrane ($r = a$) as well as its outer surface and the external medium ($r = b$), gives rise to the respective boundary conditions:

$$\varepsilon_{\mathrm{Neu}}\left(\frac{2K_B T/\langle q\rangle_{\mathrm{In}}}{2\sqrt{2}\lambda_D^{\mathrm{Neu}}}\right)\left(\frac{2a/(2\sqrt{2}\lambda_D^{\mathrm{Neu}})}{1 - [a/(2\sqrt{2}\lambda_D^{\mathrm{Neu}})]^2}\right) = \varepsilon_M \varphi_{M0}\left(\frac{1}{a}\right) \tag{9.10a}$$

$$\varepsilon_M \varphi_{M0}\left(\frac{1}{b}\right) = \frac{C_1}{\lambda_D^{\mathrm{Out}}} K_1(b/\lambda_D^{\mathrm{Out}}) \tag{9.10b}$$

since derivative of K_0 is $-K_1$. It is evident that φ_{M0}, and thus C_1, can be obtained straight forward from eqs. (9.10a) and (9.10b), respectively. Accordingly, the potential distribution is simulated assuming the conventional values of the parameters for human neuron (such as dielectric constants of the media, which is much smaller at the membrane compared to cytoplasm and the outer aqueous medium, and membrane thickness ~10 nm) and plotted in Fig. 9.3(c). It may be noted that diameter of human neuronal axon mostly vary in the range of ~100 nm to ~1 μm, but may be up to a few microns. Here, we have considered 100 nm diameter for the simulation.

9.4 Signal generation and transmission in neuron

While the equilibrium potential is maintained within the membrane as shown above, certain electrical signals, called action potential, are generated at the axon hillocks as a result of signals coming from other neurons and external stimuli as discussed earlier in this chapter. Such signals, typically of ~1 ms pulse, are transmitted through the aqueous medium containing the ions (e.g. K^+, Na^+, Ca^{2+}, and Cl^-) inside the neuronal axon toward its terminals. Here we are proposing a preliminary mathematical model to describe how the generation and transmission of such electric pulses can be possible in the aqueous medium of mobile ions. We argue here that the electric field in such an environment within the neuron is a kind of acoustic-electric field and thus a longitudinal wave along the direction of its propagation. The ions oscillate due to such electric field wave and the field propagates through such ionic oscillations. We also argue that the signal transmission is not carried out by ion flow through the membrane gates (as conventionally believed [13]), rather, such ions flow only when gate is turned open due to field amplitude over/below a threshold, and thus such ion flow only controls the field amplitude from decaying out completely before reaching the terminals or from enhancing unaddressed. The signal transmission, on the other hand, is carried out owing to electrodynamic behaviour of the medium of mobile charges inside neuron, which may be imagined to be some kind of a moderate-temperature dense plasma, where ion movements are influenced by the electrical signals and transmission of such signals are carried out by ion movements.

Thus, the motion of the mobile ions inside neuron can be described by the momentum equation of conventional fluid dynamics:

$$\left(\frac{\partial}{\partial t}+\vec{v}_j\cdot\vec{\nabla}\right)(m_jn_j\vec{v}_j)=m_jn_j[(q_j/m_j)\vec{E}+\vec{g}]-\vec{\nabla}P-\gamma_Rm_jn_j\vec{v}_j-\eta\nabla^2\vec{v}_j \tag{9.11}$$

where $q_j, m_j, n_j, \vec{v}_j$ are the charge, mass, concentration, and velocity field of jth ion, respectively. The first term in the RHS of eq. (9.12) represents the forces due to electric field and gravity, while the second term is the pressure gradient, which, considering the velocity of sound in the aqueous medium of ions to be $\beta=\sqrt{\gamma_CP/(m_jn_j)}$ ($\gamma_C=C_P/C_V$), can be written as $\vec{\nabla}P=m_j(\beta^2/\gamma_C)\vec{\nabla}n_j$. The ion oscillations in the electrodynamic field as well as the possible ion flow (net outflow) through membrane gate are incorporated as the third term, where the real part of γ_R represents escape rate of ion flow while its imaginary part describes the oscillation. The last term is for the viscosity of the fluid medium for ions to move through. Here, for simplicity of the problem, we are neglecting the effects of gravitation and viscosity, which, we although believe to be of significance for neural information transmission in reality. Thus, in the simplified model, eq. (9.11) can be written as

$$\left(\frac{\partial}{\partial t}+\vec{v}_j\cdot\vec{\nabla}\right)(n_j\vec{v}_j)=(q_jn_j/m_j)\vec{E}-(\beta^2/\gamma_C)\vec{\nabla}n_j-\gamma_Rn_j\vec{v}_j \tag{9.12}$$

Now, $(\vec{v}_j\cdot\vec{\nabla})(n_j\vec{v}_j)=1/m_j[n_j(m_j\vec{v}_j\cdot\vec{\nabla}\vec{v}_j)+(m_j\vec{v}_j\cdot\vec{v}_j)\vec{\nabla}n_j]-\vec{v}_j\times(\vec{\nabla}\times n_j\vec{v}_j)$

$$=n_j(\vec{\nabla}E_j^K/m_j)+2(E_j^K/m_j)\vec{\nabla}n_j-\vec{v}_j\times(\vec{\nabla}\times n_j\vec{v}_j)$$

If no vortex is formed in the ion medium inside neuronal axon, the last term is zero, while the ionic kinetic energy $E^K{}_j=(1/2)m_jv_j^2$ on an average of all ions gives rise to $(1/2)K_BT$. Therefore, multiplying eq. (9.12) by q_j followed by summing over 'j' for all ions, we get:

$$\frac{\partial}{\partial t}\vec{J}+(1/2)\rho\vec{\nabla}(K_BT/\langle m\rangle)+(K_BT/\langle m\rangle)\vec{\nabla}\rho=\varepsilon_{\text{Neu}}\omega_P{}^2\vec{E}-(\beta^2/\gamma_C)\vec{\nabla}\rho-\gamma_R\vec{J}$$

$$\Rightarrow\left[\frac{\partial}{\partial t}+\gamma_R\right]\vec{J}=\varepsilon_{\text{Neu}}\omega_P{}^2\vec{E}-(\beta^2/\gamma_C-K_BT/\langle m\rangle)\vec{\nabla}\rho-(1/2)(\rho/\langle m\rangle)\vec{\nabla}(K_BT) \tag{9.13}$$

where $\vec{J}=\sum_j q_jn_j\vec{v}_j$ is the current density due to ion movements and $\omega_P=\sqrt{\sum_j q_j^2n_j/\varepsilon_{\text{Neu}}m_j}$ is the equivalent plasma frequency of ionic aqueous medium inside the neuron. Thus, neglecting the temperature gradient ($\vec{\nabla}T$) within the neuron (again a crude approximation) and writing $\beta^2/\gamma_C-K_BT/\langle m\rangle=\tilde{\beta}^2$, we have:

$$\left[\frac{\partial}{\partial t}+\gamma_R\right]\vec{J}=\varepsilon_{\text{Neu}}\omega_P{}^2\vec{E}-\tilde{\beta}^2\vec{\nabla}\rho \tag{9.14}$$

Now, for an electrodynamic field, Maxwell's equations $\vec{\nabla}\times\vec{E}=-\frac{\partial}{\partial t}\vec{B}$ and $\vec{\nabla}\times\vec{B}=\mu\vec{J}+\mu\varepsilon\frac{\partial}{\partial t}\vec{E}$ leads to $\vec{\nabla}\times(\vec{\nabla}\times\vec{E})=-\mu\frac{\partial}{\partial t}\left(\vec{J}-\varepsilon\frac{\partial}{\partial t}\vec{E}\right)$, which, since $\vec{\nabla}\times\vec{E}=\vec{0}$ for a longitudinal propagating electric field, gives rise to:

$$\frac{\partial}{\partial t}(\vec{J}+\varepsilon\frac{\partial}{\partial t}\vec{E})=\vec{0} \tag{9.15}$$

$$\Rightarrow\vec{J}+\varepsilon\frac{\partial}{\partial t}\vec{E}=\vec{J}_0 \tag{9.15a}$$

Using eq. (9.15a), eq. (9.14) can be written as:

$$\left[\frac{\partial^2}{\partial t^2}+\gamma_R\frac{\partial}{\partial t}+\omega_P{}^2\right]\vec{E}-\tilde{\beta}^2\vec{\nabla}(\rho/\varepsilon_{\text{Neu}})+(\gamma_R/\varepsilon_{\text{Neu}})\vec{J}_0 \tag{9.16}$$

Now, $\vec{\nabla}(\rho/\varepsilon)=\vec{\nabla}(\vec{\nabla}\cdot\vec{E})=\nabla^2\vec{E}$ (since $\vec{\nabla}\times\vec{E}=\vec{0}$), which leads to:

$$\left[\left(\frac{\partial^2}{\partial t^2}+\gamma_R\frac{\partial}{\partial t}+\omega_P{}^2\right)-\tilde{\beta}^2\nabla^2\right]\vec{E}=(\gamma_R/\varepsilon_{\text{Neu}})\vec{J}_0 \tag{9.17}$$

Equation (9.17) is the wave equation that describes the acoustic-electric field propagation inside the neuron, where $\sim\vec{J}_0$ is the source term representing the current density that includes the sources at axon hillock while generating the signal as well as the ion inflow/outflow through the membrane gates during its propagation.

Considering the longitudinal electric field $\vec{E}$ to be propagating along Z-axis, and further considering electric field wave to be a function of the space-time coordinate $\tau=(t-z/\tilde{\beta})$, let the corresponding Green's function for eq. (9.17) is given by

$$G(\tau,\tau')=A\vartheta(\tau-\tau')e^{i\omega(\tau-\tau')} \tag{9.18}$$

where A and ω are complex numbers in general. Remembering $\frac{d\tau}{dz}=-1/\tilde{\beta}$, $\frac{d\tau}{dt}=1$, and $\frac{d}{d\tau}\vartheta(\tau-\tau')=\delta(\tau-\tau')$, we have:

$$\left[\frac{\partial^2}{\partial t^2}-\tilde{\beta}^2\nabla^2\right]G(\tau,\tau')=0 \tag{9.19a}$$

$$\left[\gamma_R\frac{\partial}{\partial t}+\omega_P{}^2\right]G(\tau,\tau')=\gamma_R A\delta(\tau-\tau')e^{i\omega(\tau-\tau')}+(i\gamma_R\omega+\omega_P{}^2)G(\tau,\tau') \tag{9.19b}$$

which suggest that $A=1/\gamma_R$ and $i\omega=-\omega_P{}^2/\gamma_R$, so that (noting $\delta(\tau-\tau')e^{i\omega(\tau-\tau')}=\delta(\tau-\tau')$):

$$\left[\left(\frac{\partial^2}{\partial t^2}+\gamma_R\frac{\partial}{\partial t}+\omega_P{}^2\right)-\tilde{\beta}^2\nabla^2\right]G(\tau,\tau')=\delta(\tau-\tau') \tag{9.20}$$

Thus, the Green's function is finally found to be

$$G(\tau,\tau')=(1/\gamma_R)\vartheta(\tau-\tau')e^{-(\omega_P{}^2/\gamma_R)(\tau-\tau')} \tag{9.21}$$

Therefore, considering the electric wave pulse being generated at $t'=0, z'=0$, the space-time spectral density of the electric field is obtained as:

$$D(t,z)=i[G(t-z/\tilde{\beta},0)-G^{+}(t-z/\tilde{\beta},0)] \tag{9.22}$$

The neuronal electrical signal, as obtained from eq. (9.22), is plotted in Fig. 9.4 illustrating how a generated pulse is transmitted through neuronal axon. It should be noted at this point that this model is based on a preliminary theoretical construct, as described above on the basis of our argument, and needs several other factors to be considered for a complete formalism. For instance, we have not considered the dipole interactions of water molecules and ion-dipole interactions in detail, whereas it is assumed to be grossly included in the average dielectric constants. Similarly, we have not incorporated any quantum phenomena here which many researchers (including Penrose, see [20]) consider to be important, especially for understanding the creativity

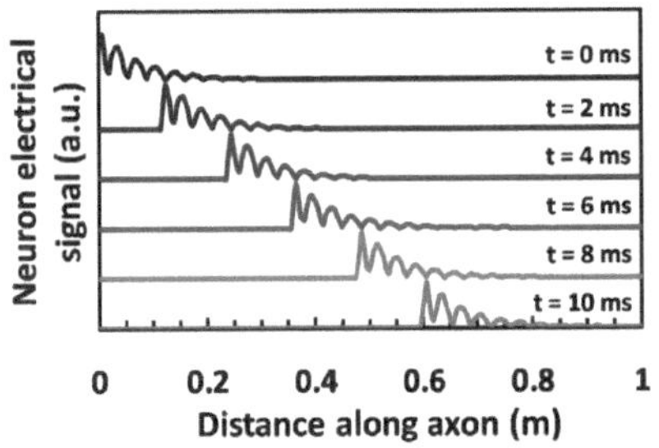

Fig. 9.4: The plots of electrical signal generation and transmission: it is assumed to be generated at $z = 0$ at time $t = 0$ and the simulated results show its transmission in 10 ms (at steps of 2 ms) traveling almost up to ~1 m.

of human brain. We have also not shown here how the neurotransmitters works in synaptic communications, which may be further developed on the basis of the present model, and may be with the help of quantum theory. However, we believe that the overall concept is quite satisfactory to understand the mechanism of signal generation and transmission in neuron, and more particularly, has been able to obtain naturally the orders of signalling speed (~ms) to pass through the entire length of neuron (~m). Finally, the characteristic retarded Green's function obtained here may have interesting implications in replicating the neuronal communication in the form of ANN as briefly illustrated below.

9.5 Learning and memorizing in neural network

The electrical signals, as shown above, are generated at the soma (on the basis of input signals received at its dendrites) by the ion charges and ion currents present in the neuron and are transmitted through the axon to its terminals. As shown above, such signal propagation from (t', z') to (t, z) can be described by the Green's function $G(\tau, \tau')$. The output signal at an axon terminal is then transmitted to the next neural dendrite as its input signal through the corresponding synaptic junction, as illustrated in Fig. 9.5. Such input signal may get a corresponding weightage depending on the synaptic strength, as discussed earlier in this chapter, and being added to other such signals, can generate and transmit new output signal. Thus, the input signals at several dendrites (let $X_j(n)$) of a post-synaptic neuron (suppose at the nth layer) are directly related to the output signals at axon terminals (let $Y_j(n-1)$) of the pre-synaptic neurons (i.e. at the $n-1$th layer) in the entire neural network of tens of billions of neurons. Considering an extensive and linear input-output relation, this can be mathematically expressed apparently as

$$Y_i(n-1) = \sum_j A_{ij} X_j(n) \tag{9.23}$$

where each element of the matrix A_{ij} represents the corresponding synaptic strength or weightage of the pre-synaptic output in the post-synaptic inputs.

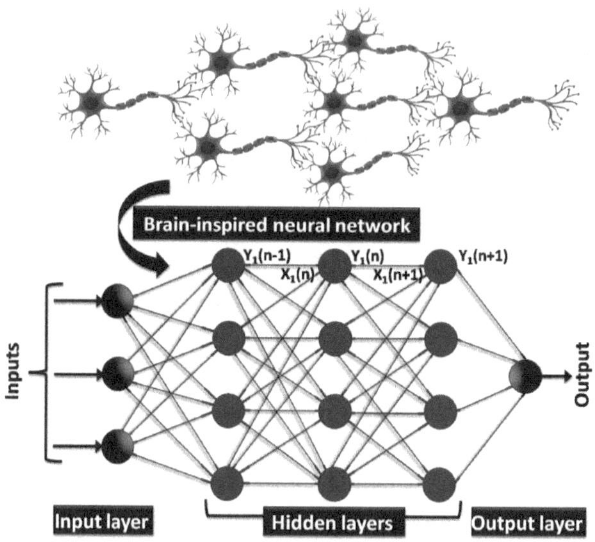

Fig. 9.5: An illustration of artificial neural network with hidden layers for deep learning inspired from biological neural circuit of human brain.

At this point, it is imperative to note that each single output signal $Y_i(n-1)$ of the pre-synaptic neuron transmits into many dendrites $X_j(n)$ of the post-synaptic neurons whereas the reverse is not true in general (i.e. a single dendrite is not connected to many axon terminals). Thus, the inverse relation $X_i(n) = \sum_j a_{ij} Y_j(n-1)$ (where $\sum_j A_{ij} a_{jk} = \delta_{ik}$) is not valid in general. Now, the rate of attenuation of the signal while transporting from one neuron to its consecutive one is given by

$$\frac{\Delta Y_i(n)}{\Delta n} = \frac{Y_i(n-1) - Y_i(n)}{(n) - (n-1)} = \sum_j A_{ij} X_j(n) - Y_i(n) \tag{9.24}$$

Considering the large numbers (i.e. tens of billions) of neurons connected to each other, we can take the limit $\Delta n \to 0$ to define:

$$\frac{d}{dn} Y_i(n) = \sum_j A_{ij} X_j(n) - Y_i(n) \tag{9.25}$$

Further, the numbers of such neurons can be assumed, to a first order approximation, to be proportional to the time required for signal transport, so that $n \approx \tau/\tau_0$. Thus, we get:

$$\left[\frac{d}{d\tau} + 1/\tau_0\right] Y_i(\tau) = (1/\tau_0) \sum_j A_{ij} X_j(\tau) \tag{9.26}$$

Therefore, now the signal transport from each neuron (let ith) to others (let jth) in the entire neural networks can be described based on the non-equilibrium Green's function, given by, $\left[\frac{d}{d\tau} + 1/\tau_0\right] G_{ij}(\tau, \tau') = \delta_{ij}\delta(\tau - \tau')$, so that the output at ith node is obtained as

$$Y_i(\tau) = Y_i(0)e^{-\tau/\tau_0} + (1/\tau_0) \sum_{j,k} \int G_{ij}(\tau, \tau') A_{jk} X_k(\tau') d\tau' \tag{9.27}$$

The first term of RHS of eq. (9.27) is mathematically the complementary function and physically represents the own residual information of a neuron, while the second term, i.e. mathematically the particular integral, shows the effect of inputs from other neurons. Now, using eq. (9.23), we can write:

$$\begin{aligned} Y_i(\tau) &= Y_i(0)e^{-\tau/\tau_0} + (1/\tau_0) \sum_j \int G_{ij}(\tau, \tau') Y_j(\tau') d\tau' \\ &= Y_i(0)e^{-\tau/\tau_0} + (1/\tau_0) \sum_j \int G_{ij}(\tau, \tau') \left[Y_j(0)e^{-\tau'/\tau_0} + (1/\tau_0) \sum_k \int G_{jk}(\tau', \tau'') Y_k(\tau'') d\tau'' \right] d\tau' \\ &= Y_i(0)e^{-\tau/\tau_0} + (1/\tau_0) \sum_j \int G_{ij}(\tau, \tau') Y_j(0) e^{-\tau'/\tau_0} d\tau' \\ &\quad + (1/\tau_0)^2 \sum_{j,k} \int\int G_{ij}(\tau, \tau') G_{jk}(\tau', \tau'') Y_k(\tau'') d\tau'' d\tau' \end{aligned}$$

and so on. Thus, the solution for a large neural network is finally found to be:

$$\begin{aligned} Y_i(\tau) &= Y_i(0)e^{-\tau/\tau_0} + (1/\tau_0) \sum_j \int G_{ij}(\tau, \tau') Y_j(0) e^{-\tau'/\tau_0} d\tau' \\ &\quad + (1/\tau_0)^2 \sum_{j,k} \int\int G_{ij}(\tau, \tau') G_{jk}(\tau', \tau'') Y_k(0) e^{-\tau''/\tau_0} d\tau'' d\tau' + \cdots \end{aligned} \tag{9.28}$$

Equation (9.28) can be regarded as the general characteristic equation of the neural network in human brain. It describes the output signal (Y_i) at any axon terminal of a neuron as a function of all the previous outputs at all its pre-synaptic nodes ($Y_j, Y_k, \ldots$) in the chains of the network in terms of the signal propagation given by several orders of the Green's functions. Interestingly, comparing eqs. (9.27) and (9.28), it may be worthy to note that such Green's functions are, in fact, representing all the previous inputs (X_i) applied to such pre-synaptic neurons and their respective synaptic strengths (A_{ij}). It can further be observed that the synaptic strength $A_{i,i'}$, corresponding to the neuron with output Y_i, that will respond to the synapses of the next neurons (at i'), is also dependent on all the previous outputs $Y_j, Y_k, \ldots$ (as well as the corresponding sets of inputs) and the associated synaptic strengths $A_{j,j'}, A_{k,k'}, \ldots$, a phenomenon known to be the synaptic plasticity, as discussed earlier in this chapter.

This is the method of learning by human brain, i.e. to provide an output on the basis of all previous information acquired from input signals and analysed through their appropriate synaptic strengths, and memorizing them in the gigantic network before they are attenuated in time (the decay time being given by τ_0). It may be interesting to note that τ_0 may be considered as the average forgetting time that depends, as discussed in the previous section of this chapter, on the ion concentrations. We argue here that this can explain why people behave in so peculiar manner when they suffer from Na +/K + imbalances. It can hamper the entire signal processing in human body and cause dysfunctional behaviours or failures of any of the organs leading to severity. Further, it may be noted that if the signals were not attenuated, they would increase in cycles of the neural network leading to diverging itself (see eq. (9.28)), which might lead to completely hazardous effects for human brain to function owing to overburden. Thus, we also argue that although memorizing is a maintained process through LTP of neural network, slowly forgetting (i.e. LTD) is an inherent feature of neurons that is compatible with human brain functioning.

It can, therefore, be noted that the neurons and synapses themselves process the relevant information tasks as well as store such information in the network. This very feature of human brain is quite different from our existing computers which have separate processing units and storage devices, and thus requires much more energy (~million times) than brain for their similar functioning. This motivates us, not only to compute in the same logical structure (i.e. through gigantic neural networks) but also to develop computing devices that will be able to process the information tasks and memorize them in the same units, similar to that brain does through neural signal generation, processing and memorizing concurrently. This will be briefly discussed in the next chapter.

9.6 Activation function for artificial neural network

The ANNs, despite their significant development, are ultimately formulated based on a simplified version of neural networks of human brain as-well-as their functioning. Up to these days, the ANNs are being used fundamentally for the computational tasks that are conventionally believed to be much complex, such as identifying different objects from an image or understanding/interpreting words (more correctly, 'concepts') of different languages. It utilizes the notion that the synaptic connections between neurons change their strength, i.e. weights of inputs, by repeatedly being fed with data to get better accuracy at the output [3, 15]. For a large neural network, such method of learning can be carried out efficiently by computing an identical gradient for all the synaptic strengths so that the performance of the entire network improves with such weights [16, 17]. The method can be understood analogically as collective learning from individual's experiences.

The transformation function for converting all such weighted inputs into the output is known to be the activation function. Despite numerous attempts to find a suitable activation function (e.g. Swish, Mish, leaky rectified linear unit, exponential linear unit, dynamic rectified linear unit (ReLU), and Gaussian error linear unit [18]), the most efficient and widely used activation function still remains to be the simple ReLU, i.e. $g(x) = \max(0, x)$. However, what the connection or analogy of such function with human neurons is, or, why it is serving so well has not been made clear by the neural network community. That is why it is difficult to modify such an activation function (except some guess works) to overcome the conventional challenges of covariate shift, inaccuracy in getting intricate information as well as leaning cost. Here, we are giving an idea on the origin of such activation function and relation with human neural networks based on the discussion in the above section.

It is apparent from eq. (9.28) that if we consider only one layer equivalent to the entire neural network that functions like human brain, then considering only the first order term the output becomes:

$$Y_i(\tau) = Y_i(0)e^{-\tau/\tau_0} + (1/\tau_0)\sum_j \int G_{ij}(\tau, \tau')Y_j(0)e^{-\tau'/\tau_0}d\tau' \tag{9.28a}$$

where the first term is residual information while the second term represents the information upgradation by learning. Now, the Green's function given by $\left[\frac{d}{d\tau} + 1/\tau_0\right]G_{ij}(\tau, \tau') = \delta_{ij}\delta(\tau - \tau')$ has a solution:

$$G_{ij}(\tau, \tau') = \delta_{ij}\vartheta(\tau - \tau')e^{-(\tau-\tau')/\tau_0} \tag{9.29}$$

where $\vartheta(\tau - \tau')$ is the unit step function that is 1 for $\tau \geq \tau'$ and 0 otherwise. Thus:

$$\int G_{ij}(\tau, \tau')Y_j(0)e^{-\tau'/\tau_0}d\tau' = \delta_{ij}Y_j(0)e^{-\tau/\tau_0}\int_0^\infty \vartheta(\tau - \tau')d\tau' = \delta_{ij}Y_j(0)e^{-\tau/\tau_0}\int_0^\tau \vartheta(\tau - \tau')d\tau'$$

since $\vartheta(\tau - \tau') = 0$ for $\tau' > \tau$. Thus, if we do not consider the residual memory (i.e. $Y_j(0)e^{-\tau/\tau_0}$), then the activation function is given by

$$g(\tau) = (1/\tau_0)\int_0^\tau \vartheta(\tau - \tau')d\tau' = \max(0, \tau/\tau_0) \tag{9.30}$$

Thus, ReLU is in fact a replication of human neural network system where one layer equivalently represents the entire network based on learning from each node, although without any residual memory. If we include the residual memory term to accommodate previous history in some training method, then the more suitable activation function would evidently be:

$$g(\tau) = e^{-\tau/\tau_0} \max(0, \tau/\tau_0) \tag{9.30a}$$

Recently, this function has also been proposed as a relevant activation function (termed as 'attenuation function') for intricate categorization [19].

At this point, it is worthy to note that, firstly, if one needs to consider two layers instead of one intermediate layer (between input and output), the second-order term of eq. (9.28) should also be considered to design the activation function, and so on for more layers; secondly, if infinite numbers of layers are considered, that would be equivalent to consider a cyclic loop thereby equating the input and output nodes, which can be confirmed by summing up all the terms in eq. (9.28) that apparently leads to $Y_i(\tau) = Y_i(0)$; and thirdly, the Green's function may have other solutions that can give rise to several other pertinent activation functions relevant for specific tasks. The method described here can explore deeper insights with broader applications in the field of the so-called AI. Nevertheless, why such intelligence cannot replace natural human intelligence can be understood by further reading the arguments of Penrose [20].

References

[1] Herculano-Houzel, S. "The remarkable, yet not extraordinary, human brain as a scaled-up primate brain and its associated cost." *Proceedings of the National Academy of Sciences* 109, no. supplement_1 (2012): 10661–10668.

[2] Marković, D., Mizrahi, A., Querlioz, D., and Grollier, J. "Physics for neuromorphic computing." *Nature Reviews Physics* 2, no. 9 (2020): 499–510.

[3] Hopfield, J. J. "Neural networks and physical systems with emergent collective computational abilities." *Proceedings of the national academy of sciences* 79, no. 8 (1982): 2554–2558.

[4] McGilchrist, I. *The master and his emissary: The divided brain and the making of the western world.* Yale University Press, 2019.

[5] Basinger, H. and Hogg, J. P. *Neuroanatomy, brainstem.* 2019.

[6] Jimsheleishvili, S. and Dididze, M. "Neuroanatomy, cerebellum." In *StatPearls [Internet]*, StatPearls Publishing, 2023.

[7] Anand, K. S. and Dhikav, V. "Hippocampus in health and disease: An overview." *Annals of Indian Academy of Neurology* 15, no. 4 (2012): 239–246.

[8] Wirtshafter, H. S. and Wilson, M. A. "Artificial intelligence insights into hippocampal processing." *Frontiers in Computational Neuroscience* 16, (2022): 1044659.

[9] Chung, K., Wallace, J., Kim, S.-Y., Kalyanasundaram, S., Andalman, A. S., Davidson, T. J., Mirzabekov, J. J., et al. "Structural and molecular interrogation of intact biological systems." *Nature* 497, no. 7449 (2013): 332–337.

[10] Renier, N., Adams, E. L., Kirst, C., Zhuhao, W., Azevedo, R., Kohl, J., Autry, A. E., et al. “Mapping of brain activity by automated volume analysis of immediate early genes.” *Cell* 165, no. 7 (2016): 1789–1802.

[11] Attarpour, A., Osmann, J., Rinaldi, A., Tianbo, Q., Lal, N., Patel, S., Rozak, M., et al. “A deep learning pipeline for three-dimensional brain-wide mapping of local neuronal ensembles in teravoxel light-sheet microscopy.” *Nature methods* 22, no. 3 (2025): 600–611.

[12] Pang, J. C., Aquino, K. M., Oldehinkel, M., Robinson, P. A., Fulcher, B. D., Breakspear, M., and Fornito, A. “Geometric constraints on human brain function.” *Nature* 618, no. 7965 (2023): 566–574.

[13] Purves, D., Augustine, G. J., Fitzpatrick, D., Katz, L. C., LaMantia, A.-S., McNamara, J. O., and Williams, S. M. “Circuits within the basal ganglia system.” In *Neuroscience, 2nd edition*. Sinauer Associates, 2001.

[14] Valero, M., Abad-Perez, P., Gallardo, A., Picco, M., García-Hernandez, R., Brotons, J., Martínez-Félix, A., Machold, R., Rudy, B., and Buzsáki, G. “Cooperative actions of interneuron families support the hippocampal spatial code.” *Science* 389, no. 6764 (2025): eadv5638.

[15] Iyer, R., Menon, V., Bulce, M., Koch, C., and Mihalas, S. “The influence of synaptic weight distribution on neuronal population dynamics.” *PLoS computational biology* 9, no. 10 (2013): e1003248.

[16] Hopfield, J. J. “Neurons with graded response have collective computational properties like those of two-state neurons.” *Proceedings of the national academy of sciences* 81, no. 10 (1984): 3088–3092.

[17] Hinton, G. “Nobel lecture: Boltzmann machines.” *Reviews of Modern Physics* 97, no. 3 (2025): 030502.

[18] Rasamoelina, A. D., Adjailia, F., and Sinčák, P. “A review of activation function for artificial neural network.” In *2020 IEEE 18th world symposium on applied machine intelligence and informatics (SAMI)*, pp. 281–286. IEEE, 2020.

[19] Jiang, W., Yuan, H., and Liu, W. “Neuron signal attenuation activation mechanism for deep learning.” *Patterns* 6, no. 1 (2025).

[20] Penrose, R. *Shadows of the Mind*. 4, Oxford: Oxford University Press, 1994.

10 Neuromorphic computing devices: fundamental concepts of memory-based processor

The term 'artificial intelligence' (AI) has brought to us, especially very recently with the emergence of ChatGPT and DeepSeek, an extreme hope to cross all the boundaries that the individual human brains are challenged by, as well as simultaneously, a serious concern that the necessity of human contributions will be completely driven out of the periphery of the civilization. Hinton, known as 'the godfather of AI', has himself expressed his anxiety(!) that it may create a 'digital god' [1], whereas Penrose, known to be 'the founding father of quantum gravity', is still hopeful(!) that AI can never 'transcend' the human brains [2]. Nevertheless, keeping the debate aside for a while, no one can deny the recent development of tremendous computing efficiencies using the artificial neural network (ANN) schemes, which began with the basic neural network theory proposed by Hopfield [3], and are now being able to provide us with the stuffs beyond even the expectation of our conventional imaginations.

As mentioned in Chapter 1, AI is fundamentally based upon the algorithmic mimicking of the learning based information processing in human brains known to be the ANN. For a comprehensive understanding on the fundamentals of ANNs and development of its various classes, the readers may see [4]. The ANN framework, in its course of evolution, has created a messy ontology tree, since the meaning of its categories, sub-categories, and the evolved counterparts are still not mature and changing with everyday developments. For instance, in regard to the direction of data transport from one layer to other in the network, ANNs can be categorized into two major classes: (1) feed-forward networks, where the data travels only in the forward direction of the network from input to output, and (2) recurrent neural networks (RNNs), which contain feedback loops in their layers leading to have internal state memory in each node. On the other hand, from a perspective of the numbers of network layers (which has been observed to have significant impact on learning ability), in contrast to the primitive single-layer perceptrons (i.e. single input-output structure), the multi-layer perceptrons (with at least one hidden layer in-between input and output) have exhibited remarkable improvement in output, thereby leading to the emergence of a more general class, known to be the deep neural networks with multiple hidden layers [5]. It is now being further advanced towards the recent emerging trend to consider time as a parameter over which the signal is studied until a threshold is reached to fire the pulse, known as spiking neural networks (SNNs) that are based, more closely, on the biological neuronal activities [6]. One other important class of ANNs is the highly specialized convolutional neural networks (CNNs), which utilizes convolution of inputs to offer a filtered inter-layer connectivity to manage the over-fitting issues during training of the system [7, 8].

As discussed in Chapter 9, the development of fundamental algorithmic methodology of ANNs has been motivated directly from the basic learning mechanism of

 | https://doi.org/10.1515/9783111545950-010

human brain by the virtue of its signal generation, transmission and memorization through the gigantic networks of neurons and synapses. Evidently, such algorithms are mostly run on the state-of-the-art highly efficient computing systems (e.g. using supercomputers or GPUs) that are built upon an architecture, known to be the von Neumann computer, which, however, is fundamentally different from the computing structure of the human brain [6, 9].

In his very short chapter, we will briefly discuss the concepts of brain-inspired computing systems constituted of memory-based processors, their fundamental categories and sub-categories, and their device-level implementations. Finally, we will show, illustrating a preliminary mathematical model to describe the basic memristor operation, how a general memristor model can be developed for understanding device physics of neuromorphic systems.

10.1 Human brain vis-à-vis computer: the need for brain-inspired computing system

The von Neumann computers, or simply the computers, which we use today, can process enormous amount of data with superfast speeds compared to individual human capacity providing remarkable accuracy as well as precision in mathematical computations and quantitative data analysis based on formal logical frameworks. It seems that, at least while dealing with heavy computational workload, a computer is not only efficient but also much more superior than a single human brain, even if it is virtually the brain of John von Neumann himself (who at the age of only six years could divide two 8-digit numbers almost instantaneously in his head).

However, it may be recalled from the previous chapter that human brain, using its neurons, synapses and the neural network as a whole, analyses, processes, and memorizes information in itself, i.e. in the same units of it. In contrast, a von Neumann computer that we use everyday, unlike human brain, is constituted of an architecturally separate processing unit and memory storage that poses the burden of huge energy consumption compared to brain. For instance, calculations have shown that the training process of ANNs for a state-of-the-art natural language processing (NLP) model using supercomputers (of von Neumann architecture) requires an amount of energy per hour of the order of ~MW that is million times higher than that a human brain needs to perform it (with only ~20 W) [9]. Moreover, the current AI algorithms, despite their unprecedented success, are limited by the requirement of large-scale training datasets for even simpler information recognition tasks and challenged by the disastrous forgetting issues.

Further, brain is distinguishingly different from the conventional von Neumann computers in performing information tasks in its core method that the latter needs instantaneous instructions at each processing steps about what is to be performed, while the former can learn from experiences. Numerous efforts have been made to physically and mathematically describe the mechanism of neural information processing that makes such learning skill possible, on the basis of the theory of complex systems, nonlinear dynamics, as well as statistical physics including the renowned Hopfield network and Boltzmann machines motivated from the spin statistical theory [9, 10]. Such studies in turn have led to the development of ANNs that are trained by relevant datasets to learn recognizing some externally unidentified patterns, in order to accurately identify similar objects, just like human brain does. However, such training process that often involves the back-propagation method (a numerical method of self-adjustment) is a heavily energy consuming course, which upholds the requirement of new-generation computing devices that can work similar to human brains, especially in terms of their energy efficiency. In this context it is worthy to note, as mentioned earlier, that emerging trends are also being observed that adopt the SNNs which operate on the principles more close to the real neural mechanisms, and thus can be built up on inherently energy efficient devices analogous to human brain.

It may also be interesting to note that human brain, in contrast to the von Neumann computers, not at all functions using an all-regulatory universal clock, based on Galilean global isochronism, for communication, but rather depends on the collective behaviour of its correlated subsystems. Alongside all such distinguishing features, the plasticity, as discussed in the previous chapter, that brain exhibits, differentiates it widely from the basic logic elements and the corresponding devices constituting the state-of-the-art electronic circuits.

In this context and course of evolution, currently the brain-inspired computing systems, that also include and emphasize on the neuromorphic computing devices, are emerging as the potential future computing technology, with the motivational ground of developing devices and circuits that are proficient to mimicking the neural/synaptic networking topology of human brain along with their plastic and non-volatile memorizing capacity. However, it is still to find its compatibility with the state-of-the-art CMOS technology, especially in terms of scaling [11]. It has been observed that the CMOS-based artificial neural systems (including transistors as neurons and memory devices as synapses) usually exhibit a physical dimension of a few micrometres even when the transistor size is only a few nanometres. Evidently, it requires quite large area (may be of the order of square metres) for incorporating millions, or in fact billions, of neurons and even more synapses, for appropriately performing the brain-alike computing tasks, while additionally at the worthless expense of energy in the interconnects of the circuit. Further to realize neural interconnectivity through tens of thousands of synaptic junctions in CMOS architectures in 2D-circuit frameworks is itself extremely challenging.

In this context, the primary challenge is to map the neural networking topology along with the neuronal/synaptic functionality of human brain on to the electronic, optoelectronic, and/or quantum device platforms, known as brain-inspired computers (BICs). They must be different from the von Neumann computers in the following basic contexts:

1) **Processing and memory:** BICs must have a device scheme that can process and memorize information in the same unit unlike von Neumann computers where processing units and storage devices are separate. It will also reduce the recall delay in the processor-memory physical communication.
2) **Event-driven computation with high plasticity:** The constituent devices of BICs need to imitate the biological neuronal/synaptic methods of functionality along with relevant plasticity to perform a task only when a spike is received, and may be repeated in some cases, known as event-driven computation with plasticity-driven learning ability, which is completely unlikely in any von Neumann computers.
3) **Inherent parallelism:** BICs must be inherently parallel in the sense that all devices representing neurons and synaptic junctions can operate simultaneously but asynchronously, where the von Neumann computers operate with only externally imposed parallel connections, if any, and that also with a universally synchronized clock.
4) **Architectural scalability:** The BICs must comprise the devices, artificially representing neurons and synapses, that are inherently scalable so as to incorporate as many as neurons/synapses into smaller chip area as well as can be easily coupled to other similar chips to enhance performances.
5) **Hardware-level stochasticity:** BICs should be capable of introducing intrinsic stochastic noise in order to mimicking real brain-aptitude, which must be the randomness accessed from its hardware not mathematical programmes.

10.2 Brain-inspired computers: categories and sub-categories

The major categories (although not all) of the brain-inspired computing device schemes (see Fig. 10.1) currently belong to three fundamentally different classes: (1) *neuromorphic computing*, (2) *reservoir computing*, and (3) *hyper-dimensional computing* (HDC). A fourth category that belongs to the overlapping region of such three basic conceptual designs is known to be the *in-memory computing* that is considered to be the most exclusive to the key agenda of brain-inspired computing. Here we are briefly discussing the approaches being adopted to realize such classes of device in practice. For a detail study on the development of BIC systems on the basis of more realistic human brain simulation as-well-as their device-level implementations, the readers may see [11–14].

10.2.1 Neuromorphic computing devices

Neuromorphic computing devices comprise the designs of hardware that is capable of imitating neural networks topology of human brain including artificially mimicking the neurons and synaptic junctions. Numerous attempts are being made to realize them by designing various materials and device architectures based on two-terminal and multi-terminal neuromorphic memristors, neuromorphic spintronic devices, chalcogenide-based neuromorphic devices, organic memristors, and other emerging systems [15].

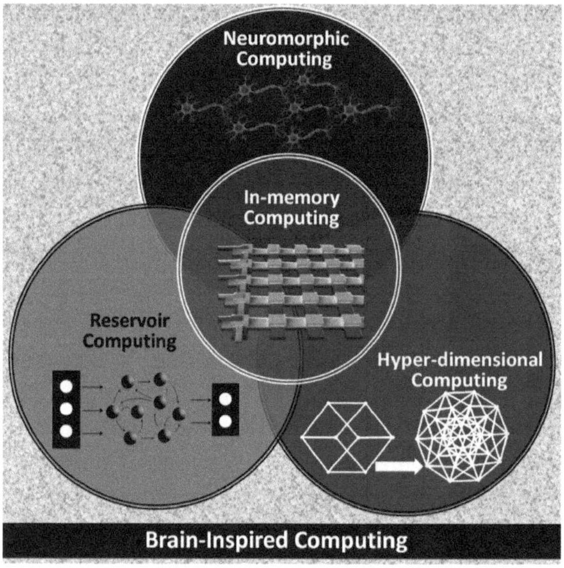

Fig. 10.1: A schematic representation of the major categories of brain-inspired computing approaches including the neuromorphic, reservoir, hyper-dimensional, and in-memory computing.

At this point, it may be worthy to note that a memristor, as can be simply understood, is an electronic device with a resistance that depends on the previous current flowed through it. From operational perspective of device, generally memristors have (at least) two resistance states, i.e. low (LRS) and high (HRS), that can be memorized and forgotten by appropriate SET and RESET biases, respectively (see the schematic of Fig. 10.2(a)). The possible existence of memristors was theoretically proposed by Chua in 1971 [16] as the fourth fundamental passive element (i.e. not a power source) of an electrical circuit (alongside inductor, capacitor, and resistor) on the basis of an assumed symmetrical property of circuits, whereas it has been experimentally achieved for the first time in 2008 by Hewlett-Packard research team. The memristors are usually classified into volatile and non-volatile depending on whether they lose the memory of latest resistance state, or not, when the power is OFF and then ON again.

A neuromorphic memristor, however, is something much more than that in the sense that it must have plasticity in its memorizing capacity, i.e. it should be able to reach the LRS in both ways: rapid and transient short-term potentiation (STP) for quick reflex; and stable long-term potentiation (LTP) for relevant training. And similarly, it should be capable of getting back to HRS through short-term depression (STD) to filter out unnecessary information and long-term depression (LTD) for important information to be memorized for long, respectively. The memristors also need to replicate the biological neurons, where the STPs are usually associated with paired-pulse facilitation (PPF) so that, not an arbitrary pulse, but a repetition of the same pulse promotes the state transition. Further, if spikes can be generated in a regulated manner in order to make it transit from HRS to the LRS by consecutive pulsed biases, it may replicate the neuronal function known to be spike-timing-dependent plasticity and/or spiking rate-dependent plasticity that apparently indicates its skill of learning. These are schematically shown in Fig. 10.2(b)–(d).

Among several neuromorphic devices, a two-terminal neuromorphic memristor is typically a metal/insulator/metal (MIM) device, as schematically shown in Fig. 10.2 (e), where initially hardly any current flows through the insulator and thus the device remains at HRS. However, application of a bias leads to a directional migration of mobile ions, usually oxygen vacancies of the insulator (e.g. TiO_2, HfO_2, and Ta_2O_3) or metal ions from electrochemically active electrodes (e.g. Ag and Cu), so that at a particular value of bias (SET) can form a random conductive path between metal electrodes. Such a pavement is commonly called the conductive filament (CF) that allows high current to flow even through the insulator leading the device to turn on from HRS to its LRS. On the other hand, an opposite bias of a specific value (i.e. RESET) can randomly rupture the CF resulting in a transition back to the HRS again. Interestingly, during the rupture of the CF (i.e. when the mobile ions are not completely randomized), a smaller bias than SET can re-form the filament and excite the device again into its LRS that is quite a mimicking of human brain to recall something forgotten on getting a small clue.

It is apparent, however, that a two-terminal memristor exhibits the so-called homosynaptic plasticity, so that only the directly activated synapses can change their own synaptic strengths, whereas biological neural networks also learn from synaptic interactions through competition and cooperation. Such heterosynaptic plasticity, thus, offers a mechanism of influencing the non-stimulated synapses by the activated groups of synapses thereby leading to a compensatory effects to modulate the strength change of all such synapses collectively. This, in turn, can regulate the strength of each of the synapses thereby curbing the risk of synaptic weight saturation, and holistically improve the computational performance of the entire network. On the device level, this is achieved by developing multi-terminal memristors, where alongside two terminals, the additional modulatory terminal(s) (see Fig. 10.2(f)) can provide improved control over the formation and rupture of CFs thereby exploring multimodal operation as-well-as offering larger dynamic range and enhanced linearity along with high-order state-variable complexity. More detail on the recent developments in such multi-terminal memristive devices can be found in [17].

Among other neuromorphic memristor candidates, the chalcogenide-based alloys (e.g. $Ge_2Sb_2Te_5$ and $Ag_5In_5Sb_{60}Te_{30}$) exhibit a crystalline LRS and an amorphous HRS where a rapid Joule heating at the latter state through high current followed by cooling makes a transition to the former state thereby leading to a phase-change memris-

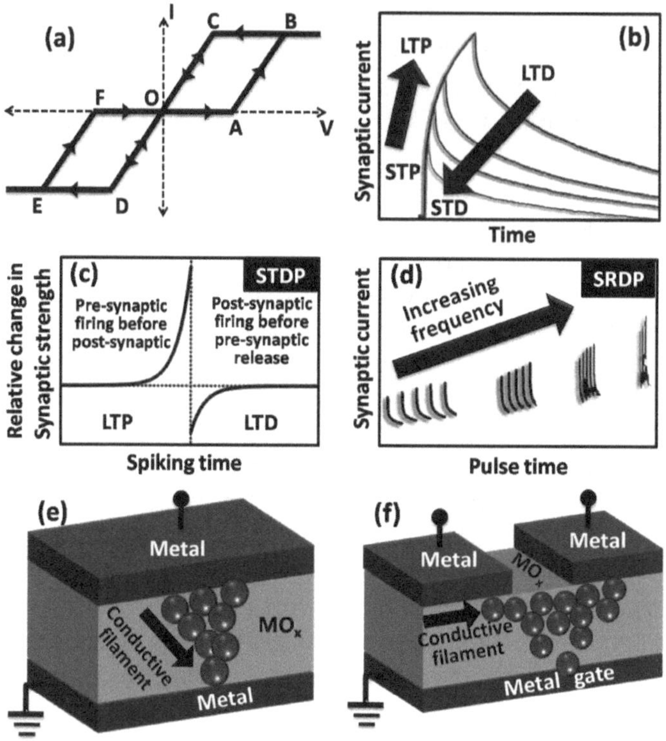

Fig. 10.2: (a) The fundamental operational output of a memristor to memorize its HRS and LRS until SET and RESET are not applied, respectively; (b)–(d) illustrations of key characteristics of neuromorphic computing devices; and (e) two-terminal and (f) multi-terminal neuromorphic devices structures forming conductive filament.

tive (PCM) behaviour. However, such devices are suffering from the challenges of energy requirement (which, for such Joule heating, is two orders higher than the conventional non-volatile memristors), variability of switching conditions, device reliability due to spontaneous structural relaxation of the amorphous phase. On the other hand, attempts are being made to develop spintronic neuromorphic devices based on *skyrmions* and magnetic tunnelling junctions owing to their static magnetoresistive effect as well as the dynamical properties like 'breathing mode' and stochastic switching behaviour, respectively. It may be briefly said that the *skyrmions* are self-localized wave packets (i.e. *solitons*) of collective spin -textures (swirl-like) that are observed in chiral magnetic medium without inversion symmetry, and are topologically protected against defects and thermal fluctuations.

10.2.2 Reservoir computing devices

As mentioned earlier in this chapter, the RNNs feed the output of a neuron to its own input or the input of other neuron of same or preceding layers, and therefore are capable of time-series analysis tasks. However, working with RNNs face challenges due to vanishing gradient issue, requirement of large numbers of recurrent units as well as training data, and the associated runtime and power budget. These problems are addressed by adopting the approach of reservoir computing, where the reservoir, constituted of the recurrent units, projects the inputs onto a higher-D space using its non-linearity owing to feedback loops, which then gives rise to the desired output at the linear read-out layer.

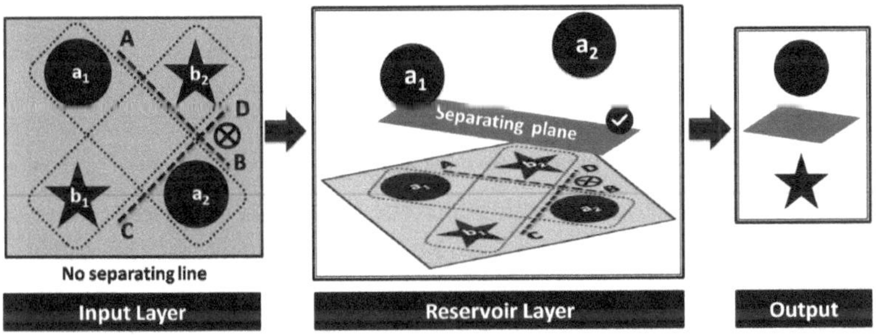

Fig. 10.3: A simple example of reservoir computing to classify two objects (e.g. circle and star) on a 2D plane, which cannot be done using any single demarcating line; however, on a reservoir layer creating two virtual objects in 3D space that projects as the circles on the 2D surface, can be easily categorized, as shown in the output.

Such a method is illustrated with a simple example in Fig. 10.3, which shows an input data consisting objects of two different classes that are inseparable on a 2D plane by any linear boundary. The reservoir, in such case, projects the data onto a 3D space so that now a linear plane is sufficient to classify such objects as linear read-outs. The reservoir in such computing approach does not require any training; rather, it utilizes the intrinsic dynamics associated with time delay of the device through recurrence process in order to perform such higher-D projections. The computing reservoirs at the device level are being developed mostly using memristors and spin-torque devices.

10.2.3 Hyper-dimensional computing devices

Mimicking human brain functioning through biological neural networks on device platforms is challenging majorly due to the burden of energy consumption, as discussed earlier in this chapter and Chapter 1. The concept of HDC addresses this key issue by focussing on replicating the neural circuits at device level instead of fabricating artefacts of individual neurons. Such approach replicates human brain functioning, e.g. memory, perception, and cognition, onto thousands of *bits* forming the so-called hyper-vectors.

In HDC approach, information is distributed uniformly over an entire hyper-vector through all of its independent and identically distributed components so that the failure of one or a few bits does not lead to instant loss of the memory, thereby upholding its robustness. Interestingly, it is apparent that any two randomly generated vectors in such high-dimensional space ($d \geq 1,000$) with binary (0, 1) components are highly probable to be mutually orthogonal, i.e. the inner product will be negligibly small. Such quasi-orthogonality offers that numerous distinct information can be stored in the same hyper-dimensional space without significant cross-talks. Moreover, the cognitive operations, i.e. binding (analogous to vector product), bundling (equivalent to vector addition), and permutation may lead to the generation of new hyper-vectors that can represent novel and complex information out of the simpler known set of information. A detail on basic approach and recent progresses in HDC can be found in [18].

Further, it may be noted that HDC offers a remarkable advantage of learning from one or a very few example(s) as training in comparison to the large numbers of iteration needed in support vector machines or CNN models, thereby further reducing energy burden on the processors. At device level, such HDC tasks are being performed using reduced-instruction-set computing architectures, field-programmable gate arrays as well as wearable sensors, however, in-memory computing devices are showing most prominence in terms of energy efficiency.

At this point, it is worthy to mention that the in-memory computing is carried out on several devices, such as using SRAM, DRAM, flash memory, as well -as memristors like RRAM and PCMs [19]. Usually, the memristors-based in-memory computing removes the steps of accessing memory in order to assign relevant synaptic-equivalent weights to inputs, where the applied voltage inputs to bit lines, being multiplied with the conductance of each memristor results in the corresponding current. Then using analogue-to-digital converter, it gives rise to the binary outputs thereby significantly improving the entire circuit efficiency. It is evident that parallelism of the circuitry and memorization of conductance by the memristors in such approach significantly curbs the energy budget of computation.

10.3 A basic mathematical model of generic memristors

The above discussions on the tremendous attraction towards developing the memristive devices with different on-demand properties relevant to brain-inspired computing on device platforms uphold the requirement of detail and general understanding of the underlying physics of such mechanism. However, theoretical propositions to describe the memristor functionalities have been empirical in nature, and thus, several memristive devices and associated materials are usually mathematically modelled in a case-specific manner [20]. It is therefore crucial to develop a generic mathematical model that can describe different materials and device schemes for various classes of memristors. Here we are exploring such a general model, however, at a very preliminary level to describe the basic memristor operation that can be widely extended for complex and higher order devices.

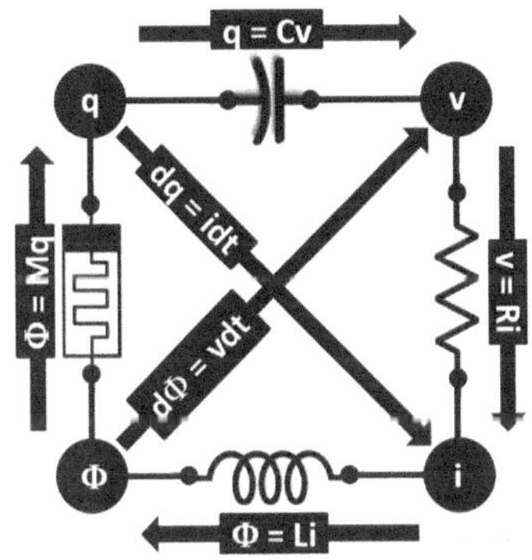

Fig. 10.4: The symmetry in electrical circuit in terms of its fundamental passive elements, i.e. resistors, inductors, capacitors, and memristors, owing to the relations between charge, voltage, current, and flux.

The operational principle of a memristor, as described by Chua [16], is based on the relation between the charge and flux of a device in the circuit, which cannot be described by the other fundamental passive elements, i.e. resistors, inductors, or capacitors, of an electric circuit. The possible existence of such a device was proposed theoretically in search of a missing element to complete the symmetry of electrical circuit, quite analogous to the symmetric nature of Maxwell's equations. As schematically illustrated in Fig. 10.4, the output of a capacitor is the accumulated charge for an input voltage (given by $q(t) = Cv(t)$), which, on the other hand, may be the output across a resistor if an input current passes through it (i.e. $v(t) = Ri(t)$), and such a current can be output of an inductor due to a flux input ($i(t) = (1/L)\Phi(t)$). Then, is it possible to find another fundamental circuit element that completes such a symmetric loop, so that an output flux is obtained from an input charge, represented by $\Phi(t) = Mq(t)$? Such element is known to be the memristor, i.e. a resistor with a memory, given by $V(t) = MI(t)$, owing to the time-derivatives $V(t) = d\Phi(t)/dt$ and $I(t) = dq(t)/dt$, differing from $v(t) = Ri(t)$ since the latter is associated with the time dynamics of the circuit resulting in a memory effect.

Conventionally, an ideal memristor is defined as a two-terminal resistive device that exhibits an instantaneous current-voltage relationship following the state-dependent Ohm's law. It is worthy to note that a memristor is fundamentally different from a typical storage device in this particular sense of the state-dependence of its conduction behaviour. Such a memristive device can be either current-controlled or voltage-controlled, respectively given by

$$V(t) = M(x_j)I(t) \tag{10.1a}$$

$$\frac{d}{dt}x_j = f_j(x_k, I) \tag{10.1b}$$

and

$$I(t) = W(x_j)V(t) \tag{10.2a}$$

$$\frac{d}{dt}x_j = g_j(x_k, V) \tag{10.2b}$$

where $M(x_j)$ is known to be the memristance and $W(x_j)$ is the memductance, that are dependent on internal state defined by the set of variables $\{x_j\}$. The time dynamics of such devices are described by the functions $f_j(x_k, I)$ or $g_j(x_k, V)$, and thereby resulting in a dependence on the history of the device, e.g. the current passed through just before.

In order to describe such devices mathematically as a general framework, here we are considering the latter case, i.e. the voltage-controlled memristor, where voltage is increased and decreased in forward and backward directions monotonically with time, for instance, at equal steps per unit time, so that dV/dt is positive or negative constant, respectively. Apparently, the same method can be applied for current-controlled memristors as well.

The memductance now can be expanded as

$$W(x_j) = W(0) + \sum_k \left.\frac{\partial W}{\partial x_k}\right|_{x_k=0} x_k + \frac{1}{2}\sum_{r,s} \left.\frac{\partial^2 W}{\partial x_r \partial x_s}\right|_{x_r=0,\, x_{sr}=0} x_r x_s + \cdots \tag{10.3}$$

so that its time-derivative is given by:

$$\frac{d}{dt}W(x_j) = \sum_k \left.\frac{\partial W}{\partial x_k}\right|_{x_k=0} \frac{dx_k}{dt} + \frac{1}{2}\sum_{r,s} \left.\frac{\partial^2 W}{\partial x_r \partial x_s}\right|_{x_r=0,\, x_{sr}=0} \left(x_r\frac{dx_s}{dt} + \frac{dx_r}{dt}x_s\right) + \cdots \tag{10.4}$$

Now, differentiating eq. (10.2a) with respect to time, and using eqs. (10.3) and (10.4), we have

$$\begin{aligned}\frac{d}{dt}I(t) = &\left[W(0) + \sum_k \left.\frac{\partial W}{\partial x_k}\right|_{x_k=0} x_k + \frac{1}{2}\sum_{r,s} \left.\frac{\partial^2 W}{\partial x_r \partial x_s}\right|_{x_r=0,\, x_{sr}=0} x_r x_s + \cdots\right]\left(\frac{d}{dt}V(t)\right)\\ &+ \left(\sum_k \left.\frac{\partial W}{\partial x_k}\right|_{x_k=0} \frac{dx_k}{dt} + \frac{1}{2}\sum_{r,s} \left.\frac{\partial^2 W}{\partial x_r \partial x_s}\right|_{x_r=0,\, x_{sr}=0} \left(x_r\frac{dx_s}{dt} + \frac{dx_r}{dt}x_s\right) + \cdots\right)V(t)\end{aligned}$$

and then since eq. (10.2b), we get:

$$\frac{d}{dt}I(t) = \left[W(0) + \sum_k \frac{\partial W}{\partial x_k}\bigg|_{x_k=0} x_k + \frac{1}{2}\sum_{r,s}\frac{\partial^2 W}{\partial x_r \partial x_s}\bigg|_{x_r=0,\, x_{sr}=0} x_r x_s + \cdots\right]\left(\frac{d}{dt}V(t)\right)$$
$$+\left(\sum_k \frac{\partial W}{\partial x_k}\bigg|_{x_k=0} g_k(x_l, V) + \frac{1}{2}\sum_{r,s}\frac{\partial^2 W}{\partial x_r \partial x_s}\bigg|_{x_r=0, x_{sr}=0}(x_r g_s(x_l, V) + x_s g_r(x_l, V)) + \cdots\right)V(t) \tag{10.5}$$

Equation (10.5) is the generic equation for a voltage-controlled memristor (that for a current-controlled device can be obtained similarly). Now the set of state variables $\{x_j\}$ are in general different for different material mechanisms and device architectures (for detail see [21]).

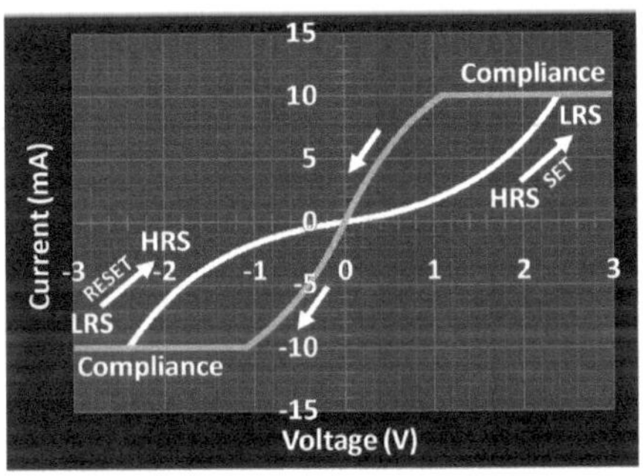

Fig. 10.5: The plot of current with voltage in a memristive circuit over an entire loop, obtained using the above model, assuming a constant voltage change rate in both forward and reverse directions.

Here, we are illustrating a simple case, where we assume the current passed through the device just before (i.e. in principle the existing resistance state) to be the determining state of the memductance, so that the only state-variable is $x(t) = I(t)$. For further simplicity of the solution of eq. (10.5), let us consider the first-order approximation to get (writing $g(V(t))V(t) = \tilde{g}(t)$):

$$\frac{d}{dt}I(t) = \left[W(0) + W'(0)I(t)\right]\left(\frac{d}{dt}V(t)\right) + W'(0)\tilde{g}(t)$$
$$\Rightarrow \left[\frac{d}{dt} - \left(W'(0)\frac{dV}{dt}\right)\right]I(t) = \left[W(0)\frac{dV}{dt}\right] + W'(0)\tilde{g}(t) \tag{10.6}$$

Assuming at $t = 0$, $V = 0$ and thus $I = 0$, the solution of eq. (10.6) is obtained as:

$$I(t) = \int G(t,t')\left[W(0)\frac{dV}{dt} + W'(0)\tilde{g}(t')\right]dt' \tag{10.7}$$

where the Green's function is the solution of $\left[\frac{d}{dt} - \left(W'(0)\frac{dV}{dt}\right)\right]G(t,t') = \delta(t-t')$, and is apparently obtained as:

$$G(t,t') = \vartheta(t-t')e^{\left(W'(0)\frac{dV}{dt}\right)(t-t')} \tag{10.8}$$

Equation (10.7) is solved using the Green's function of eq. (10.8) and the result for the first-order approximation is plotted in Fig. 10.5. Here, we have considered a compliance current of 10 mA, as usually done in practical cases. It may be noted that such a device, once (due to applying SET) reaches the compliance current, memorize the path determined by the LRS, and needs a RESET voltage to return to the HRS. A more rigorous result can be obtained using the above equations with suitable approximations or numerical solutions for different physical mechanisms and device architectures of memristive devices in general.

References

[1] Bengio, Y., Bucknall, B., Clare, S., Prunkl, C., Andriushchenko, M., Fox, P., Tiancheng, H. U., et al. "International AI safety report: First key update capabilities and risk implications." *SuperIntelligence-Robotics-Safety & Alignment* 2, no. 6 (2025).

[2] Penrose, R. "The emperor's new mind." *RSA Journal* 139, no. 5420 (1991): 506–514.

[3] Hopfield, J. J. "Neural networks and physical systems with emergent collective computational abilities." *Proceedings of the national academy of sciences* 79, no. 8 (1982): 2554–2558.

[4] Haykin, S. *Neural networks: A comprehensive foundation*. Prentice hall PTR, 1994.

[5] LeCun, Y., Bengio, Y., and Hinton, G. "Deep learning." *nature* 521, no. 7553 (2015): 436–444.

[6] Schuman, C. D., Kulkarni, S. R., Maryam Parsa, J. P. M., Date, P., and Kay, B. "Opportunities for neuromorphic computing algorithms and applications." *Nature Computational Science* 2, no. 1 (2022): 10–19.

[7] Zhao, X., Wang, L., Zhang, Y., Han, X., Deveci, M., and Parmar, M. "A review of convolutional neural networks in computer vision." *Artificial Intelligence Review* 57, no. 4 (2024): 99.

[8] Krizhevsky, A., Sutskever, I., and Hinton, G. E. "Imagenet classification with deep convolutional neural networks." In *Advances in neural information processing systems*, 25, 2012.

[9] Marković, D., Mizrahi, A., Querlioz, D., and Grollier, J. "Physics for neuromorphic computing." *Nature Reviews Physics* 2, no. 9 (2020): 499–510.

[10] Hinton, G. "Nobel lecture: Boltzmann machines." *Reviews of Modern Physics* 97, no. 3 (2025): 030502.

[11] Zolfagharinejad, M., Alegre-Ibarra, U., Chen, T., Kinge, S., and Van der Wiel, W. G. "Brain-inspired computing systems: A systematic literature review." *The European Physical Journal B* 97, no. 6 (2024): 70.

[12] Amunts, K. "Workflows for ultra-high resolution 3D models of the human brain on massively parallel supercomputers." In *Brain-Inspired Computing: Second International Workshop, BrainComp 2015, Cetraro, Italy, July 6–10, 2015, Revised Selected Papers*, 10087, p. 15. Springer, 2016.

[13] Christensen, D. V., Dittmann, R., Linares-Barranco, B., Sebastian, A., Gallo, M. L., Redaelli, A., Slesazeck, S., et al. "2022 roadmap on neuromorphic computing and engineering." *Neuromorphic Computing and Engineering* 2, no. 2 (2022): 022501.

[14] Jin, B., Wang, Z., Wang, T., and Meng, J. "Memristor-based artificial neural networks for hardware neuromorphic computing." *Research* (2025).

[15] Wan, Q. and Shi, Y., eds. *Neuromorphic Devices for Brain-inspired Computing: Artificial Intelligence, Perception, and Robotics.* John Wiley & Sons, 2022.

[16] Chua, L. "Memristor-The missing circuit element." *IEEE Transactions on Circuit Theory* 18, no. 5 (1971): 507–519.

[17] Leng, Y., Zhang, Y., Ziyu, L., Wang, J., Xie, T., Zhu, S., Qin, J., Runze, X., Zhou, Y., and Han, S. "Recent progress in multiterminal memristors for neuromorphic applications." *Advanced Electronic Materials* 9, no. 6 (2023): 2300108.

[18] Lulu, G. and Parhi, K. K. "Classification using hyperdimensional computing: A review." *IEEE Circuits and Systems Magazine* 20, no. 2 (2020): 30–47.

[19] Sebastian, A., Gallo, M. L., Khaddam-Aljameh, R., and Eleftheriou, E. "Memory devices and applications for in-memory computing." *Nature nanotechnology* 15, no. 7 (2020): 529–544.

[20] Radwan, A. G. and Mohammed, E. *Fouda. On the mathematical modeling of memristor, memcapacitor, and meminductor*. Cham, Switzerland: Springer International Publishing, 26, 2015.

[21] Yang, K., Yang, J. J., Huang, R., and Yang, Y. "Nonlinearity in memristors for neuromorphic dynamic systems." *Small Science* 2, no. 1 (2022): 2100049.

11 How an NEGF simulator can be constructed?

In the previous chapters of this book, particularly from Chapters 3–10, we have illustrated the beauty and the strength of non-equilibrium Green's function (NEGF) formalism in terms of mathematically describing a wide range of physical phenomena. It includes the quantum behaviour of nature and its dynamics with the classical world observed in several aspects of various systems such as electronic transport in ultra-small devices, quantized lattice vibrations, optical, and photonic systems, spin-based properties, double-quantum-dot qubits as well as signalling in human brain along with its algorithmic and device replications. If we have been a little successful, over whatever small scale it may be, in attracting the readers toward the immense potential and possible prospect of this mathematical technique, then the natural question arises as to whether, and if so then how, one can construct a NEGF simulator in an efficiently computable manner. In this chapter, we will illustrate with some examples of such programs how to compute some of the previously discussed physical processes.

11.1 Electron in a 1D-confined rectangular potential well: double-gate MOS

Before going to illustrate any NEGF-based simulating problem, let us first try to understand the complexity of a much (so-called) simpler problem in reality. For instance, we will start from investigating the motion of an electron confined in a 1D-rectangular potential well, which is quite trivial (as described in any preliminary textbook) if we consider the well depth to be infinite, i.e. strong confinement of electron leading to no wave function tail within the barriers. It is apparent that the solutions of the Schrödinger equation, with a corresponding boundary condition that the wave function be zero at both the barrier edges (well width being a), are $\psi_n(x) = \sqrt{2/a}\sin(n\pi x/a)$ which are infinitely many since $n = 1, 2, 3, \ldots$ leading to the corresponding energy eigenvalues to be $\varepsilon_n = n^2(\hbar\pi/a)^2/2m_e$ (where m_e is mass of electron). Evidently, the probability of the electron along the well is given by $P_0(x) = \sum_n |\psi_n(x)|^2$.

Such a 1D-rectangular well can be physically realized in a double-gate MOS device where a semiconductor thin film (of thickness a) is sandwiched between two insulators that are connected to two metal electrodes. We are not considering any voltage applied at the two gates and assuming the tunnelling leakage to be almost zero. In such a condition the device should behave ideally as a 1D-rectangular well exhibiting the wave functions and energy eigenvalues given above. But NO! The material world has many other dialectical constraints. In fact, we have previously discarded two physical realities that are not ignorable within materials, not even theoretically in the ideal cases: 1) the potential well for the electron is representative of its electrostatic

 | https://doi.org/10.1515/9783111545950-011

interaction with all other charges, and 2) the electron is free to move in the plane (say YZ-plane) perpendicular to the MOS-axis (i.e. the X-axis) thereby possessing the statistical properties like temperature and Fermi level. The former requires Poisson's equation to be coupled to the Schrödinger equation and the latter needs to be modelled by incorporating the corresponding Fermi-Dirac statistical distribution. The infinite barrier may be replaced with a finite barrier to get more accuracy; for instance, such barrier for a Si/SiO_2-system is ~3.16 eV while the energy gaps of the quantized states are of the order of a few meV.

Thus to take into account all such unavoidable physical properties, the mathematical method to describe an electron confined in a 1D-rectangular well needs to pursue the following steps:

1. Start with the Schrödinger equation assuming a 1D-rectangular well with an infinite barrier and get the preliminary solutions, i.e. $\psi_n^{(0)}(x) = \sqrt{2/a}\sin(n\pi x/a)$ and $\varepsilon_n^{(0)} = n^2(\hbar\pi/a)^2/2m_e$.
2. Assuming the electron to be free to move in the YZ-plane, its 1D-probability density along the MOS direction, i.e. the X-axis, is given by, $P(x) = \sum_{n,\vec{k}_{yz}} |\psi_n^{(0)}(x)|^2 f_{FD}(\varepsilon_n^{(0)} + \varepsilon_{\vec{k}_{yz}} - E_F)$, where $f_{FD}(\varepsilon_n^{(0)} + \varepsilon_{\vec{k}_{yz}} - E_F)$ is the Fermi-Dirac distribution function for all possible wave vectors $\vec{k}_{yz}$ in the YZ-plane. On replacing the sum over $\vec{k}_{yz}$ by integral over $A_{yz}2\pi kdk$ (A_{yz} being the film area), it gives rise to $P(x) = A_{yz}(2\pi m_e K_B T/\hbar^2)\sum_n \ln(1 + e^{-\varepsilon_n^{(0)}/K_B T})|\psi_n^{(0)}(x)|^2$. It is interesting to note that such a probability density of the electron is an ideal example of a 'mixed state' in contrast to the previous one given only by $P_0(x) = \sum_n |\psi_n^{(0)}(x)|^2$ indicating a 'pure state'. Furthermore, the probability density $P(x)$ is dependent on temperature and shows lesser contribution of the higher energy states (due to the factor $e^{-\varepsilon_n^{(0)}/K_B T}$) unlike $P_0(x)$.
3. Such 1D-probability density of an electron gives rise to the 3D charge density to be $\rho(x) = (-e)P(x)/A_{yz}$, so that, the 1D-Poisson's equation (i.e. $\frac{d^2}{dx^2}\phi(x) = -\rho(x)/\varepsilon_0 k$ where $\phi(x)$ is potential, ε_0 is permittivity of vacuum, and k is the dielectric constant of well material), to find the potential variation along the well that was initially assumed to be flat, i.e. $\phi^{(0)}(x) = 0$, is now $\frac{d^2}{dx^2}\phi^{(1)}(x) = \frac{e}{\varepsilon_0 k}(2\pi m_e K_B T/\hbar^2)\sum_n \ln(1 + e^{-\varepsilon_n^{(0)}/K_B T})|\psi_n^{(0)}(x)|^2$.
4. The solution $\phi^{(1)}(x)$ of Poisson's equation is then to be put into 1D-Schrödinger equation $\left[-\frac{\hbar^2}{2m_e}\frac{d^2}{dx^2} + \phi^{(1)}(x)\right]\psi_n^{(1)}(x) = \varepsilon_n^{(1)}\psi_n^{(1)}(x)$ to get the modified solutions $\psi_n^{(1)}(x)$ and $\varepsilon_n^{(1)}$.
5. Same steps from 4 are now repeated to get $\phi^{(2)}(x)$, $\psi_n^{(2)}(x)$ and $\varepsilon_n^{(2)}$ in the second iteration loop, and such iterations are continued until the difference between $(\phi^{(r)}(x), \psi_n^{(r)}(x), \varepsilon_n^{(r)})$ and $(\phi^{(r-1)}(x), \psi_n^{(r-1)}(x), \varepsilon_n^{(r-1)})$ become negligibly small (i.e. set out by optimizing the desired accuracy and program runtime).

We are illustrating below the corresponding program codes written in MATLAB and the simulated results. While carrying out the above simulation numerically, in order to write both Schrödinger and Poisson's equations, the first step is to define a second-order derive matrix D_2, and a unity matrix I_0 to express the potential $U=-\phi$ (in eV). The well width is assumed in the following codes to be divided into N numbers of grids of length Δ:

```
% 2nd derivative matrix
for jc1=1:1:N
  for jc2=1:1:N
    if jc2==jc1
    D2(jc1,jc2)=-2;
    elseif jc2==jc1+1
    D2(jc1,jc2)=1;
    elseif jc2==jc1-1
    D2(jc1,jc2)=1;
    else
    D2(jc1,jc2)=0;
    end
  end
end
```

```
% Unity matrix
for jc1=1:1:N
  for jc2=1:1:N
    if jc2==jc1
    I0(jc1,jc2)=1;
    else
    I0(jc1,jc2)=0;
    end
  end
end
```

```
% rectangular potential well matrix
U=0*I0;
U(1,1)=phiB;
U(N,N)=phiB;
```

The next step is to write the Hamiltonian matrix with above matrices and predefined constants and find the eigenvectors and eigenvalues up to a relevant order (n_0), e.g. less than the barrier:

```
% Hamiltonian matrix, eigenvectors & eigenvalues
t=(1/e)*((hcut^2)/(2*mef))*(1/(del^2));
H=-t*D2 +U;
[Vec,Val]=eig(H);
EV=diag(Val)
```

and to calculate the electron density from normalized quantum and statistical probabilities:

```
% calculation of electron density
P0=0;
for jn=1:1:n0
  PQ=((Vec(:,jn)).^2)/trapz(Dis,((Vec(:,jn)).^2));
  P=log(1 +exp(-beta*(EV(jn)-EF)))*PQ;
  P0=P0+P;
end
```

The subsequent step is to solve Poisson's equation with boundary conditions (i.e. barrier at the edges of well):

```
% solving Poisson's equation
t=(1/e)*((hcut^2)/(2*mef))*(1/(del^2));
te=(e/(beta*epsS))*(t*D2)^-1;
phi=te*P0;
U0=-phi';
U(1,1)=phiB;
U(N,N)=phiB;
```

Now, to achieve self-consistency between Schrödinger and Poisson's equations, the above codes are written in a feedback loop:

```
% rectangular potential well matrix
U=0*I0;
U(1,1)=phiB;
U(N,N)=phiB;

t=(1/e)*((hcut^2)/(2*mef))*(1/(del^2));
te=(e/(beta*epsS))*(t*D2)^-1;

for con=1:1:15

% Hamiltonian matrix, eigenvectors & eigenvalues
        H=-t*D2 +U;
        [Vec,Val]=eig(H);
        EV=diag(Val);

% calculation of electron density
        P0=0;
        for jn=1:1:n0
        PQ=((Vec(:,jn)).^2)/trapz(Dis,((Vec(:,jn)).^2));
        P=log(1+exp(-beta*(EV(jn)-EF)))*PQ;
        P0=P0+P;
        end

% solving Poisson's equation
        phi=te*P0;
        U0=-phi';
        U=diag(U0-U0(1));
end
```

Once the gap between the potential values obtained from successive loops is negligibly small, the iteration can be stopped and self-consistency is considered to be achieved, as shown in the plot of the potential gap at the middle of the well in successive loops plotted in Fig. 11.1(a). Figure 11.1(b) plots the initially assumed rectangular potential well and the corresponding ideal eigenvalues $\varepsilon_n^{(0)} = n^2(\hbar\pi/a)^2/2m_e$ (solutions of the Schrödinger equation) as 'dotted' lines and the Schrödinger-Poisson self-consistent solutions and the obtained potential (which is curved instead of being flat as presumed) as 'continuous' lines. The energy eigenvalues corresponding to the confined eigenstates in the well of specified materials and dimensions (see Fig. 11.1(b)) are thus modified as:

Schrödinger eigenvalue (eV)	0.077	0.307	0.690	1.225	1.908	2.738
Schrödinger-Poisson self-consistent eigenvalue (eV)	0.096	0.320	0.698	1.227	1.903	2.724

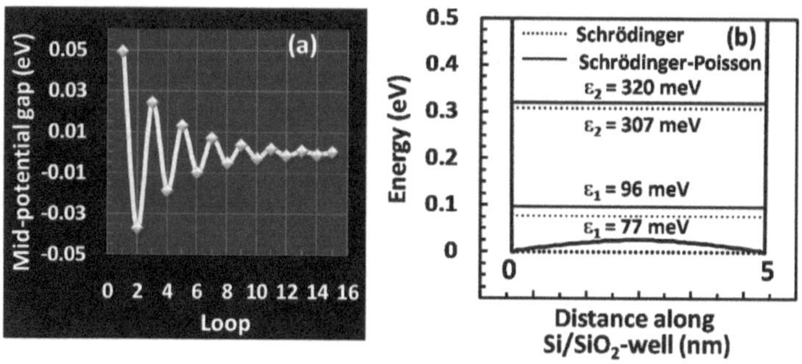

Fig. 11.1: (a) The plot of the potential gap at mid-well for successive iteration loops showing convergence, and (b) the Schrödinger-Poisson self-consistent potential that is modified from a rectangular well thereby improving the energy eigenstates.

Thus, the confined eigenstates of an electron in a real system cannot be obtained accurately by solving only the Schrödinger equation, but it requires a Schrödinger-Poisson self-consistent solution. It may be interesting to note at this point that the above results show that there can be no rectangular potential well for electrons, indeed, since the quantum-electrostatic (or mass/charge) dialectics makes the flat surface a curved one. Can it be somehow compared with the inertial-mass/gravitational-mass equivalence principle!

Nevertheless, now the same Schrödinger equation in the rectangular potential well can be solved using the NEGF framework by writing the local Green's function as $G(E,x) = Lt_{\Sigma\to 0}\left[E - H(x) - i\Sigma\right]^{-1}$ where $H(x) = \left(-\frac{\hbar^2}{2m_e}\frac{d^2}{dx^2} + \phi(x)\right)$. Consequently, the local density of states (LDOS) in per energy unit, i.e. the spectral density, is given by $D(E,x) = i(G(E,x) - G^+(E,x))$. Numerically it can be obtained using the following illustrative code (note that I_0 and H are already defined):

```
j0=1;
for E=0:0.01:6
        G=(E*I0 -H -1i*0.01*I0)^-1;
        D=1i*(G-G');
        D0=diag(D);
   S=size(D0);
   for j1=1:1:S(1,1)
        ZD(j0,j1)=abs(D0(j1,1));
        XX(j1)=(j1-1)/(N-1)*(N*del);
   end
YE(j0)=E;
j0=j0+1;
End
contourf(XX,YE,ZD)
```

The direct solutions ($\varepsilon_n, \psi_n(x)$) of Schrödinger equation in the rectangular well and the spectral density obtained from NEGF model using the above code are plotted in Fig. 11.2(a) and (b), respectively. It is imperative to note that spectral density represents the energy eigenvalues as well as eigenfunctions together as can be observed from Fig. 11.2(b) (compare with Fig. 11.2(a)). The maxima of spectral density along the energy axis depict the eigenvalues while their variations along the spatial axis refer to the corresponding probabilities (i.e. norm of the wave functions). Thus, the NEGF framework can replace the method of finding direct solutions of the Schrödinger equation in the Schrödinger-Poisson self-consistent solver following the same above-mentioned steps. Further, it may be worthy to note that the broadening of the eigenvalues (as can be seen in Fig. 11.2(b)) are due to the self-energy (Σ), which depends on the coupling of electron in the well to its surroundings in practical devices leading to associated leakages.

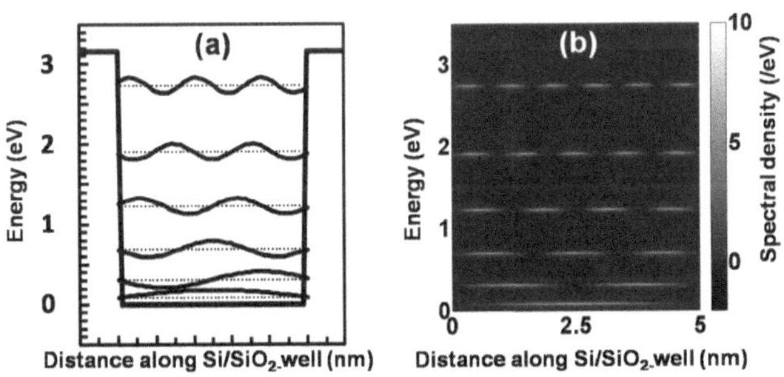

Fig. 11.2: (a) Wave functions corresponding to the eigenstates of a rectangular potential well obtained by solving the Schrödinger equation, (b) plots of spectral density obtained from Green's function formalism showing the same wave function distribution, however, along with the respective state broadening.

Here, we have not considered, and in the following sections will not be considering (unless mentioned) the effect of Poisson's equation on the nature of potential, as described earlier, with the aim of focusing on the NEGF part of the whole formulation. However, it is imperative to note that consistency between such NEGF part and Poisson's equation is necessary for accurate and realistic estimation of the device characteristics.

11.2 1D-rectangular potential barrier: gate-leakage in MOS devices

In order to find out the leakage of electrons from a system to its surroundings, for instance in a leaky MOS capacitor, the tunnelling can be obtained by considering the barrier (e.g. oxide/semiconductor and oxide/metal barrier) potential. As an illustration, electron tunnelling through the barrier (when energy is lower than potential) or scattering by the barrier (when energy is higher than potential) can be modelled in the NEGF framework considering a rectangular potential barrier for understanding the concept and method of simulating. Now the Green's function is modified to $G(E,x) = Lt_{\Sigma\to 0}\left[E - H(x) - i\Sigma_1(1) - i\Sigma_2(N)\right]^{-1}$, where the self-energies corresponding to the semiconductor and metal are given by $\Sigma_{1/2} = -\frac{\hbar^2}{2me}\frac{1}{\Delta^2}e^{ik\Delta}I_{1/2}$, with $k = \sqrt{2m_eE}/\hbar$ assuming a flat-band condition (i.e. metal/semiconductor work function to be equal), and $I_{1/2}$ is the matrix with non-zero unit value only at $(1,1)$ or (N,N) positions, respectively. The transmission coefficient for the electron to transport through or overcoming the barrier potential, for instance from semiconductor to the metal-lead through the oxide, is given by $T(E) = Trace([i(\Sigma_1 - \Sigma_1{}^+)][G(E,x)][i(\Sigma_2 - \Sigma_2{}^+)][G^+(E,x)])$. The corresponding program code is as follows:

```
j0=1;
for E=0:0.01:4.05
  k=(sqrt(2*mef*(e*E)))/hcut;
  sig1=(-t)*exp(1i*k*del)*I1;
  sig2=(-t)*exp(1i*k*del)*I2;
  G=(E*I0 -H -(sig1+sig2))^-1;
  D=1i*(G-G');
  D0=diag(D);
  S=size(D0);
  for j1=1:1:S(1,1)
    YD(j0,j1)=abs(D0(j1,1));
    XX(j1)=(j1-1)/(N-1)*(a0+a1);
  end
  T=trace((1i*(sig1-sig1'))*G*(1i*(sig2-sig2'))*G');
  XE(j0)=E;
  YT(j0)=T;
  j0=j0+1;
End
```

The results of simulation are plotted in Fig. 11.3(a) and (b) representing the LDOS and transmission coefficient. It can be observed from the plots that tunnelling probability, i.e. the transmission at energy below the barrier value, is very low for the considered material properties (e.g. effective mass, barrier value) and dimensions (e.g. barrier thickness). However, for energy higher than the barrier leading to scattering, the transmission shows peaks of value 1 at several energy values thereby indicating resonance scattering phenomena. It may also be observed that such peaks are broadened with increasing order of energy suggesting a higher rate of scattering as per the energy-time uncertainty principle as discussed in Chapter 3.

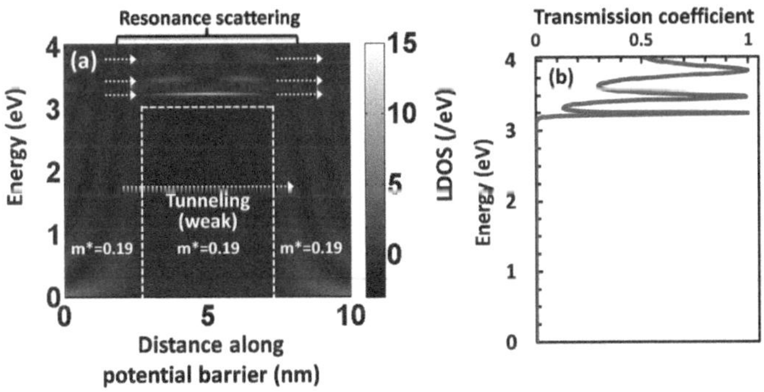

Fig. 11.3: Transport of an electron through a rectangular potential barrier showing resonance scattering: (a) the variation of LDOS with energy along the barrier, and (b) transmission coefficient with energy showing resonance peaks.

It is worthy to note at this point that the electron's effective mass is considered above to be identical throughout the entire device, which is unlikely most of the time in reality. Firstly, effective mass of an electron is a second rank tensor with at least 6 independent components, and secondly, it is different in different materials. For instance, in a conventional MOS device made up of Si/SiO_2/Al, the effective mass (m^*) of the electron varies accordingly as ~0.19 (along [100] crystal direction), 0.42 and 0.4. In order to model such real devices with more complex material properties, the Hamiltonian can be modified by the effective-mass approximation to be $H(x) = \left(-\frac{\hbar^2}{2m_e}\left[\frac{d}{dx}\frac{1}{m^*(x)}\frac{d}{dx}\right] + \phi(x)\right)$. The differential operator in the code is modified accordingly with a defined effective-mass row matrix:

```
% 2nd derivative matrix with effective-mass correction
for jc1=1:1:N
  for jc2=1:1:N
    if jc2==jc1
    D2(jc1,jc2)=-(1/m0(jc2) +1/m0(jc2+1));
    elseif jc2==jc1+1
    D2(jc1,jc2)=1/m0(jc2);
    elseif jc2==jc1-1
    D2(jc1,jc2)=1/m0(jc2);
    else
    D2(jc1,jc2)=0;
    end
  end
end
```

The simulated results are accordingly plotted in Fig. 11.4(a) and (b). The effective-mass mismatch between materials constituting the device brings more resonant scattering states closer to each other within the same energy range, however, this reduces the transmission probability significantly (by almost 50% in the present case). The result indicates that hot-electron transport can be modelled and designed accordingly by effective-mass engineering in semiconductor devices on the basis of such a concept.

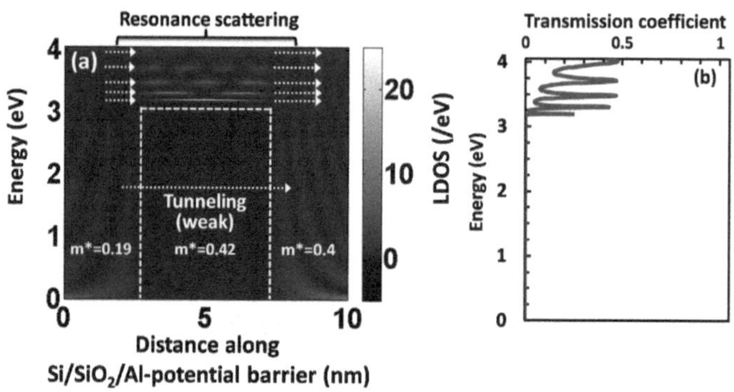

Fig. 11.4: Same as Fig. 11.3, however, considering the variation of effective mass for different materials to simulate practical devices.

11.3 1D-rectangular potential well coupled to leads: leaky DG MOS

We started with illustrating the simplest case of a 1D-rectangular potential well assuming it to confine the electrons completely inside it that can be realized experimentally in a DG MOS device with negligible gate tunnelling leakage. However, now if we include the possible gate leakage in the model, then the coupling with two leads can be incorporated by the corresponding self-energies (as in the above section). The respective code (without considering the effective mass mismatch) is given by

```
% potential row matrix
for nn=1:1:10
  U(nn,nn)=0;
end
for nn=11:1:18
  U(nn,nn)=phiB;
end
for nn=19:1:38
  U(nn,nn)=0;
end
for nn=39:1:46
  U(nn,nn)=phiB;
end
for nn=47:1:56
  U(nn,nn)=0;
End

%Hamiltonian matrix
t=(1/e)*((hcut^2)/(2*mef))*(1/(del^
2));
H=-t*D2 +U;
```

```
% LDOS and T

j0=1;
for E=0:0.01:4.05
  k=(sqrt(2*mef*(e*E)))/hcut;
  sig1=(-t)*exp(1i*k*del)*I1;
  sig2=(-t)*exp(1i*k*del)*I2;
  G=(E*I0 -H -(sig1+sig2))^-1;
  D=1i*(G-G');
  D0=diag(D);
  S=size(D0);
  for j1=1:1:S(1,1)
    YD(j0,j1)=abs(D0(j1,1));
    XX(j1)=(j1-1)/(N-1)*(a0+a1);
  end
  T=trace((1i*(sig1-sig1'))*G*(1i*(sig2-
sig2'))*G'),
  XE(j0)=E;
  YT(j0)=T;
  j0=j0+1;
end
```

The simulated LDOS and transmission coefficient are plotted in Fig. 11.5(a) and (b) which show the possibility of sharp tunnelling from one lead to the other through the well, e.g. one gate to the other of a DG MOS, with increasing probability for higher quantized energy states. It may be noted that for the considered values of material and dimensional parameters, the tunnelling probability is very low (<0.1%), which can however be much higher for different values. It is also important to note that if a voltage difference is introduced between the two leads, the same NEGF model can ac-

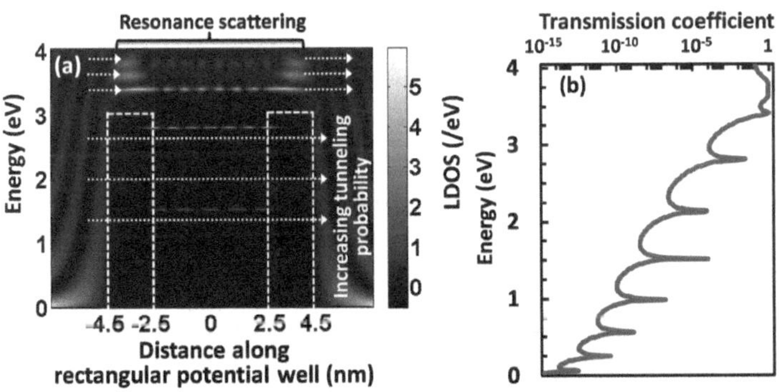

Fig. 11.5: Transport of electron through rectangular potential well separated from metallic leads by two barriers, showing tunnelling and resonance scattering: (a) the variation of LDOS with energy along the well; and (b) the transmission coefficient with energy showing tunnelling and resonance peaks.

commodate the conventionally known Fowler-Nordheim (FN) tunnelling and conduction-to-valence band tunnelling as well as the trap-assisted hopping if the corresponding discrete defect states within the barrier (i.e. physically the oxide) are considered.

11.4 Two coupled 1D-rectangular potential wells connected to leads: resonant tunnelling devices

If two adjacent potential wells separated by a barrier are now considered that are connected to the metallic leads through respective barriers, then electrons (or holes) can tunnel back and forth through the separating wall. And particular suitable voltages at one lead with respect to the other, where the eigenstates of two wells feedback each other positively, can drive the electrons completely resulting in high-peak resonant currents. Resonant tunnelling diodes, transistors and tunnel field-effect transistors (FETs) as well as cascading systems and gate-defined or embedded coupled quantum dots are practical examples of such device. As an example, a uniform 1 eV barrier leads to the following simulated results plotted in Fig. 11.6(a) and (b):

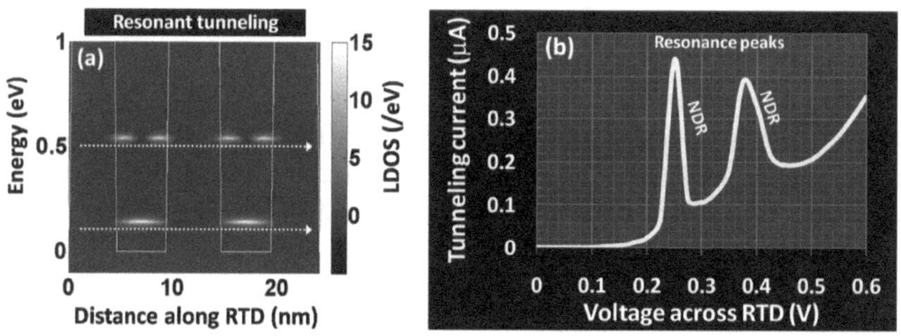

Fig. 11.6: Transport of electron through two adjacent rectangular potential wells showing resonance tunnelling: (a) the variation of LDOS with energy along the wells, and (b) tunnelling current with voltage across the resonant tunnelling device (RTD) that exhibits current peaks and negative differential resistances (NDR).

11.5 NEGF simulation of nanowire FET: effect of scaling the diameter

In order to illustrate the simulation of nanowire FET devices with ultra-scaled diameters (≤5 nm), as is being done and projected in the state-of-the-art, industrial-level fabrication (e.g. TSMC, Samsung, and Intel), we have considered a Si/SiO_2 core-shell nanowire with the channel to be ideally ballistic with a length of 10 nm, and all the material parameters are chosen accordingly. In the following simulations, we have not incorporated the contact resistances associated with source/drain electron effective mass and phonon scattering as described in Chapters 4 and 5, respectively, which have been illustrated later while simulating an experimental result. We are illustrating below the relevant parts of the codes of a complete program along with the corresponding simulated results. At this point, it is worthy to mention that in all the simulations performed and the results plotted here, the operating temperature of the reservoirs is assumed to be room temperature, i.e. 300 K. No impurity scattering centres or doping in the nanowire channel and no fixed oxide charges are considered. Further, the flat-band condition is assumed to be preserved by choosing an appropriate gate.

1. **Transverse sub-band energy:** The effective mass components and grid spacing are pre-defined to write the kinetic energy part of the Hamiltonian as:

```
%Transverse Hamiltonian part-1
for r=1:1:Nc
  for p=(r-1)*Nc+1:1:r*Nc
    for q=(r-1)*Nc+1:1:r*Nc
      if q==p
        O(p,q)=-(1/mx1+1/mx2)*(((hcut^2)/2)/(ac^2))-(1/my1+1/my2)*(((hcut^2)/2)/(bc^2));
      elseif q==p-1
        O(p,q)=(1/mx1)*(((hcut^2)/2)/(ac^2));
      elseif q==p+1
        O(p,q)=(1/mx2)*(((hcut^2)/2)/(ac^2));
      else
        O(p,q)=0;
      end
    end
  end
end
```

```
%Transverse Hamiltonian part-2
for r=2:1:Nc
  for p=(r-1)*Nc+1:1:r*Nc
    for q=(r-2)*Nc+1:1:(r-1)*Nc
      if q==p-Nc
      O(p,q)=(1/my1)*(((hcut^2)/2)/(bc^2));
      else
      O(p,q)=0;
      end
    end
  end
end
```

```
%Transverse Hamiltonian part-3
for r=2:1:Nc
  for p=(r-2)*Nc+1:1:(r-1)*Nc
    for q=(r-1)*Nc+1:1:r*Nc
      if q==p+Nc
      O(p,q)=(1/my2)*(((hcut^2)/2)/(bc^2));
      else
      O(p,q)=0;
      end
    end
  end
end
```

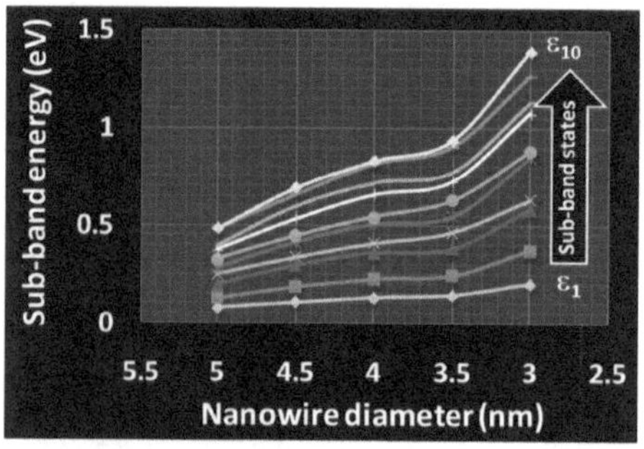

Fig. 11.7: The plots of simulated sub-band energies (for 10 transverse modes) of a nanowire FET while downscaling the diameter.

The obtained energy eigenvalues for 10 transverse sub-bands of the cylindrical Si-nanowire for scaling down the diameter from 5 nm to 3 nm plotted in Fig. 11.7 clearly suggest the increase in the energy value and energy gap between consecutive sub-bands with miniaturization.

2. **Transmission coefficient:** After writing the Green's function as described in a step-by-step manner as code, the transmission coefficient can be written as

```
%Transmission coefficient
jE=1;
for E=(-1.5:3/100:1.5);
        T1=trace(TS*fS*i*(G-G'))-trace(TS*G*(TS*fS+TD*fD)*G');
        T2=trace(TD*fD*i*(G-G'))-trace(TD*G*(TS*fS+TD*fD)*G');
        Tr(jE)=(T1-T2)/(2*(fS-fD));
        XE(jE)=E;
        jE=jE+1;
End
```

As can be seen from the simulated results plotted in Fig. 11.8(a)-(f) for different nanowire diameters and drain biases, the transmission coefficient shows unit steps at each energy sub-band state, which, we argue, is owing to the 1D transport through the nanowire FET with two quantized transverse dimensions (see Chapter 12 for details). Such transmission steps are observed to considerably shift with drain bias for higher diameters compared to the smaller ones. This is attributed to the larger band-bending due to the longitudinal field leading to an average shift of sub-bands towards lower energy. It also upholds that lowering the diameter for strong confinement reduces the transverse-longitudinal correlation (described in Chapter 4) thereby leading to achieving ideal 1D transport.

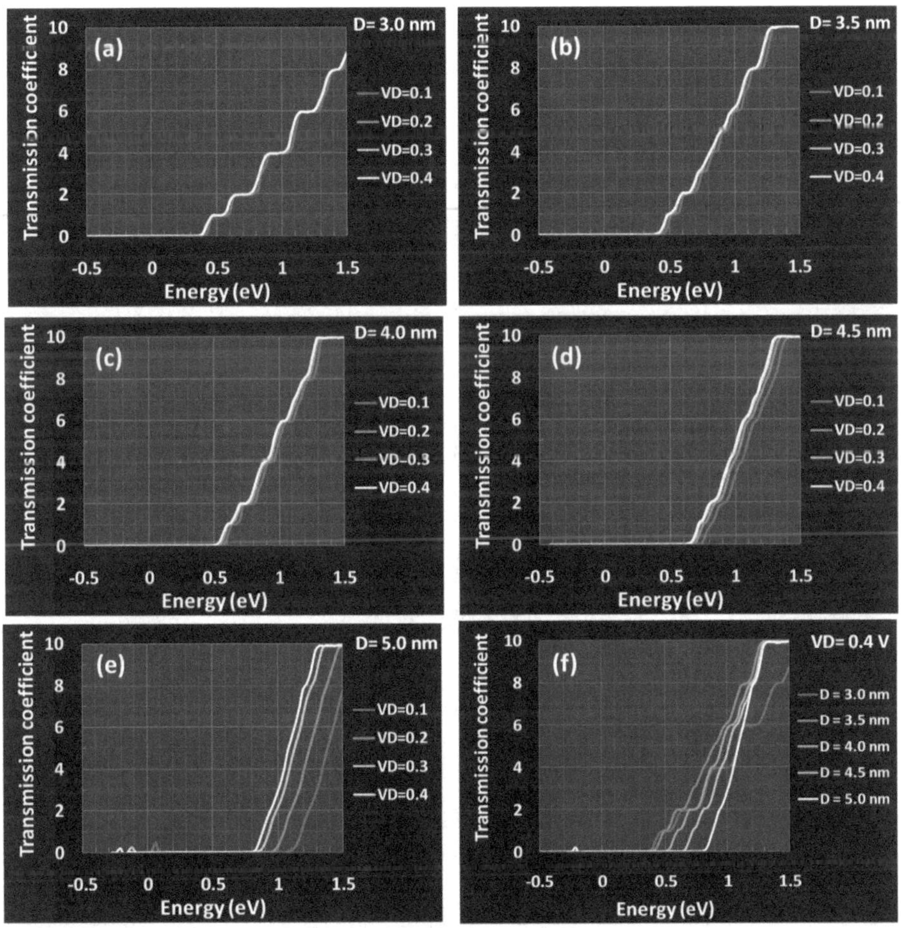

Fig. 11.8: Plots of transmission coefficient with energy for different drain voltages at a gate voltage of 0.4 V, the nanowire diameter being: (a) 3 nm; (b) 3.5 nm; (c) 4 nm; (d) 4.5 nm; and (e) 5 nm; (f) for all such diameters at drain a bias of 0.4 V.

3. **Local density of states (LDOS):** At any particular spatial point in the nanowire channel, the LDOS is given by,

```
% LDOS
jE=1;
for E=(-1.5:3/100:1.5);
        DS=(1/(2*pi*a))*G*TS*G';
        DD=(1/(2*pi*a))*G*TD*G';
        D(jE)=DS+DD;
        XE(jE)=E;
        jE=jE+1;
End
```

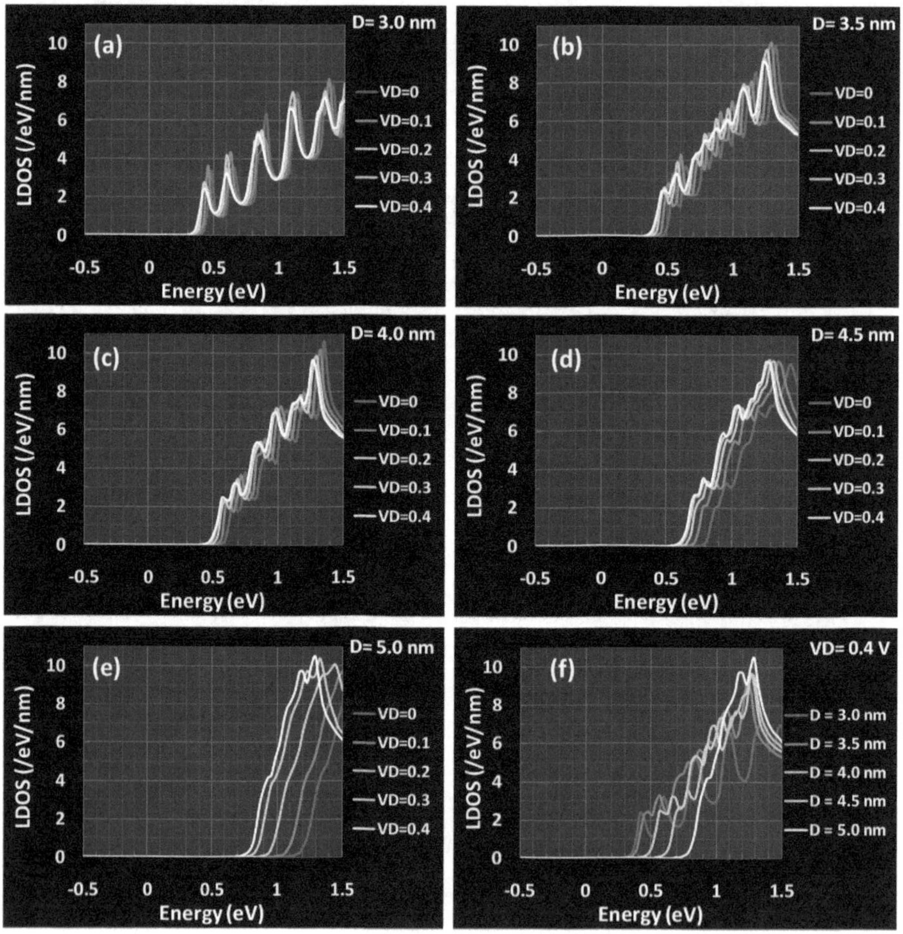

Fig. 11.9: Plots of LDOS at the mid-channel of a nanowire FET with energy for different drain voltages at a gate voltage of 0.4 V, the nanowire diameter being: (a) 3 nm; (b) 3.5 nm; (c) 4 nm; (d) 4.5 nm; and (e) 5 nm; (f) for all such diameters at drain a bias of 0.4 V.

Here, we have plotted in Fig. 11.9 the LDOS at the centre of the channel. However, it can be calculated (and accordingly plotted) for any spatial point in the channel and also separately for the contribution from the source and drain.

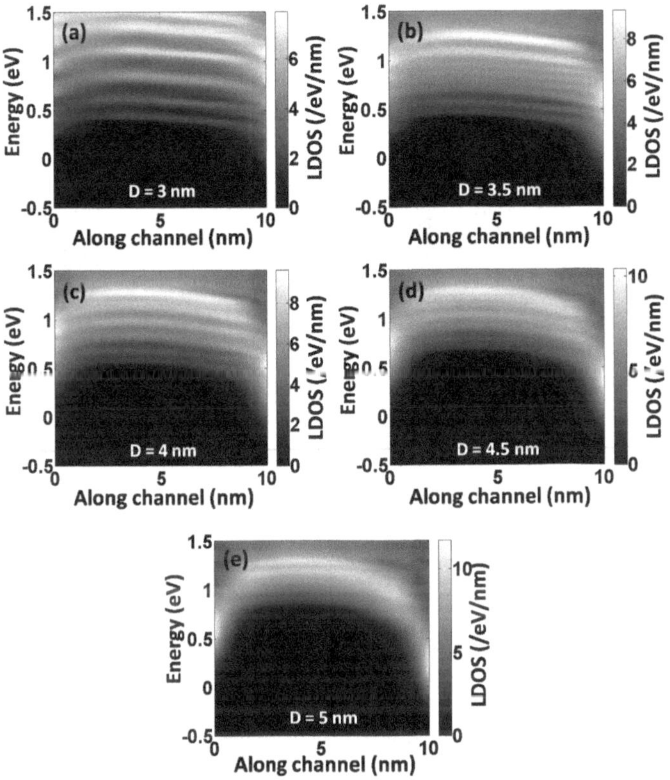

Fig. 11.10: Plots of LDOS with energy along the nanowire FET channel showing sub-band variation for different drain voltages at a gate voltage of 0.4 V, the nanowire diameter being: (a) 3 nm; (b) 3.5 nm; (c) 4 nm; (d) 4.5 nm; and (e) 5 nm; (f) for all such diameters at a drain bias of 0.4 V.

Its behaviour with nanowire diameter and drain bias is similar to that of the transmission, the difference only being in the functional form. LDOS shows peaks at the sub-band minima and then decay, which is expected for 1D transport (i.e. $\mathrm{DOSH}_{1D}(E) \propto = E^{-1/2}$). However, it will be shown in Chapter 12 that this widely accepted DOS formula derived from Green's function, in fact, does not result in such a functional form, and thus, may need modification. Nevertheless, the variation of LDOS over the channel with energy is shown in Fig. 11.10 that depict the corresponding sub-band profiles for various diameters.

4. **Conduction band alignment and carrier density:** The spatial variation of LDOS or sub-bands, on the other hand, qualitatively represents the variation of the conduc-

tion band profile along the nanowire channel, as can be seen in the plots of Fig. 11.11 for different nanowire diameters and drain biases. However, it is worth noting that, while LDOS is obtained from the Green's function solution of the modified Schrödinger equation, band potential is obtained by solving Poisson's equation; and their similar nature shows the Schrödinger-Poisson self-consistency. Interestingly, it can be observed from the plots of Fig. 11.11 that lower voltages at higher diameters do not show consistent results here (see Fig. 11.11 (e)) due to the non-convergence of the Schrödinger-Poisson coupled equations, which will be elaborated later (in Fig. 11.13).

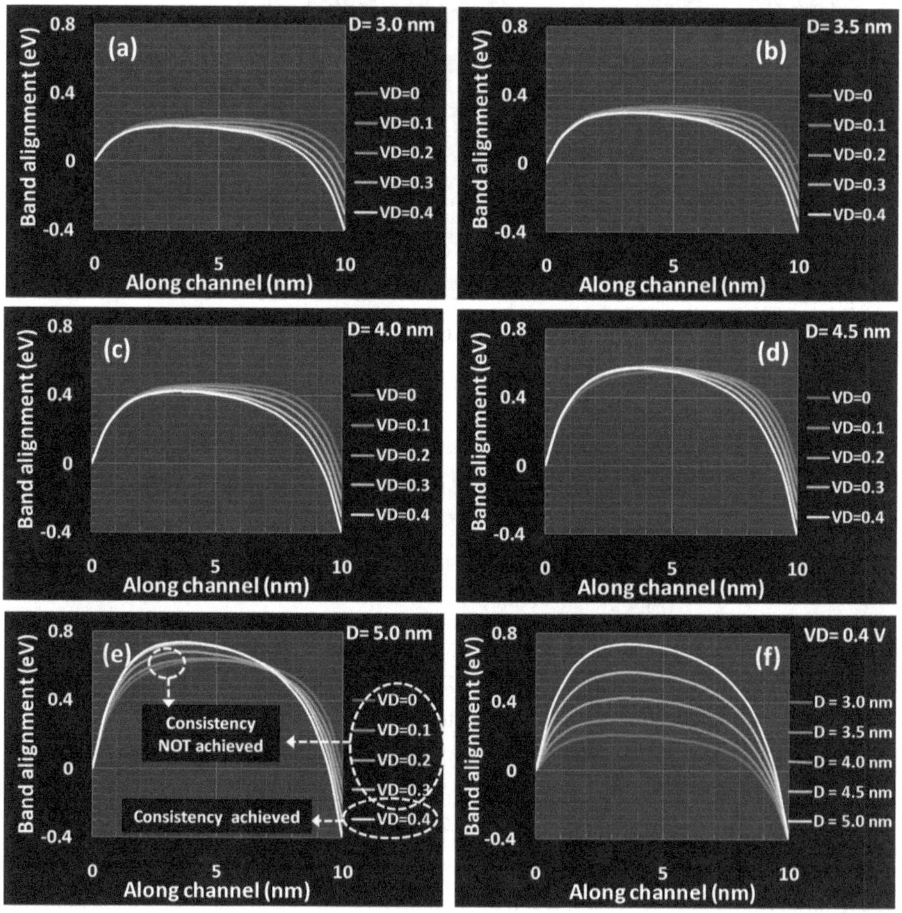

Fig. 11.11: The conduction band profile along the nanowire FET channel for different drain voltages at a gate voltage of 0.4 V, the nanowire diameter being: (a) 3 nm; (b) 3.5 nm; (c) 4 nm; (d) 4.5 nm; and (e) 5 nm; (f) for all such diameters at a drain bias of 0.4 V.

The program code for such potential calculation is trivial (see Chapter 4) with the predefined second-order derivative matrix as well as material and device parameters like permittivity and thickness of oxide:

```
% band potential calculation

D=D2-((2/R^2)*(kOx/kSm)*(1/log(1+tOx/R))*)*I;
In=inv(D);

pot=((e/(pi*R^2*f*epsilon0*kSm))*In*n1D');
pot=pot'+Vg;
```

where the charge density associated with 1D carrier concentration is obtained from LDOS and source/drain Fermi distributions as

```
% carrier concentration
for Nz=1:1:N
    for g=1:1:jE-1
            A11=e*XE(g);
            f1=1/(exp(beta*(A11-EfS))+1);%
            f2=1/(exp(beta*((A11-(EfD-e*VD))))+1);

            Sdos=DS(g,Nz);
            Ddos=DD(g,Nz);

            ft1(g)=Sdos*f1+Ddos*f2;
            E1(g)=A11/e;
    end
 n1D(Nz)=trapz(E1,ft1);
end
```

The 1D carrier density along the channel simulated by running the program is plotted in Fig. 11.12 for the assumed nanowire diameters and drain voltages. It is apparent from such plots that, at any non-zero drain voltage, the carrier density shows asymmetry around the mid-channel owing to the applied bias at drain, the source being assumed to be grounded. It also suggests that such carrier density represents a dynamic charge concentration obtained from a retarded function-based NEGF formalism in contrast to what the Schrödinger or Poisson's equation alone can offer us, i.e. to provide static particle distributions. Nevertheless, the same consistency problem of the program, as faced while calculating potential, is observed in the carrier profile for higher diameter at low voltages.

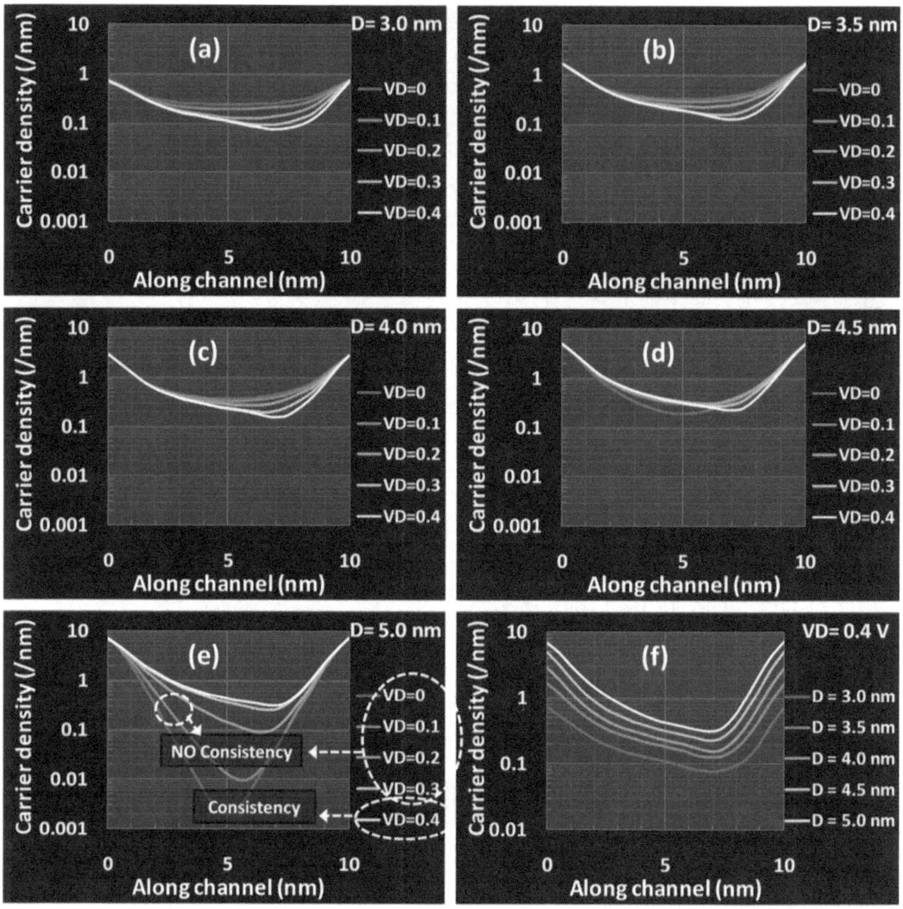

Fig. 11.12: The variation of 1D carrier density along the nanowire FET channel for different drain voltages at a gate voltage of 0.4 V, the nanowire diameter being: (a) 3 nm; (b) 3.5 nm; (c) 4 nm; (d) 4.5 nm; and (e) 5 nm; (f) for all such diameters at a drain bias of 0.4 V.

5. **Consistency of Schrodinger-Poisson solver and convergence of program:** As described earlier in this chapter and Chapter 4, the consistency gap for successive iteration loops is plotted for all such cases in Fig. 11.13. It clearly shows that the program has not converged in 10 loops for higher diameters of the nanowire FET. It is owing to the fact that we have considered only 10 sub-bands as coupled modes, and with increasing diameter, such sub-bands get closely spaced requiring more states to be incorporated within the same energy range. It clearly suggests that for higher diameters, more sub-band states need to be considered and it may be necessary to iterate over more loops to get convergence and respective self-consistency.

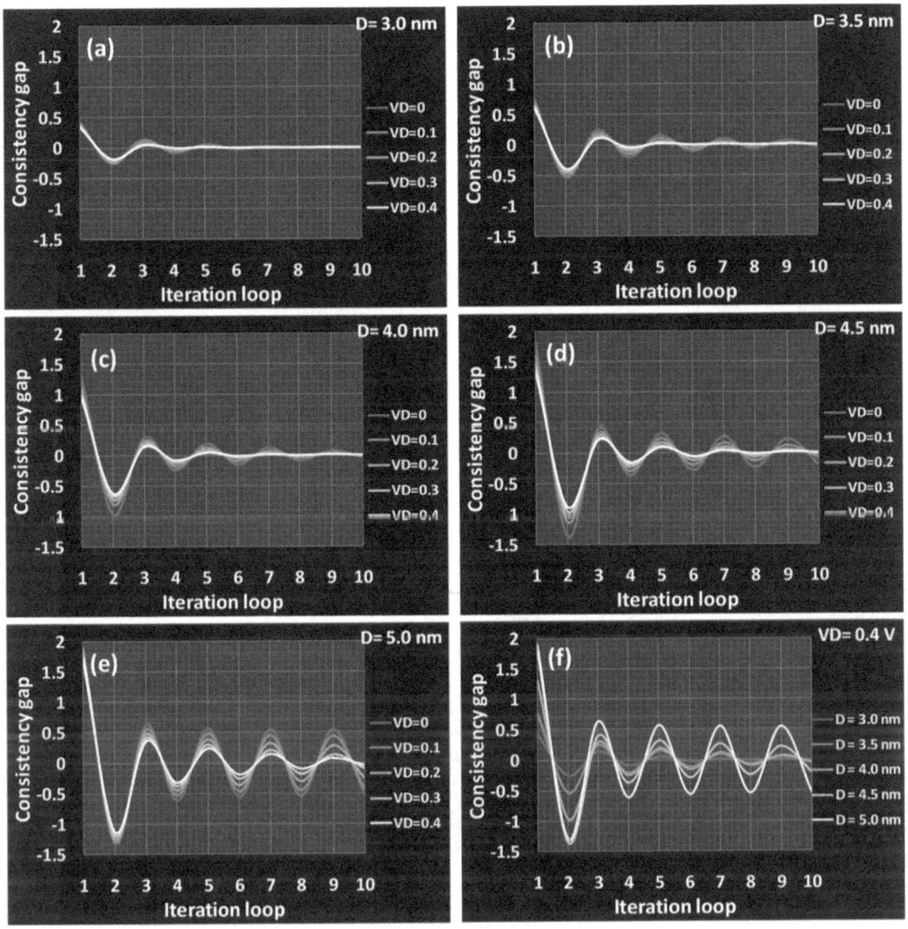

Fig. 11.13: Plots of the consistency (mid-channel potential) gap with program iteration loops for different drain voltages across the nanowire FET channel at a gate voltage of 0.4 V, the nanowire diameter being: (a) 3 nm; (b) 3.5 nm; (c) 4 nm; (d) 4.5 nm; and (e) 5 nm; (f) for all such diameters at a drain bias of 0.4 V.

6. **Output current simulation:** Once the final self-consistent transmission coefficient is obtained, the current can be calculated using the Landauer formula:

```
% Current calculation
for g=1:1:jE-1
    A11=e*XE(g);
    f1=1/(exp(beta*(A11-EfS))+1);
    f2=1/(exp(beta*((A11-(EfD-e*Vd))))+1);

    T0=Tr(g);
    ft(g)=T0*(f1-f2);
    E1(g)=A11/e;
end
  F=trapz(E1,ft);
  ID=(e^2)*(1/(2*pi*hcut))*F;
```

The simulated output characteristics, i.e. drain current against drain bias, and mutual or transfer characteristics, i.e. drain current with gate voltage, are plotted in Fig. 11.14 and Fig. 11.15, respectively, for different diameters.

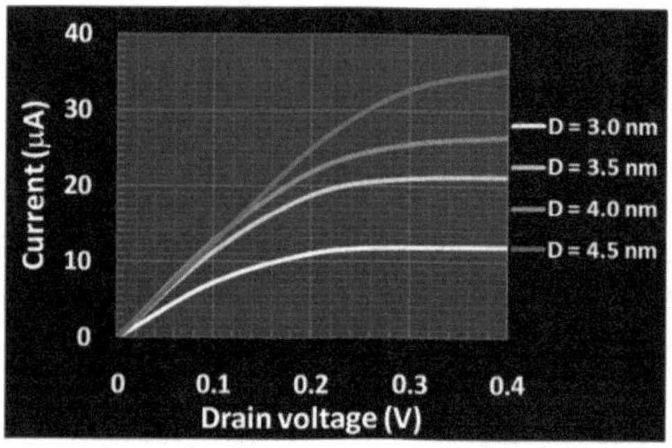

Fig. 11.14: Simulated self-consistent output characteristics of a nanowire FET showing the variation of drain current with drain bias at a gate voltage of 0.4 V, for different nanowire diameters.

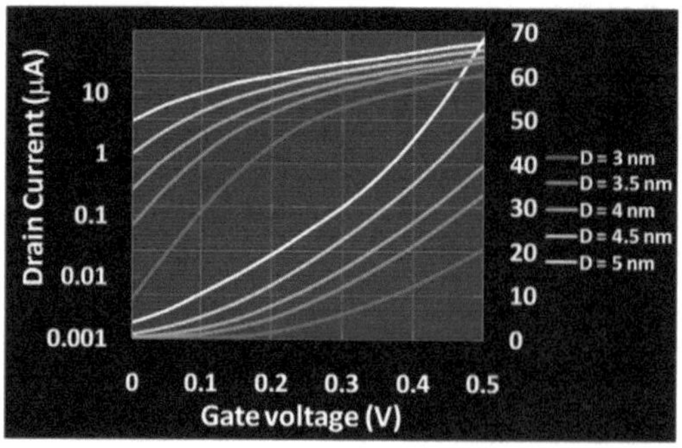

Fig. 11.15: Simulated self-consistent mutual (transfer) characteristics of a nanowire FET showing the variation of drain current with gate voltage at a drain bias of 0.4 V, for different nanowire diameters.

The device parameters of significance, like threshold voltage, sub-threshold swing, mutual conductance, etc., can be extracted from such data. In order to examine the drain-induced barrier lowering (DIBL), however, mutual characteristics need to be calculated for low drain bias, as shown in Fig. 11.16. At this point, it is worth mentioning that the NEGF model offers a more in-depth analysis of such short-channel effects like DIBL from an internal device physics perspective. For instance, the LDOS at the barrier maxima due to the source and drain can be obtained separately, as shown earlier. DIBL is originated due to the shift of the LDOS peak position due to the drain compared to that for the source at higher drain biases. On the other hand, here also the higher diameters for lower drain bias show inconsistent results due to lack of program convergence.

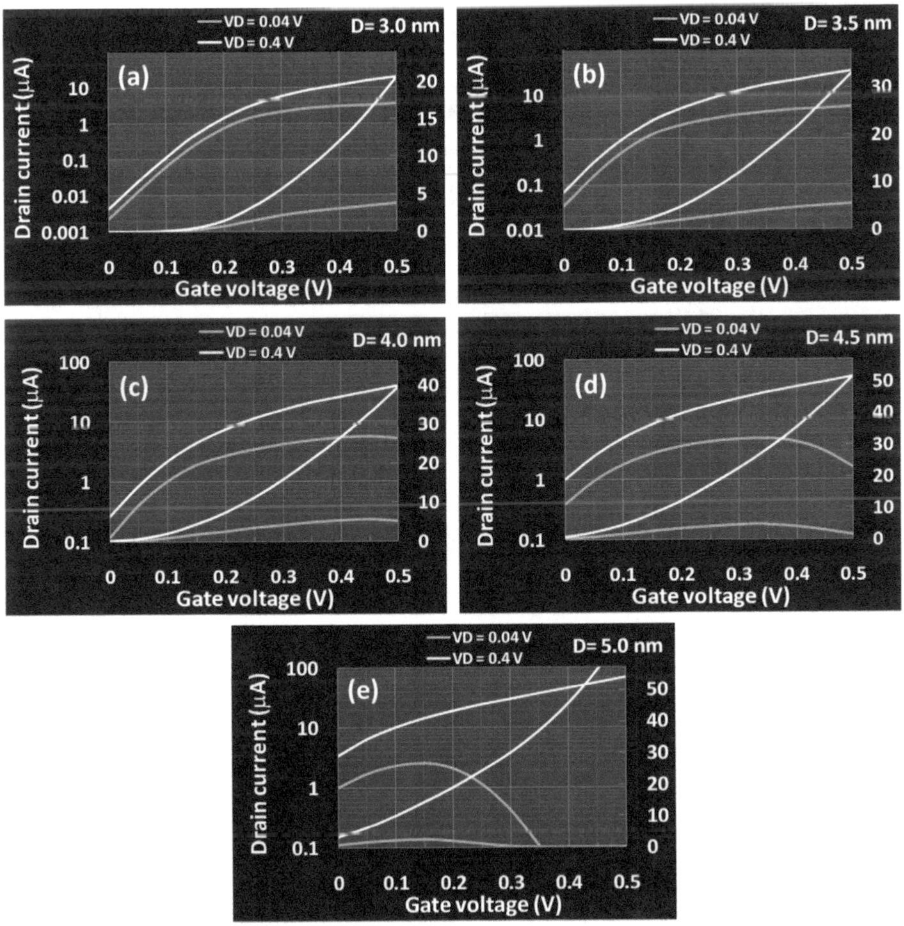

Fig. 11.16: Simulated drain current plots with gate voltage at a drain bias of 0.4 V and 0.04 V, for different nanowire diameters: (a) 3 nm; (b) 3.5 nm; (c) 4 nm; (d) 4.5 nm; and (e) 5 nm.

7. **Simulating experimental device performance:** In order to simulate practical devices on the basis of programs illustrated above, two important factors need to be incorporated: contact resistance, as described in Chapter 4, and electron-phonon scattering, as elaborated in Chapter 5. Introducing such effects into our NEGF-based program, an experimental nanowire FET with a Ω-gate structure and state-of-the-art device dimensions (see ref-[4] of Chapter 4) has been simulated and the results are plotted in Fig. 11.17. Such agreement of the experimental data and simulated result upholds the effectiveness of a complete NEGF program simulator.

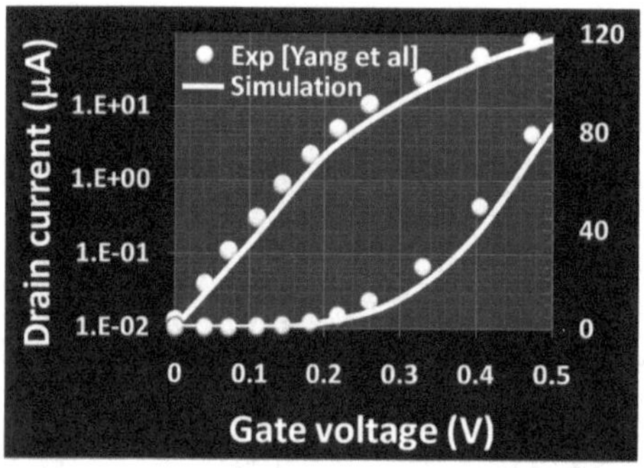

Fig. 11.17: NEGF simulation of the transfer characteristics of experimental performance of a nanowire Ω-gate fin-FET with state-of-the-art device dimensions (see ref [4] of Chapter 4) and the comparative plots of simulated (continuous line) and experimental (dots) data.

Now the entire scheme of such simulation of a nanostructure electronic and optoelectronic devices, including but not limited to the state-of-the-art COMS-compatible FETs, while operating in the ideal ballistic limit, or in the presence of phonon scattering, as well as photodevices including emitter, absorber/detector and solar cells, is diagrammatically represented in Fig. 11.18, so that such a complete NEGF simulator can be constructed.

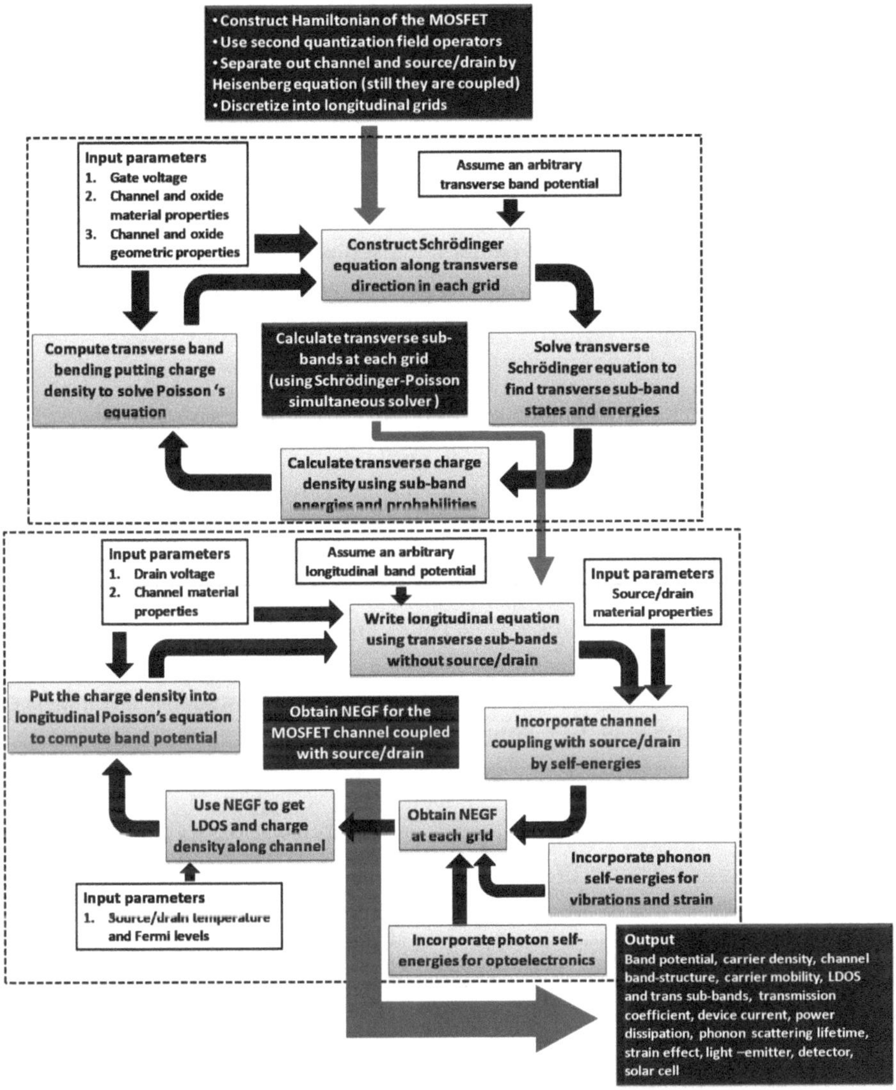

Fig. 11.18: The systematic workflow for developing the theoretical model of charge transport in a quantum scale electronic and optoelectronic devices including (planar or nanostructure) MOSFETs operating even in the ballistic limit with or without phonon scattering and in absence or presence of light, on the basis of NEGF formalism accompanying the quantum-electrostatic self-consistent framework that offers all the physical and electrical/optical outputs on basis of material and geometric properties as inputs.

12 Beyond NEGF formalism: exploring fractional dimensions

The physical dimension of an object or a system is conventionally defined as the minimum number of independent coordinates that are necessary to locate any point of such object/system. The coordinates can be generalized to the components of any physical quantity describing the system, for instance, any observable in case of a quantum system, or space-time for a classical relativistic system. It is apparent that the number of such coordinates must be a positive integer. Can it be a fractional number ever?

In fact, a mathematical concept of fractional dimension was introduced by Hausdorff in 1918, and later developed by Besicovitch, which has been applied to model the exciton dynamics in anisotropic solids by He [1] that corroborates with experimental results in quantum systems [2–4]. Such a model is based upon the mathematical equivalence of the excitons in an integral-dimensional anisotropic solid with those in a fractional-dimensional isotropic system, the dimensional fraction representing the degree of anisotropy. The concept has further been applied to analyse the electromagnetic wave propagation and scattering in fractal media [5, 6]. However, here we are not dealing with any such analogous concept of virtual equivalence, but rather we are questioning whether the real physical space (in terms of positional coordinates) of a system can have a fractional dimension.

Conventionally, the spatial dimension of an object is considered in a relative sense: if the spatial extents of the object along three mutually orthogonal directions are comparable to each other, then we call it a 3D structure; whereas if one of the spatial extent is much smaller than the other two, then it is regarded as a 2D construction; and when two of them are negligibly small compared to the third one, it is considered to be 1D. For instance, a book is 3D, a sheet of paper is 2D, and a pen can be assumed to be an 1D object (see Fig. 12.1(a)). In this sense, a 0D object is something with all three spatial extents being arbitrarily small, but this time in comparison to the spatial extents of some other object(s). However, such perception-based definition raises the question how much the dimension(s) has to be smaller (i.e. in what ratio) than the rest one(s). Can such relative smallness determine the 'dimension' of a system? Or does there exist any universal standard number to define such a ratio? Even if we get such a universal standard ratio, then also, for instance, a 2D object will have a finite thickness, whatever small it may be, and just rotating the axes of the previously assumed dimensional directions, such ratio(s) will be changed (see Fig. 12.1(b)). Thus, now the dimension of the object depends also on our choice of the axial directions! A complete mess, isn't it!

Thus, since 'dimension' cannot be defined uniquely in terms of any such universal standard value, even if any, that can ideally compare between spatial extents along independent directions, it has to be determined on the basis of a measurable

 | https://doi.org/10.1515/9783111545950-012

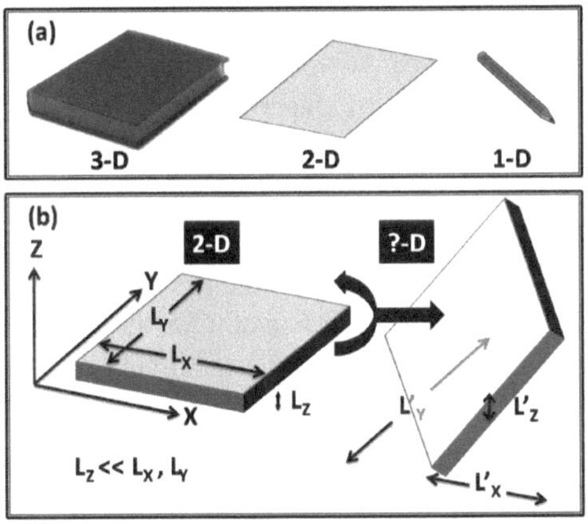

Fig. 12.1: (a) Examples of conventionally considered 3D, 2D, and 1D objects and (b) change in the spatial dimensions of a so-called 2D sheet while rotating it.

physical property of the systems that has distinguishing one-to-one behavioural correspondence with 0, 1, 2, and 3D. Interestingly, the advent of quantum theory, exploration of the fundamentals of material science, and the remarkable progress of nanotechnology (i.e. nanofabrication and nanocharacterization) has opened up a window in this context. As a result of such development, now we know that the spectral density function (i.e. the functional form of density of states in terms of energy) for a 3D, 2D, 1D, and 0D system (e.g. non-relativistic) varies, as shown schematically in Fig. 12.2, with energy (E) as $(E-\varepsilon_0)^{1/2}$, $\vartheta(E-\varepsilon_0)$, $(E-\varepsilon_0)^{-1/2}$ and $\delta(E-\varepsilon_0)$, respectively, where ε_0 is an energy eigenvalue or band minima (for 3D).

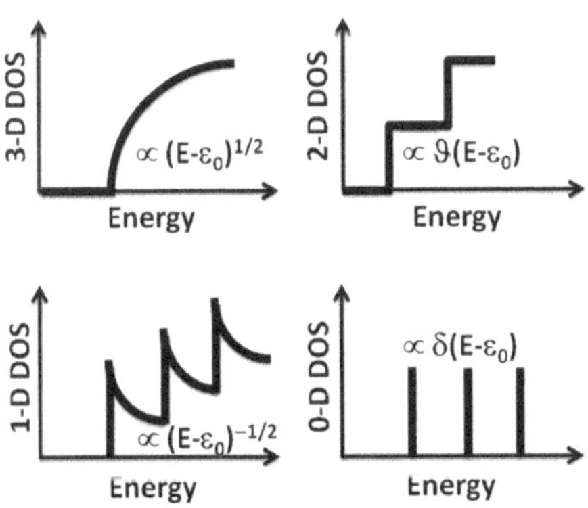

Fig. 12.2: Schematic illustration of DOS or spectral density function of 3D, 2D, 1D, and 0D systems.

Furthermore, alongside the 3D substrates of various materials, we can now fabricate nanosheets, nanowires, and quantum dots, correspondingly as 2D, 1D, and 0D systems, so that their spectral density can be measured using differential current/capacitance, scanning tunnelling microscopy and/or angle-resolved photoemission spectroscopy [7–10]. We argue here that, conversely, we can propose an objective and measurable definition of the spatial dimension in terms of such spectral density as:

> The spatial dimension of any object/system is 0, 1, 2, or 3 as defined by whether its spectral density varies with energy as $\delta(E-\varepsilon_0)$, $(E-\varepsilon_0)^{-1/2}$, $\vartheta(E-\varepsilon_0)$, or $(E-\varepsilon_0)^{1/2}$, respectively.

This can be considered as an objective definition of spatial dimension for the systems exhibiting non-relativistic internal dynamics (the functions will be modified for relativistic case), since such natures of spectral densities originate from the fundamental uncertainty principle ($\Delta x_j \Delta p_j \geq \hbar$) that restricts the phase-space volume corresponding to one spatial dimension to be not less than Planck's constant (h). Thus, the possible number of quantum states is proportional to $d^N\vec{p}$ for a N–D system, which, since $p^2 \propto (E-\varepsilon_0)$, gives rise to the above functional forms. At this point, it is worthy to note that a system is usually made less than 3D by geometrically thinning one or more spatial dimensions (e.g. nanosheet, nanowire, and quantum dot), thereby confining the motion of its constituent particles (e.g. electrons and holes in an electronic system) along such direction(s). However, such a conversion of the spatial dimension of a system from '3' to '2' or '1', or even '0', can be carried out, not only by geometrically reducing its corresponding spatial extents (e.g. by nanofabrication), but also by other means. For instance, the 3D motion of charges can be confined into 1D by applying external magnetic field (known to be Landau quantization), as described in Chapter 7, which has been realized experimentally in study of quantum Hall effect [11, 12]. Similarly, external electric field applied on the gate-defined quantum dots (quite familiar as spin/charge qubits [13]) or voltage-tunable quantum dots (explored by the present author with others [14]) can quantize two or one spatial dimension(s), respectively, to create even 0D system.

Therefore, in contrast to the concept of gravity where interaction was replaced by geometry, here we preliminarily argue that, to define spatial dimension of a system, it is necessary to replace geometry by interaction of the system with its surroundings, at least from materials point of view (although we are not discarding the geometric interpretation of gravity). For analogy, as shown in Fig. 12.3, consider the cases: (1) a woman is walking through a lane bounded by two large walls, (2) a few women running randomly on an open ground, and (3) a woman is (may be virtually) flying. It is apparent that the spatial extent of such woman (women) may be regarded as 1D, 2D, and 3D, respectively (we are not illustrating a caged woman with a spatial extent of 0D). Now, in case 1, the woman can only cross the walls if she can manage to expend an energy more than the so-called gravitational potential (i.e. simply mgH, H being the height of the wall) to get a 2D space, whereas in case 2, the women are free to move in any horizontal direction, but only keeping the footsteps on the ground, where the so-called gravitational attraction and electromagnetic repulsion (the so-called normal reaction of ground) are creating potential minima to restrict their motion vertically. On the other hand, in case 3, the woman, may be a superwoman, has somehow managed to balance the gravity and buoyancy to make herself free to fly in any direction. Evidently, the woman of case 1 cannot touch anyone outside the walls, and that of case 2 cannot touch anyone above a particular height, whereas the case 3 woman can touch anyone, sky being the limit. It clearly suggests that spatial dimension is nothing but a degree (i.e. a kind of strength) of how much a system can interact with its surroundings, and *vice versa*, '3' being the highest value of such degree/strength indicating an unrestricted communication.

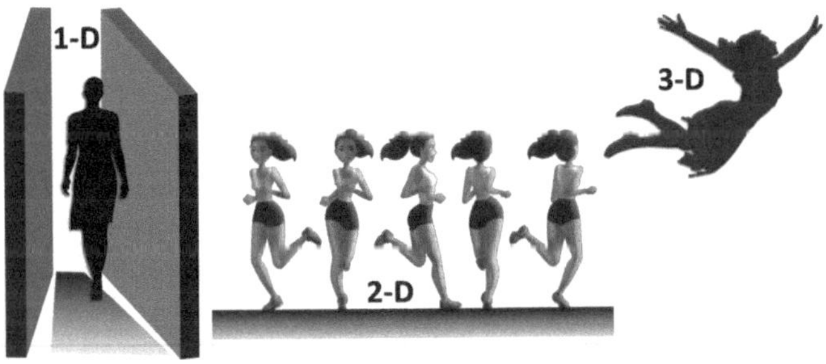

Fig. 12.3: An analogical depiction of the correlation between spatial dimension and the restrictedness of motion.

The same behaviour can be thought of for any system such as an electronic system as mentioned above. An electron in the conduction band of a bulk semiconductor (at the interior, not at its surface) is free to move in any direction (averaging out the effect of scattering), while, in a 2D material or a nanosheet, is confined along the thickness, not due to its smallness, but by the surface potentials (almost equal to electron affinity of the material). Such surface potential originates due to the interaction of charges in the material lattice potential near the surface with the vacuum states. Similarly, in both, the embedded quantum dots within insulator or gate-defined/voltage-tunable quantum dots, the electrostatic interaction of electrons inside dot and the charges outside insulator leads to the quantum confinement thereby creating the 0D structures. The thickness of nanosheet or the size of dot (along with effective mass of electron) only determines the energy gaps between quantum states, but not the nature of spectral density, and thus, we argue, not the spatial dimension. This can be more clearly visualized in case of Landau quantization, where, without any so-called geometrical boundary, two of the dimensions are quantized through interacting with the system by applying magnetic field to create a 1D electronic system.

Being such concept (if) accepted, that the spatial dimension of a system is defined by its strength of interaction with its surroundings, now the question arises: is such degree/strength of interaction, i.e. the spatial dimension, mandatorily be an integer, i.e. 0, 1, 2, or 3? While quantizing one or more spatial dimensions (or retrieving it back) by interaction (e.g. electric or magnetic field), is such a transition from a higher spatial dimension to a lower one, or *vice versa*, a single-step jump, or there may exist intermediate spatial dimensions between two integral ordered ones, namely the 'fractional dimensions'?

For instance, the nature of spectral density in quantum Hall effect or gate-defined (or voltage-tunable) quantum dots does not switch from $(E-\varepsilon_0)^{1/2}$ to $(E-\varepsilon_0)^{-1/2}$, or from $\vartheta(E-\varepsilon_0)$ (or $(E-\varepsilon_0)^{-1/2}$) to $\delta(E-\varepsilon_0)$, when, respectively, the magnetic field or gate voltage is applied with slowly increasing strength from zero. It cannot be understood in terms of how many distinct states have appeared (*C-numbers*, as discussed in Chapter 1 and 7) but on the basis of whether, and if so then how, the broadening of such states are changing (*V-numbers*) with the strength of interaction. The changes in such broadening of states with magnetic field or gate voltage or measuring apparatus parameters are observed in experiments [7, 9, 15].

It is imperative to note at this point that although the spectral density function for 0D, 1D, 2D, and 3D seems to be quite different in nature, if we look closely, a continuous broadening of the function $\delta(E-\varepsilon_0)$ results in $(E-\varepsilon_0)^{-1/2}$, $\vartheta(E-\varepsilon_0)$, and $(E-\varepsilon_0)^{1/2}$ at certain values of broadening. Such a broadening of spectral density, as described in Chapter 3 and illustrated in the subsequent chapters, appears due to coupling of the system with other systems, e.g. what we are considering here as surroundings as a whole. Thus, a system has 0-spatial dimension if it is not interacting with any surrounding system, or we can say, completely isolated from its surroundings. And depending on the strength of interaction with other surrounding systems, it may be 1D, 2D, or 3D as increasing order of such strength. Therefore, such strength may also be something in between that for 0D and 1D, or 1D and 2D, and so on. Furthermore, from mathematical point of view, it clearly indicates that the spectral density functions corresponding to 0D, 1D, 2D, and 3D systems are special cases of a general function, and thus, we can also have the intermediate natures between such particular seemingly different functions as spectral densities, apparently representing the intermediate (between integers) spatial dimensions, i.e. *fractional dimensions*.

Before trying to find such spectral densities, we need to modify a little the previously discussed NEGF formalism. It is worthy to recall from Chapter 3 that spectral density of a system while interacting with other system(s) (which may be considered as surroundings) can be described mathematically using non-equilibrium Green's function (NEGF) formalism as:

$$A(E) = i(G(E) - G^{+}(E)) \tag{12.1}$$

where $A(E)$ is the spectral density of the system characterized by the NEGF given by:

$$G^{S}(E) = [E - \varepsilon^{S} - \Sigma_{\mathrm{Sur}}(E)]^{-1} \tag{12.2}$$

where $\Sigma_{\mathrm{Sur}}(E)$ is the self-energy of the system corresponding to its interaction with the surroundings. Thus, a 0D system, i.e. a system with 0-spatial dimension, is some system that is not interacting with any surrounding system, so that $\Sigma_{Sur}(E) \to 0$. Then it is apparent that:

$$A(E) = \underset{\Sigma_{Sur}(E)\to 0}{Lt} i(G(E) - G^{+}(E)) = 2\pi\delta(E - \varepsilon^{S}) \qquad \text{(using Poisson kernel).}$$

Now, for an ideal uniform 1D system, such self-energy $\Sigma_{Sur}(E)$ is non-zero and finite, and, since the discussion of Chapter 3, can be expressed as $\Sigma_{Sur}(E) = 2\tau g\tau^{+}$ (for neighbours at two sides), where τ is the coupling between system and surroundings, and owing to the translational symmetry g is given by

$$\tau g\tau^{+} - (E - \varepsilon^{S})g + 1 = 0 \tag{12.3}$$

which leads to the solutions $\Sigma_{Sur}(E) = (E - \varepsilon^{S}) \pm \sqrt{(E - \varepsilon^{S})^2 - 4\tau\tau^{+}}$, so that the Green's function results to be $G(E) = [\mp\sqrt{(E - \varepsilon^{S})^2 - 4\tau\tau^{+}}]^{-1}$. Accordingly, the spectral density is obtained to be:

$$A(E) = \begin{cases} 2\left[4\tau\tau^{+} - (E - \varepsilon^{S})^2\right]^{-1/2} & \text{for } (E - \varepsilon^{S})^2 < 4\tau\tau^{+} \\ 0 & \text{else where} \end{cases} \tag{12.4}$$

It is apparent from eq. (12.4) that the widely accepted conventional NEGF formalism, as also elaborated in all the previous chapters, results in the spectral density (proportional to $[4\tau\tau^{+} - (E - \varepsilon^{S})^2]^{-1/2}$) for 1D system that is contradictory to its obvious functional form proportional to $(E - \varepsilon^{S})^{-1/2}$.

In order to find out the origin of such problem, let us start from the expression of spectral density of a 0D, i.e. isolated, system, which is evidently $A_{0\text{-}D} \equiv \delta(E - \varepsilon^{S})$. As mentioned earlier, using Poisson kernel, it may be written as:

$$\begin{aligned} A_{0\text{-}D} &\equiv \delta(E - \varepsilon^{S}) \\ &= \frac{1}{\pi} \underset{\Sigma\to 0}{Lt} \left[\frac{\Sigma}{(E - \varepsilon^{S})^2 + \Sigma^2}\right] \\ &= \frac{1}{2\pi} \underset{\Sigma\to 0}{Lt}\, i \left[\frac{1}{(E - \varepsilon^{S}) + i\Sigma} - \frac{1}{(E - \varepsilon^{S}) - i\Sigma}\right] \\ &= (1/2\pi) \underset{\Sigma\to 0}{Lt}\, i[G(E) - G^{+}(E)] \end{aligned}$$

Thus, the delta function is expressed in the limiting form of Poisson kernel with an arbitrarily small element $\Sigma \to 0$, where such generic element Σ is interpreted as the interaction between the system with its surroundings, which tends to zero for an isolated system leading to 0D. Thus, for > 0D systems, i.e. where such interaction is non-zero, the spectral density can be defined in general as $A(E) \equiv i[G(E) - G^{+}(E)]$. However, as shown above, such a formulation of spectral density, which is widely accepted, unfortunately leads us to contradictory results.

In this context, we argue here that from physical point of view $\Sigma_{Sur}(E)\rightarrow 0$ can have meaning only in comparison to other energies associated with the system and surroundings since the interaction potential always needs a reference. Thus, we may rather consider the transmission coefficient ($T(E)|_{E\rightarrow\varepsilon^S}$) per energy eigenstate (see Chapter 4) as an absolute (not relative) measure of communication between a system and its surroundings which is absolute zero for an isolated system, i.e. 0D. Furthermore, the transmission is also inherently probabilistic in nature that has close resemblance with the fundamental definition of information, as described in Chapter 1. Thus, we propose that the spatial dimension of the system can be obtained simply in trams of system-surrounding communication as $N = T(E)|_{E\rightarrow\varepsilon^S}$. For instance, in the above case of an ideal 1D system, the transmission coefficient is $T(E) = G(E)(\Gamma_{Sur}/2)G^{+}(E)(\Gamma_{Sur}/2)$ (see Chapter 4), where for each of the two neighbours as considered above, $\Gamma_{Sur}/2 = i(\Sigma_{Sur} - \Sigma_{Sur}{}^{+})/2$. It is apparent that $\Gamma_{Sur}\neq 0$ only for $(E-\varepsilon^S)^2 < 4\tau\tau^{+}$, which evidently results in $T(E) = 1$ for each of such eigenstates.

We further argue here that, mathematically, the expression of the delta function as a limiting form has many other solutions (i.e. other than the Poisson kernel); more pertinently, such a form is given by $A_{0\text{-}D}(E) \equiv \delta(E-\varepsilon^S) = Lt_{\eta\rightarrow 0}\frac{\eta}{2}\left|E-\varepsilon^S - i\Sigma_{Sur}\right|^{\eta-1}$. Thus, comparing with the standard functional forms of spectral density for 1D, 2D, and 3D systems, it can be generalized for any system with any spatial dimension as:

$$A(E) \equiv \frac{\eta}{2}|G(E)|^{1-\eta} \tag{12.5}$$

where $\eta = N/2$. It may be interesting to further note that since the fact that spectral density is proportional to $d^N\vec{p}$ for an N-D system and owing to the non-relativistic relation $p \propto (E-\varepsilon_0)^{1/2}$, we can propose that confining one spatial-dimension through some kind of interaction is equivalent to reduce half order of elementary energy, e.g. 1/2-order fractional derivative of 3D spectral density results in that for 2D,. and so on, so that

$$A_{(N-f)\text{-}D}(E) \equiv \frac{d^{f/2}}{dE^{f/2}}A_{N\text{-}D}(E) \tag{12.6}$$

which is apparently consistent with eq. (12.5). Now, such spectral densities for any dimensional (including all the fractions $0 \leq f \leq 3$) system can be obtained by fractional derivative of the 3D spectral density. This has been illustrated in Fig. 12.4(a)–(d).

It can be observed from the plots of Fig. 12.4(a) – (c) that, following eq. (12.6), the spectral density (for 3D→2D, 2D→1D, and 1D→0D) changes its functional form as $(E-\varepsilon_0)^{1/2} \rightarrow \vartheta(E-\varepsilon_0) \rightarrow (E-\varepsilon_0)^{-1/2} \rightarrow \delta(E-\varepsilon_0)$ with intermediate functional forms with a continuously reducing broadening of the state finally resulting in a delta function with zero broadening. This is more apparent from the plots of Fig. 12.4(d) which shows a continuum of the spectral density over dimension from 3D to 0D.

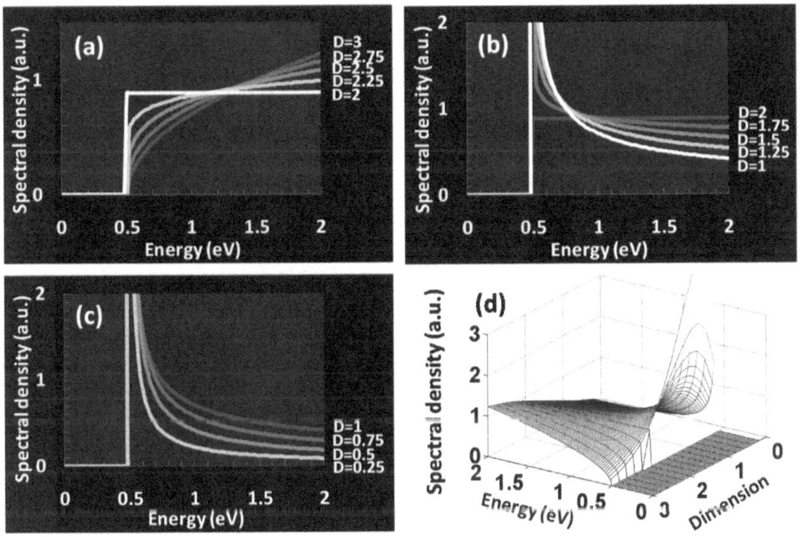

Fig. 12.4: The plots of spectral density for various spatial dimensions calculated from fractional derivative of that for 3D: (a) $f = 3 \rightarrow 2$; (b) $f = 2 \rightarrow 1$; (c) $f = 1 \rightarrow 0$ and (d) continuously varying $f = 3 \rightarrow 0$.

Such a continuum is consistent with the continuity in coupling strength and the corresponding transmission between a system and its surroundings. Thus, any (including all the fractions $0 \leq f \leq 3$) spatial dimension can now be directly correlated in a one-to-one manner with any such system-surrounding coupling with the respective strength. It can therefore be asked now, in retrospect, whether any such system-surrounding interaction can be replaced by spatial dimension or geometry in general. Can then the quantum theory and gravitation be described in a more general fractional dimensional geometry, so that the dimensions are relevant mapping of interactions on physical space for quantum case, whereas they themselves have appropriate mapping on to the equivalent gravitational interaction? Can quantum and gravity be coupled together, may be as complementary to each other, in terms of fractional dimensional geometry and interaction? Let's keep it an open question for now.

References

[1] Xing-Fei, H. "Excitons in anisotropic solids: The model of fractional-dimensional space." *Physical Review B* 43, no. 3 (1991): 2063.

[2] Birkedal, D., Jai Singh, V. G. L., Erland, J., and Marcher Hvam, J. "Binding of quasi-two-dimensional biexcitons." *Physical review letters* 76, no. 4 (1996): 672.

[3] Wagner, M., Schneider, H., Stehr, D., Winnerl, S., Andrews, A. M., Schartner, S., Strasser, G., and Helm, M. "Observation of the intraexciton autler-townes effect in GaAs/AlGaAs<? format?> semiconductor quantum wells." *Physical Review Letters* 105, no. 16 (2010): 167401.

[4] Yiling, Y., Yifei, Y., Cai, Y., Wei, L., Gurarslan, A., Peelaers, H., Aspnes, D. E., et al. "Exciton-dominated dielectric function of atomically thin MoS2 films." *Scientific reports* 5, no. 1 (2015): 16996.

[5] Zubair, M., Mughal, M. J., and Naqvi, Q. A. "Electromagnetic wave propagation in fractional space." In *Electromagnetic fields and waves in fractional dimensional space*, pp. 27–60. Berlin, Heidelberg: Springer Berlin Heidelberg, 2012.

[6] Shalaev, V. M. and Stockman, M. I. "Optical properties of fractal clusters (susceptibility, surface enhanced Raman scattering by impurities)." *Sov. Phys. JETP* 65, (1987): 509–522.

[7] Liljeroth, P., Pedro, A., Van Emmichoven, Z., Hickey, S. G., Weller, H., Grandidier, B., Allan, G., and Vanmaekelbergh, D. "Density of states measured by scanning-tunneling spectroscopy sheds new light<? format?> on the optical transitions in PbSe Nanocrystals." *Physical review letters* 95, no. 8 (2005): 086801.

[8] Hussain, Z., Brouet, V., Yang, W., Xingjiang Zhou, R. G. M., He, R., Lu, D. H., et al. "ARPES study of the evolution of band structure and charge density wave properties in RTe3 (R= Y, La, Ce,Sm, Gd,Tb, and Dy)." *Physical Review B* 77, no. LBNL-831E (2008).

[9] Fujita, K., Drozdov, I., Du, Z., Li, H., Joo, S.-H., Lee, J., Gu, G., Johnson, P. D., and Valla, T. "Combined spectroscopic imaging STM and ARPES study of different gaps measured in the cuprate phase diagram." *Physical Review B* 101, no. 4 (2020): 045136.

[10] Sohn, B., Kim, J. R., Kim, C. H., Lee, S., Hahn, S., Kim, Y., Huh, S., et al. "Observation of metallic electronic structure in a single-atomic-layer oxide." *Nature Communications* 12, no. 1 (2021): 6171.

[11] Cheng, P., Song, C., Zhang, T., Zhang, Y., Wang, Y., Jia, J.-F., Wang, J., et al. "Landau quantization of topological surface states in Bi 2 Se 3." *Physical review letters* 105, no. 7 (2010): 076801.

[12] Wang, P., Guo, Y., Jia, Y., Michael Onyszczak, F. A. C., Lei, S., Klemenz, S., et al. "Landau quantization and highly mobile fermions in an insulator." *Nature* 589, no. 7841 (2021): 225–229.

[13] Kouwenhoven, L. P., Marcus, C. M., McEuen, P. L., Tarucha, S., Westervelt, R. M., and Wingreen, N. S. "Electron transport in quantum dots." In *Mesoscopic electron transport*, pp. 105–214. Dordrecht: Springer Netherlands, 1997.

[14] Sikdar, S., Chowdhury, B. N., Saha, R., and Chattopadhyay, S. "Voltage-tunable quantum-dot array by patterned Ge-nanowire-based metal-oxide-semiconductor devices." *Physical Review Applied* 15, no. 5 (2021): 054060.

[15] Zhang, Y., Tan, Y.-W., Stormer, H. L., and Kim, P. "Experimental observation of the quantum Hall effect and Berry's phase in graphene." *nature* 438, no. 7065 (2005): 201–204.

Appendix

Relativistic quantum theory and topological materials

A.1 The basic concept of special theory of relativity

The extraordinarily famous special theory of relativity has been the most remarkable discovery (by Einstein, possibly not required to mention) of the last century and is still fascinating people due to its success in all branches of physics. Its basic postulate that all motions are relative and there can be no existing standard reference frame, but even then all laws of physics are the same in all inertial reference frames, which upholds the subjectivity of facts leading to the objective truths. Mathematically such a concept is represented by the invariance of the physical laws in tensor form under Lorentz transformation.

The concept of relativity is also fundamentally based upon the equivalence and contrast, together, between space and time, so that in a moving reference frame both are changed, but in an opposite way, i.e. length is contracted while time is dilated. Also the Lorentz invariant length of space-time $(\vec{r}, t)$ is given by $s^2 = c^2t^2 - r^2$, c being the speed of light in vacuum, where $\vec{r}$ and ct have equal footings, however, with a (–) sign in between, constraining each other. Interestingly, the concept of subjective facts and objective truth has also been upheld by considering symmetry in the relative motion of all objects, however, with the speed of light in vacuum to be absolute.

For instance, suppose an object A has a velocity v_{12} w.r.t. B, which has a velocity v_{23} (in the same direction) w.r.t. C. Then the velocity of C w.r.t. A is v_{31} that completes a symmetric loop so that we can write a function $f(v_{12}, v_{23}.v_{31}) = 0$. Now, considering a linearity in their symmetric relation, we can write:

$$p + q(v_{12} + v_{23} + v_{31}) + r(v_{12}v_{23} + v_{23}v_{31} + v_{31}v_{12}) + sv_{12}v_{23}v_{31} = 0 \tag{A.1}$$

where p, q, r, s are constants. It is apparent that if A is stationary w.r.t. B and so is B w.r.t. C (i.e. $v_{12} = 0 = v_{23}$), then C must be stationary w.r.t. A (i.e. $v_{31} = 0$), leading to $p = 0$. Now, if A is stationary w.r.t. B, then again, due to symmetry, A and B are both equivalent w.r.t. C so that $v_{23} = v_{13}$, or we have $v_{23} = -v_{31}$. Putting this in eq. (A.1), we get, $r = 0$. Finally, since speed of light in vacuum is independent of any reference frame, if $v_{12} = c$, then $v_{13} = c$ or $v_{31} = -c$, that results in $q = sc^2$, so that eq. (A.1) can now be written as

$$c^2(v_{12} + v_{23} + v_{31}) + v_{12}v_{23}v_{31} = 0$$

$$\Rightarrow v_{31} = -\frac{(v_{12} + v_{23})}{1 + v_{12}v_{23}/c^2}$$

 | https://doi.org/10.1515/9783111545950-013

$$\Rightarrow v_{13} = \frac{(v_{12} + v_{23})}{1 + v_{12}v_{23}/c^2} \tag{A.2}$$

which is the velocity transformation formula for relativistic objects that approximates to ordinary Galilean relative motion if the velocities are much smaller than that of light in vacuum. It is well known that such relativistic transformation relation leads to the relativistic mass of an object (with rest mass m_0 and speed v) to be $m = m_0/\sqrt{1 - v^2/c^2}$, which results in the energy-momentum (i.e. $E - p$) relation to be

$$E^2 = p^2c^2 + (m_0c^2)^2 \tag{A.3}$$

A.2 The basic formalism of Dirac equation

In quantum mechanics, any physical quantity of a system in a state is defined by its measurement and its possible measured values in terms of application of an operator and its eigenvalues, respectively. This is why the physical quantities are termed here as 'observables'. Thus, the energy and momentum values are associated with the corresponding measuring operators $i\hbar\partial/\partial t$ and $-i\hbar\vec{\nabla}$. Then, can such operators be put into eq. (A-3) to describe the motion of relativistic particles like electrons?

NO. We need Dirac equation for that. The speciality of Dirac equation is that it considers, as its core base, the linearity of special relativity as shown above. Thus it starts from writing a linear equation of the form

$$E\underline{\underline{I}} = m_0c^2\underline{\underline{\beta}} + c\sum_{j=1}^{3}\underline{\underline{\alpha}}_j p_j \tag{A.4}$$

where $\underline{\underline{\alpha}}_j, \underline{\underline{\beta}}$ are matrices that anti-commute with each other and square of each of them is unity matrix $\underline{\underline{I}}$. Such a linear equation evidently satisfies eq. (A.3), and thus now the Hamiltonian operator of a free electron, or a fermion in general, is given by

$$\hat{H}_{3\text{-D}} = m_0c^2\underline{\underline{\beta}} + c\sum_{j=1}^{3}\underline{\underline{\alpha}}_j\hat{p}_j \tag{A.5}$$

It should be noted that the choice of $\underline{\underline{\alpha}}_j, \underline{\underline{\beta}}$ matrices are not unique, one most common choice (known as Dirac-Pauli representation) being $\underline{\underline{\alpha}}_j \equiv \begin{pmatrix} \underline{\underline{0}} & \underline{\underline{\sigma}}_j \\ \underline{\underline{\sigma}}_j & \underline{\underline{0}} \end{pmatrix}$ and $\underline{\underline{\beta}} \equiv \begin{pmatrix} \underline{\underline{I}} & \underline{\underline{0}} \\ \underline{\underline{0}} & -\underline{\underline{I}} \end{pmatrix}$, where $\underline{\underline{\sigma}}_j$ are Pauli matrices given by $\underline{\underline{\sigma}}_1 \equiv \begin{pmatrix} 0 & 1 \\ 1 & 0 \end{pmatrix}$, $\underline{\underline{\sigma}}_2 \equiv \begin{pmatrix} 0 & -i \\ i & 0 \end{pmatrix}$ and $\underline{\underline{\sigma}}_3 \equiv \begin{pmatrix} 1 & 0 \\ 0 & -1 \end{pmatrix}$, each having eigenvalues ±1, so that $\underline{\underline{\sigma}}_i\underline{\underline{\sigma}}_j = \delta_{ij}\underline{\underline{I}} + i\sum_k \varepsilon_{ijk}\underline{\underline{\sigma}}_k$ (where

δ_{ij} is Kronecker delta and ε_{ijk} id Levi-Civita). In case of 2D motion, the choices of such matrices may be much simpler to write,

$$\hat{H}_{2D} = m_0 c^2 \underline{\underline{\sigma}}_3 + c \sum_{j=1}^{2} \underline{\underline{\sigma}}_j \hat{p}_j \tag{A.6}$$

A.2.1 Spin and helicity

Now, it may be interesting to note that the angular momentum operator $\hat{L}_i = (\vec{r} \times \hat{\vec{p}})_i = \sum_{j,k} \varepsilon_{ijk} x_j \hat{p}_k$ does not commute with the Hamiltonian, and owing to $[x_j, \hat{p}_k] = i\hbar\delta_{jk}$ gives rise to

$$[\hat{L}_i, \hat{H}_{3\text{-D}}] = [\sum_{j,k} \varepsilon_{ijk} x_j \hat{p}_k, c \sum_{l} \underline{\underline{\alpha}}_l \hat{p}_l] = \sum_{j,k,l} \varepsilon_{ijk} c \underline{\underline{\alpha}}_l [x_j, \hat{p}_l] \hat{p}_k = i\hbar c (\underline{\underline{\vec{\alpha}}} \times \hat{\vec{p}})_i \tag{A.7a}$$

Interestingly, if we consider a matrix $\underline{\underline{\Sigma}}_j = (\hbar/2) \begin{pmatrix} \underline{\underline{\sigma}}_j & \underline{\underline{0}} \\ \underline{\underline{0}} & \underline{\underline{\sigma}}_j \end{pmatrix}$, we get

$$[\underline{\underline{\Sigma}}_i, \hat{H}_{3\text{-D}}] = -i\hbar c (\underline{\underline{\vec{\alpha}}} \times \hat{\vec{p}})_i \tag{A.7b}$$

so that, since Heisenberg equation, it results to be

$$i\hbar \frac{d}{dt} (\hat{\vec{L}} + \underline{\underline{\vec{\Sigma}}}) = [(\hat{\vec{L}} + \underline{\underline{\vec{\Sigma}}}), \hat{H}_{3\text{-}D}] = 0 \tag{A.8}$$

Equation (A.7a) shows that the mechanical (or in an atom, orbital) angular momentum does not commute with the Hamiltonian and thus, since Heisenberg equation, is not a conserved quantity. However, as can be seen from eq. (A.8), the conservation of an equivalent angular momentum $(\hat{\vec{L}} + \underline{\underline{\vec{\Sigma}}})$, in turn, reveals that electrons, or in general, fermions, have an intrinsic angular momentum $\underline{\underline{\vec{\Sigma}}}$ (which does not depend on any condition) with the eigenvalues of each of its components being $\pm(\hbar/2)$. This is called the *spin* of an electron, or in general, a fermion. Further, assuming $\hat{n}_{\vec{p}}$ to be the unit vector along momentum, eq. (A.7b) simply leads to

$$[\underline{\underline{\vec{\Sigma}}} \cdot \hat{n}_{\vec{p}}, \hat{H}_{3\text{-}D}] = -i\hbar c (\underline{\underline{\vec{\alpha}}} \times \hat{\vec{p}}) \cdot \hat{n}_{\vec{p}} = 0$$

$$\Rightarrow i\hbar \frac{d}{dt} (\underline{\underline{\vec{\Sigma}}} \cdot \hat{n}_{\vec{p}}) = 0 \tag{A.9}$$

which shows that the component of spin along momentum, called ***helicity***, is also a conserved quantity, and thus, a precisely measurable observable.

A.2.2 Anti-particle and Majorana particle

Now, writing the 4-vectors $\underline{\underline{\gamma}}^\mu \equiv (\underline{\underline{\gamma}}_0, \underline{\underline{\gamma}}_1, \underline{\underline{\gamma}}_2, \underline{\underline{\gamma}}_3)$ so that $\underline{\underline{\gamma}}_0 = \underline{\underline{\beta}}, \underline{\underline{\gamma}}_j = \underline{\underline{\beta}}\,\underline{\underline{\alpha}}_j$, and $p^\mu \equiv (E/c, p_1, p_2, p_3)$, Dirac equation can be written in a covariant form as

$$(\underline{\underline{\gamma}}_\mu \hat{p}^\mu - m_0 c)\underline{\psi}(p) = 0 \tag{A.10}$$

which in the presence of an electromagnetic field (A^μ) leads to

$$[\underline{\underline{\gamma}}_\mu (\hat{p}^\mu - eA^\mu) - m_0 c]\underline{\psi}(p) = 0 \tag{A.10a}$$

Noting that $-p$ also satisfies Dirac equation, we must have another equation

$$[\underline{\underline{\gamma}}_\mu (-\hat{p}^\mu - eA^\mu) - m_0 c]\underline{\psi}(-p) = 0$$

$$[\underline{\underline{\gamma}}_\mu (\hat{p}^\mu + eA^\mu) + m_0 c]\underline{\phi}(p) = 0 \tag{A.11}$$

which indicates a negative rest mass that has no physical meaning. However, noting that $\hat{p}^\mu * = -\hat{p}^\mu$, taking the complex conjugate of eq. (A.10a), and then multiplying it by a matrix $-\underline{\underline{C}}\underline{\underline{\gamma}}_0$ (where $(-\underline{\underline{C}}\underline{\underline{\gamma}}_0)\underline{\underline{\gamma}}_\mu * = \underline{\underline{\gamma}}_\mu(\underline{\underline{C}}\underline{\underline{\gamma}}_0)$) and writing $\underline{\psi}_C(p) = \underline{\underline{C}}\underline{\underline{\gamma}}_0\underline{\psi} * (p)$, we get

$$[\underline{\underline{\gamma}}_\mu (\hat{p}^\mu + eA^\mu) - m_0 c]\underline{\psi}_C(p) = 0 \tag{A.11a}$$

which indicates a particle with the same positive rest mass, but an opposite charge to that described by eq. (A.10a); thus, this is known to be its anti-particle, such as positron for electron. Since $\hat{p}^\mu * = -\hat{p}^\mu$ satisfies eq. (A.11a), the anti-particle can be assumed as if the respective particle is moving not only in opposite direction but also backward in time. This concept was utilized by Feynman in his diagrammatic concept as described in Chapter 2. At this point, it can be said simply that if for some particle, the charge conjugate state is equal to the state itself, i.e. $\underline{\psi}_C(p) = \underline{\underline{C}}\underline{\underline{\gamma}}_0\underline{\psi} * (p) = \underline{\psi}(p)$, then such a particle can be considered to be its own anti-particle, and is known to be a Majorana particle. Unfortunately, no such fundamental particle in free state has still been observed experimentally. However, recently tremendous efforts are being made to find them in topological materials as Majorana bound state (MBS).

A.2.3 Chirality and Weyl particle

Considering the matrix $\underline{\underline{\gamma}}_5 = i\underline{\underline{\gamma}}_0\underline{\underline{\gamma}}_1\underline{\underline{\gamma}}_2\underline{\underline{\gamma}}_3$, it can be observed that $\underline{\underline{\gamma}}_5$ commutes with $\underline{\underline{\alpha}}_j$ but not with $\underline{\underline{\beta}}$, and thus does not commute with the Hamiltonian $\hat{H}_{3\text{-D}}$ unless the rest mass is zero, which is known as Weyl particle. Interestingly, we can define two operators

$\hat{P}_{R/L} = (1/2)(\underline{\underline{I}} \pm \underline{\underline{\gamma}}_5)$ so that owing to the property $\underline{\underline{\gamma}}_5^{\,2} = \underline{\underline{I}}$, we have $\hat{P}_{R/L}^{\,2} = \hat{P}_{R/L}$, $\hat{P}_R + \hat{P}_L = 1$, and $\hat{P}_R\hat{P}_L = 0$. Thus, $\hat{P}_R = (1/2)(\underline{\underline{I}} + \underline{\underline{\gamma}}_5)$ and $\hat{P}_L = (1/2)(\underline{\underline{I}} - \underline{\underline{\gamma}}_5)$ can be considered as two mutually orthogonal projection operators, usually called right- and left-handed projection operators, and therefore $\underline{\underline{\gamma}}_5$ is known to be the chirality. Now, for a Weyl particle of zero rest mass (i.e. $E = pc$), eq. (A.4) gives rise to $E\underline{\psi} = c\sum_{j=1}^{3} \underline{\underline{\alpha}}_j p_j \underline{\psi}$, which since $\underline{\underline{\gamma}}_5 \underline{\underline{\alpha}}_j = \underline{\underline{\Sigma}}_j$ leads to

$$\underline{\underline{\gamma}}_5 \underline{\psi} = \vec{\underline{\underline{\Sigma}}}_j \cdot \hat{n}_{\vec{p}} \underline{\psi} \tag{A.12}$$

Thus, chirality is identical to helicity for a Weyl particle. Unfortunately, again no fundamental Weyl particle in free state has been found in experiments. However, in particular topological materials called Weyl semimetal, such states have been experimentally observed.

A.3 Relativistic electrons in topological materials

Before going to describe the motion of relativistic electrons in some topological classes of materials, let us start with understanding the physics of the matrices $\underline{\underline{\alpha}}_j, \underline{\underline{\beta}}$ appearing in Dirac Hamiltonian given by eq. (A.5). We have already observed that the ±1 eigenvalues of $\underline{\underline{\beta}}$ correspond to two solutions with rest mass $\pm m_0$ that represent the Dirac particle and its anti-particle, respectively. Now, in the following we can see that the commutation of position with Hamiltonian results in $[x_k, \hat{H}_{3\text{-D}}] = c\sum_{j=1}^{3} \underline{\underline{\alpha}}_j [x_k, \hat{p}_j] = i\hbar c \sum_{j=1}^{3} \underline{\underline{\alpha}}_j \delta_{kj} = i\hbar c \underline{\underline{\alpha}}_k$, which, since Heisenberg equation, gives rise to

$$\frac{d}{dt} x_k = c\underline{\underline{\alpha}}_k \tag{A.13}$$

Thus, firstly, the matrices $\underline{\underline{\alpha}}_j$ physically represent the velocity components (in units of c) of a Dirac particle/anti-particle (e.g. electron/positron), and secondly, the eigenvalue of each of such velocity components are $\pm c$! Then the total velocity will anyhow exceed the speed of light in vacuum! However, this is not a contradiction to the theory of relativity since $\underline{\underline{\alpha}}_j$'s neither commute with each other nor with the Dirac Hamiltonian. Thus, neither any two (or all three) components of velocity can be measured at the same state nor can any one of them be measured precisely. The velocity operator only commutes with the chiral matrix ($\underline{\underline{\gamma}}_5$) that, indeed, has significance in topological classes of materials. At this point, we want to point out that astonishingly $\sum_{j=1}^{3} \underline{\underline{\alpha}}_j^{\,2} = 3$, and since it commutes with the Dirac Hamiltonian, it indicates that the rms velocity of any Dirac particle is $v_{\text{rms}} = \sqrt{3}c$, i.e. greater than the speed of light, and we argue, at least in principle, if an experiment can be arranged to measure it, we can verify it.

Nevertheless, in a 2D material that allows electrons to move with relativistic speed, since eq. (A.6), the Hamiltonian in such material can now be written as

$$\hat{H} = m^* c^2 \underline{\underline{\sigma}}_3 + v_x \underline{\underline{\sigma}}_1 \hat{p}_x + v_y \underline{\underline{\sigma}}_2 \hat{p}_y \tag{A.14}$$

where $m^* = m_x + m_y$, indicating the corresponding electron effective mass components, and $v_x^2 p_x^2 + v_y^2 p_y^2 = c^2 p^2$. It is imperative to note here that the X-axis is considered along the thickness of the material so that the wave function decays towards interior from its surface, while Y-axis is along the motion of electron in the plane of its surface. Thus, we can have a solution of eq. (A.14) to be $\underline{\psi}(x, y) = \underline{u}^{(s)} \exp(\kappa x + iKy)$, where (s) represents the spin state and (κ, K) are real. Considering the energy associated with decaying wave function along the material thickness to be reference, we can thus write

$$[m_x c^2 \underline{\underline{\sigma}}_3 + v_x \underline{\underline{\sigma}}_1 \hat{p}_x] \underline{u}^{(s)} \exp(\kappa x + iKy) = 0 \tag{A.15a}$$

$$[m_y c^2 \underline{\underline{\sigma}}_3 + v_y \underline{\underline{\sigma}}_2 \hat{p}_y] \underline{u}^{(s)} \exp(\kappa x + iKy) = E \underline{u}^{(s)} \exp(\kappa x + iKy) \tag{A.15b}$$

which lead, respectively, to

$$[m_x c^2 \underline{\underline{\sigma}}_3 - i v_x \hbar \kappa \underline{\underline{\sigma}}_1] \underline{u}^{(s)} = 0 \tag{A.16a}$$

$$[m_y c^2 \underline{\underline{\sigma}}_3 + v_y \hbar K \underline{\underline{\sigma}}_2] \underline{u}^{(s)} = E \underline{u}^{(s)} \tag{A.16b}$$

Thus the non-trivial solution of eq. (A.16a) is given by $\det[m_x c^2 \underline{\underline{\sigma}}_3 - i v_x \hbar \kappa \underline{\underline{\sigma}}_1] = 0$ leading to

$$\kappa = \pm m_x c^2 / \hbar v_x \tag{A.17}$$

with the eigenstates

$$\underline{u}\Big|^{(s)}_{\kappa = \pm \frac{m_x c}{\hbar}} \equiv \frac{1}{\sqrt{2}} \begin{pmatrix} 1 \\ \mp i \end{pmatrix} \tag{A.18}$$

It is interesting to note that such states are also eigenstates of the spin component along Y-axis, i.e. $\underline{\underline{\sigma}}_2 \underline{u}\Big|^{(\downarrow\uparrow)}_{\kappa = \pm \frac{m_x c}{\hbar}} = \mp \underline{u}\Big|^{(\downarrow\uparrow)}_{\kappa = \pm \frac{m_x c}{\hbar}}$, so that $|\kappa = +m_x c/\hbar\rangle$ and $|\kappa = -m_x c/\hbar\rangle$ are one-to-one correlated with the down Y-spin $|\downarrow\rangle$ and up Y-spin $|\uparrow\rangle$ states, respectively. Thus, eq. (A.16b) gives rise to

$$E = m_y c^2 + (\mp)^{\downarrow\uparrow}_{\kappa = \pm m_x c/\hbar} v_y \hbar K \tag{A.19}$$

$$\Rightarrow \frac{1}{\hbar} \frac{\partial E}{\partial K} = (\mp)^{\downarrow\uparrow}_{\kappa = \pm m_x c/\hbar} v_y \tag{A.20}$$

Equation (A.20) shows that v_y is indeed the band velocity on the plane of the surface of such a 2D material so that its directions for ‘up’ and ‘down’ spin along v_y are opposite to each other. Further, since the wave function varies as $\sim \exp(\kappa x)$, it is apparent from eq. (A.17) that $\kappa = +m_x c^2/\hbar v_x$ is for $-x$ and $\kappa = -m_x c^2/\hbar v_x$ is for $+\times$ indicating them to be defined for the upper and lower surfaces of the material, respectively. Now the fact that $|\kappa = +m_x c/\hbar\rangle$ and $|\kappa = -m_x c/\hbar\rangle$ are one-to-one correlated with $|\downarrow\rangle$ and $|\uparrow\rangle$, respectively, suggests from eq. (A.19) that the direction of momentum is locked with spin along it, and thus the upper and lower surfaces will exhibit opposite spin currents.

At this point, it is worthy to note from eq. (A.19) that, firstly, the energy-momentum dispersion relation is linear; and secondly, the energy is positive or negative as per $m_y c^2 > |v_y \hbar K|$ or $m_y c^2 < |v_y \hbar K|$ accordingly, with band touching at $m_y c^2 - |v_y \hbar K| = 0$, indicating it to be a Dirac semimetal. However, a material for which $m_y = 0$, such positive and negative energy bands exhibit a gap of $2|v_y \hbar K|$, known as Weyl semimetal. Interestingly, for such a Weyl material in the presence of a magnetic field is described by (since eq. (A.19)) $E = (\mp)^{\downarrow\uparrow}_{\kappa = \pm m_x c/\hbar} v_y(\hbar K - eA)$, which putting $E \to -E$; $K \to -K$ leads to $E = (\mp)^{\downarrow\uparrow}_{\kappa = \pm m_x c/\hbar} v_y(\hbar K + eA)$, indicating MBSs. It may be noted that $\kappa x + iKy$ as the phase indicates that $-ip_x + p_y$ is the origin of all such topological materials.

Index

 | https://doi.org/10.1515/9783111545950-014

De Gruyter Series in Mathematics and Life Sciences

Volume 15
Leonid Hanin
Modeling Cancer Progression. Mathematical, Statistical
and Computational Methodologies with Clinical Perspectives
ISBN 978-3-11-133332-8, e-ISBN (PDF) 978-3-11-133403-5, e-ISBN (EPUB) 978-3-11-133425-7

Volume 14
Jochen Mau, Sergey I. Mukhin, Guanyu Wang, Shuhua Xu, Željko Čupić (Eds.)
BIOKYBERNETIKA II. Thermodynamics, Functional Physiology, Process Control, and Society Impact
ISBN 978-3-11-914480-3, e-ISBN (PDF) 978-3-11-221662-0, e-ISBN (EPUB) 978-3-11-221681-1

Volume 13
Jules J. Berman
Fundamentals of Reality. Emergent Systems Development from Vacuum and Waves
ISBN 978-3-11-149814-0, e-ISBN (PDF) 978-3-11-149869-0, e-ISBN (EPUB) 978-3-11-149841-3

Volume 12
Olga Vasieva
Origins of Biocomplexity. The Fundamental Principles of Life in Terms of Physics
ISBN 978-3-11-133472-1, ISBN (PDF) 978-3-11-133479-0, ISBN (EPUB) 978-3-11-133483-7

Volume 11
Apul Dev, Sayantan Guha (Eds.)
Applied Engineering Mathematics. Fluid and Solid Mechanics in Life Science
ISBN 978-3-11-163853-9, ISBN (PDF) 978-3-11-163878-2, ISBN (EPUB) 978-3-11-163868-3

Volume 10
Fateh Mebarek-Oudina (Ed.)
CFD Simulation. Thermo-Fluids and Nanofluids in Engineering and Biomedicine, 2025
ISBN 978-3-11-140454-7, ISBN (PDF) 978-3-11-140509-4, ISBN (EPUB) 978-3-11-140523-0

Volume 9
Jochen Mau, Sergey Mukhin, Guanyu Wang, Shuhua Xu (Eds.)
BIOKYBERNETIKA. Mathematics for Theory and Control in the Human and in Society, 2025
ISBN 978-3-11-134197-2, ISBN (PDF) 978-3-11-134199-6, ISBN (EPUB) 978-3-11-134204-7

Volume 8/2
Igor M. Rouzine
Mathematical Modeling of Evolution. Volume 2: Fitness Landscape, Red Queen,
Evolutionary Enigmas, and Applications to Virology, 2023
ISBN 978-3-11-069731-5, ISBN (PDF) 978-3-11-069738-4, ISBN (EPUB) 978-3-11-069746-9

www.degruyterbrill.com

Volume 8/1
Igor M. Rouzine
Mathematical Modeling of Evolution. Volume 1: One-Locus and Multi-Locus Theory and Recombination, 2020
ISBN 978-3-11-060789-5, ISBN (PDF) 978-3-11-061545-6, ISBN (EPUB) 978-3-11-060819-9

Volume 7
George Dassios, Athanassios S. Fokas
Electroencephalography and Magnetoencephalography, 2020
ISBN 978-3-11-054583-8, ISBN (PDF) 978-3-11-054753-5, ISBN (EPUB) 978-3-11-054578-4

Volume 6
Piotr Biler
Singularities of Solutions to Chemotaxis Systems, 2019
ISBN 978-3-11-059789-9, ISBN (PDF) 978-3-11-059953-4, ISBN (EPUB) 978-3-11-059862-9

Volume 5
S. V. Masiuk, A. G. Kukush, S. V. Shklyar, M. I. Chepurny, I. A. Likhtarov
Radiation Risk Estimation. Based on Measurement Error Models, 2016
ISBN 978-3-11-044180-2, ISBN (PDF) 978-3-11-043366-1, ISBN (EPUB) 978-3-11-043347-0

Volume 4
Sergey Vakulenko
Complexity and Evolution of Dissipative Systems: An Analytical Approach, 2013
ISBN 978-3-11-026648-1, e-ISBN 978-3-11-026828-7

Volume 3
Zoran Nikoloski, Sergio Grimbs
Network-based Molecular Biology: Data-driven Modeling and Analysis, 2013
ISBN 978-3-11-026256-8, e-ISBN 978-3-11-026266-7

Volume 2
Shair Ahmad, Ivanka M. Stamova (Eds.)
Lotka-Volterra and Related Systems: Recent Developments in Population Dynamics, 2013
ISBN 978-3-11-026951-2, e-ISBN 978-3-11-026984-0

Volume 1
Alexandra V. Antoniouk, Roderick V. N. Melnik (Eds.)
Mathematics and Life Sciences, 2012
ISBN 978-3-11-027372-4, e-ISBN 978-3-11-028853-7

FSC
www.fsc.org
MIX
Papier | Fördert gute Waldnutzung
FSC® C083411

Kolibri 360 GmbH
Ferdinand-Jühlke-Straße 7
99095 Erfurt, Deutschland
produktsicherheit@kolibri360.de